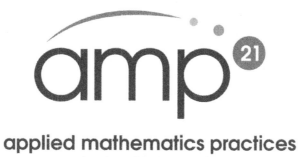

applied mathematics practices
for the 21st century

MATHEMATICAL MODELING WITH ALGEBRA

Using Authentic Problem Contexts

Thomas Edwards, Ph.D.
Kenneth Chelst, Ph.D.

applied mathematics practices
for the 21st century

Copyright © 2017 Thomas Edwards and Kenneth Chelst.
All rights reserved.

This book or any portion there of may not be reproduced or used in any manner whatsoever without the expressed written permission of the authors except for the use of brief quotations in a book review.

Applied Mathematics Practices for the 21st Century (AMP21) is an affiliation of professors and mathematics educators who share a common desire to bring relevance to K-12 mathematics by using authentic problem contexts in teaching and developing mathematics concepts and skills. The lead faculty are Kenneth Chelst (College of Engineering) and Thomas Edwards (College of Education) who are both professors at Wayne State University in Detroit, MI.

AMP21 is a Non-Profit developer and provider of curriculum that is aligned with the eight Standards for Mathematical Practice in the Common Core State Standards. Our team has published two textbooks for high school 1) Algebra and Mathematical Modeling and 2) Probability and Mathematical Modeling. Our new curriculum development focuses on middle school topics central to proportional reasoning: percentages, rates, ratios and proportions. Everything is presented in authentic problem contexts that involve making decisions. AMP21 offers professional development workshops in conjunction with local universities to help teachers and schools develop the programs needed to enable students to succeed in the global economy. In addition, this curriculum can form a basis for project based learning in mathematics.

First Printing, 2017
ISBN -13: 978-1979305655

Applied Mathematics Practices (AMP21)
In conjunction with Wayne State University
www.appliedmathpractices.com

Contact: Kenneth Chelst
4815 Fourth St. (Room 2017)
Detroit, MI 48201
kchelst@wayne.edu

Acknowledgements:

This work was developed initially with funding from the National Science Foundation (NSF Directorate for Education and Human Resources Project # DRL-0733137). The grant was a partnership between North Carolina State University, University of North Carolina – Charlotte, and Wayne State University. The following people contributed to the initial project, known as Project MINDSET:

Wayne State University

Dr. Thomas Edwards
Dr. Kenneth Chelst
Mr. Jay Johnson

North Carolina State University

Dr. Karen Keene
Dr. Karen Norwood
Dr. Robert Young
Dr. Molly Purser
Dr. Amy Craig-Reamer

University of North Carolina Charlotte

Dr. David Royster
Dr. David Pugalee
Ms. Sarah Johnson

Researchers

Dr. Saman Alaniazar
Dr. Hatice Ucar
Dr. Richelle Dietz
Dr. Krista Holstein
Mr. Will Hall
Ms. Zeyneb Yurtseven

Teachers

Derek Blackwelder, Jamie Blanchard James Bruckman, Kevin Busfield, Robert Butler, Molly Charles, Suzanne Christopher, Kristina Clayton, Krystle Corbin, Dean DiBasio (teacher writer), Robin Dixon, Caroline Geel, Michael Gumpp, Breanna Harrill, Jack Hunter, Amy Johnson, Jennifer Johnson, Jenise Jones (teacher writer), Anna Kamphaus, Chris Kennedy, George Lancaster (teacher writer), Craig Lazarski, Jeffrey Loewen, Charles Ludwick, Margaret Lumsden, Leon Martin, Cora McMillan, Kristen Meck (teacher writer), Karen Mullins, Erica Nelson, Bethany Peters, Joseph Price, Angela Principato (teacher writer), John Pritchett, Tyler Pulis, Paul Rallo, Tonya Raye, Kym Roberts, William Ruff, Amy Setzer, Rob Sharpe, Christy Simpson, Catherine Sitek, Cheryl Smith, Julia Smith, Wendy Carol Srinivasan, Susan Sutton, Greg Tucker, Laura Turner, Thad Wilhelm (teacher writer), Tracey Walden, Tracey Weigold, Lanette Wood

Cover Design: Susan Marie Dials

Significant contribution to the 2017 second edition revision: Thad Wilhelm

When Will I Ever Use This?
Volume I: Algebraic Modeling
Thomas Edwards and Kenneth Chelst

Table of Contents

Introduction

Chapter 1: Make Hard Decisions—Multi-Criteria Decision Making (MCDM)	**1**
▪ cellphone plan ▪ college ▪ used car	
Chapter 2: Optimize Product Mix: Profit Maximization with Linear Programming	**49**
▪ computer assembly ▪ skateboards ▪ sports shoes	
Chapter 3: Analyze Optimal Solutions—Sensitivity Analysis	**109**
▪ computer assembly ▪ skateboards ▪ sports shoes	
Chapter 4: Minimize Calories or Cost with Linear Programming	**161**
▪ Malawi diet ▪ Wisconsin watershed pollution ▪ gasoline blending	
Chapter 5: Optimize Effectiveness or Cost with Integer Programming	**205**
▪ political advertising ▪ workforce in pizza palace ▪ trucking oranges	
Chapter 6: Optimize Selection with Binary Programming	**253**
▪ flip houses ▪ apply to colleges ▪ form swim teams ▪ form project teams	
Chapter 7: Find Optimal Locations with Algorithms	**319**
▪ hot dog stand ▪ strawberry warehouse ▪ smoothie store ▪ disaster response facility	
Chapter 8: Waiting in Line with Polynomials – Non-linear functions and Queueing	**349**
▪ ticket box office ▪ post offices ▪ airport security screening	

Chapter 1: Make Hard Decisions—Multi-Criteria Decision Making (MCDM)

We all face decisions in our jobs, in our communities, and in our personal lives. For example,
- Where should a new airport, manufacturing plant, power plant, or health care clinic be located?
- Which college should I attend, or which job should I accept?
- Which car, house, computer, stereo, or health insurance plan should I buy?
- Which supplier or building contractor should I hire?

Multi-criteria decision making (MCDM) is used when one needs to make a hard decision with many criteria. The method introduced in this chapter is a structured methodology designed to handle the tradeoffs among multiple criteria. MCDM is a systematic approach to quantify an individual's preferences. Measures of interest are rescaled to numerical values on a 0–1 scale, with 0 representing the worst value of the measure and 1 representing the best. The decision maker assigns weights to each criterion to reflect the relative value of each criterion to the decision maker. This allows the direct comparison of many diverse measures. In other words, with the right tool, it really is possible to compare apples to oranges! The result of this analysis is a rank ordering of the alternatives that reflects the decision makers' preferences.

This chapter uses basic mathematics skills in an intellectually challenging environment. It was designed to provide all students with an appreciation that they can successfully apply their math skills to sophisticated decisions.

Chapter 2: Optimize Product Mix with Linear Programming (Maximization)

Chapter 2 is the introductory chapter for the optimization topics in chapters 2-7. While the succeeding chapters will extend this topic to minimization, integer programming, and binary programming, the skills in this maximization of linear programming (LP) chapter are the basic tools that will be used throughout. There is an introductory activity that involves using Lego to model the decision of making chairs and tables. This provides the students with a concrete understanding of decision variables and constraints. The problem context for this first substantive problem involves assembling two types of computers with different profit margins and labor requirements. Students are led through a graphical solution to a two decision variable problem involving two constraints. This is called the corner point principle. Lastly, the problem is expanded to include more decision variables, x_3, and x_4, to represent two additional configurations of computers. Once there are more than two decision variables, the problem cannot be solved graphically. In the next section of the text, students learn how to use SOLVER, a standard add-in to EXCEL, to solve mathematical programming problems. From this point on, Excel is a critical element of each chapter.

Chapter 3: Analyze Optimal Solutions—Sensitivity Analysis

This chapter revisits the three examples from chapter 2. In addition to solving problems, analysts are often interested in learning how sensitive their solutions are to changes in the parameters of the problem. Consider the computer assembly problem in the chapter 2. How sensitive is the solution to changes in the amount of profit that is made on each type of computer? What would be the effect of increasing the amount of available installation time or testing time? Questions such as these are part of what is called **sensitivity analysis**. This chapter is designed around the sensitivity analysis report that is part of Solver. Students learn how to explore and answer diverse what if questions.

Chapter 4: Minimize Calories or Cost with Linear Programming

This chapter presents three decision contexts in which the goal is to minimize the objective function. The first example was taken from a UN report and involves designing a nutritious diet for children in Malawi. The number of constraints and wide range of units of measures adds to the complexity of the problem. The second decision context motivates the need to use double subscripted variables to represent a pollution control decision in the Wisconsin watershed. The final decision involves more complex constraints that require students to manipulate equations to convert them to a standard form required for the Excel implementation of the model.

Chapter 5: Optimize Effectiveness or Cost with Integer Programming

In the previous chapters on linear programming, the decision variables were things that can be measured continuously such as production rates, grams of a food source, or 100-gallons of gasoline. However, in some contexts, the decision variable could be restricted. For example, a manager might need to know how many of each type of worker to hire. In this chapter, we will discuss how to solve mathematical programming problems in which the decision variables must be restricted to integer values. We will also discuss why such a restriction makes a difference in how the problem is solved and how its solution is analyzed. The chapter begins by investigating integer and non-integer solutions to linear equations in the context of purchasing advertisements to support a political campaign. The second example involves scheduling workers and supervisors at a fast food restaurant. The final decision context involves planning the shipment of truckloads of oranges to different markets in the Midwest. It illustrates the basic concepts of logistics planning. This chapter reinforces the core algebraic modeling skills of defining decision variables, framing the objective function and structuring the constraints.

Chapter 6: Optimize Selection with Binary Programming

Binary Integer Programming (BIP) problems have the same basic features as other mathematical programming problems: a set of decision variables, an objective function, a system of constraints. The distinguishing feature of BIP problems is that the possible values of the decision variables are limited to zero and one. These are called *binary decision variables*. Chapter 6 opens with a simple investment with just two decision variables and a single constraint. Students are asked to consider different values for the RHS of the constraint and investigate the implications of those changes on the optimal solution of the BIP problem. The first substantive example involves flipping houses. This example uses a complex spreadsheet to capture all of the information required to select which houses to buy and fix up.

Assignment problems are a special case of binary integer programming. Assignment problems involve matching a number of agents (e.g., athletes, students, machines) to a number of tasks (e.g., events, teams, jobs). This information is represented using matrices. If there are m agents and n tasks, then an $m \times n$ matrix of binary decision variables assigns agents to tasks. Assignment problems have a second $m \times n$ matrix, called a *cost matrix* that shows the "cost" (e.g., time, distance, monetary cost) associated with each agent performing each task.

Chapter 7: Find Optimal Locations with Algorithms

Amy Craig, Ph.D. - UNC Wilmington

This chapter focuses on different types of location problems. Location decisions arise in many contexts. All fast food companies, oil companies, drugstore chains or other retail outlets routinely evaluate locations for new facilities. Similar decisions are made in the public sector with regard to the location of libraries, fire stations, school buildings, and health care clinics. In some instances a simple measure of travel distance suffices to guide the decision. In other instances multiple criterion are used as in chapter 1 of this text.

The chapter starts with a simplified example involving two smoothie stands located along a single stretch of road. This example introduces the concept of minimizing the average distance traveled. Next we explore where to locate a small warehouse to store excess inventory for a downtown store. The third decision involves the location of a warehouse along a major interstate. Trucks from this warehouse are to make deliveries to different cities along the interstate. We then move to two dimensions as the decision involves locating a food stand in a downtown area. In some contexts, the preferred measure involves ensuring that all potential users of a service are within a fixed distance of the nearest facility. Any set of users that are within the prescribed distance are said to be *covered* by that facility. The final example utilizes binary integer programming to find the optimal locations for disaster response facilities to provide coverage of the region.

Chapter 8: Waiting in Line with Polynomials – Non-linear functions and Queues

There are relatively few broadly relevant applications of non-linear functions in high school mathematics curriculum. Mathematical models of waiting in line provide a rich array of decision contexts that utilize non-linear functions to calculate critical performance measures. Queues is the British word for waiting lines, and they are an ever present element of all societies. The chapter's first context explores how to reduce the average waiting time for people standing in line to buy tickets to a show. The second example assesses the impact of merging small rural post offices. The final example evaluates two alternative layouts of airport security screening stations. The nonlinear functions presented in this chapter include higher order polynomials as well complex exponential functions.

Applied Mathematics Practices for the 21st Century AMP21

The Common Core State Standards Initiative is the result of an effort at the state level coordinated by the National Governors Association Center for Best Practices (NGA Center) and the Council of Chief State School Officers (CCSSO). The most provocative aspect of the document is the Standards for Mathematical Practice. Six of the eight standards for Mathematical Practice require that students:

1. Make sense of problems and persevere in solving them.
2. Reason abstractly and quantitatively.
3. Construct viable arguments and critique the reasoning of others.
4. Model with mathematics.
5. Use appropriate tools strategically.
6. Look for and make use of structure.

Making sense of problems and solving them is the essential characteristic of this textbook. Mathematical modeling is typically the vehicle by which OR problems are conceptualized and explored. Moreover, OR practitioners use technological tools to obtain solutions to problems and perform sensitivity analyses. Finally, sensitivity analysis, the quintessential form of explaining and interpreting results, is a clear demonstration of understanding both the nature of the problem *and* its solution.

A New Way to Learn Mathematics

We created AMP21 to develop mathematical curriculum that directly address the above standards. On the pages that follow, you will be introduced to a wide range of real world problems that operations researchers solve every day. While real world problems can be difficult to solve due to their enormous size, in many cases, their solution depends on mathematics that you have already learned in your previous high school mathematics courses. So how, you may wonder, will this course be different?

This new course in the mathematics of operations research for high school students is different because it is applications-based and problem-driven. This means that the applications of the mathematics will be upfront, not at the end of the chapter and that the key ideas will be developed within the sorts of problems that gave them birth. The applications you will see focus around making decisions in business, government, or your own personal life. In addition, mathematics is not a spectator sport. Therefore, in the text, many questions are asked but not answered. Instead, you will provide the answers. Studying mathematics in this way may require you to develop a new mindset about what mathematics is, how it can be used, and the best way to learn it.

What is Operations Research and Business Analytics?

Operations research (OR) is a scientific way to analyze problems, make decisions, and improve processes. OR professionals try to provide a sound basis for decision-making. These decisions may focus on day-to-day operations that arise in a manufacturing plant. Or, they may involve long-range issues such as designing new environmental regulations or establishing minimum prison sentence guidelines.

Operations researchers attempt to understand the structure of complex situations. They develop mathematical and computer models of a system of people, machines, and procedures. If you have ever played the game Sim City, you have manipulated a computer model. Operations researchers often use numerical, algebraic, and statistical techniques to model the decision context. Then they manipulate their models to study the behavior of the system. They use this understanding to predict how the system will behave under different rules and policies to improve system performance.

Unlike most disciplines, we can point to specific events that mark the birth of operations research. OR was born in the years just prior to World War II. The British anticipated an air war with Germany. In 1937, they began to test radar. By 1938, they were studying how to use the information radar provided to direct the operations of their fighter planes.

Until this time, the word experiment usually meant a scientist carrying out a controlled experiment in a laboratory. In contrast, this radar-fighter plane project used a multidisciplinary team of scientists. They studied actual operating conditions in the field instead of in the laboratory. They then designed experiments in the field of operations, and the new term "operations research" was born. Their goal was to understand the operation of the complete system of equipment, people, and environmental conditions (e.g. weather, nighttime). Then they tried to improve the total system's performance. Their work was an important factor in winning the air war in Battle of Britain. OR eventually spread to all of the military services. Several of the leaders of this effort eventually won Nobel Prizes in their original fields of study.

All branches of the US Armed Forces during WWII formed similar groups of interdisciplinary scientists. These groups worked to protect naval convoys, search for enemy convoys, enhance anti-submarine warfare and improve the effectiveness of bombers. To do so, they collected data by directly observing operations. Then they built a mathematical model of the system. Next, they used the model to recommend improvements. Finally, they obtained feedback on the impact of the changes. Today, every branch of the military has its own operations research group. These OR groups include both military and civilian personnel. They play a key role in long-term strategy and weapons development. They also direct the operation of actions such as Operation Desert Storm. In addition, the National Security Agency has its own Center for Operations Research.

In the 1950s, national professional organizations were formed. These organizations published research journals and universities added OR departments. All of this raised operations research to the level of a profession. The leading professional organization is INFORMS (Institute for Operations Research and the Management Sciences). There are also operations research societies all across the globe. With regard to formal education, operations research became and remains one of the core competencies of the field of Industrial Engineering (IE). At the graduate level there is significant overlap between IE and OR. In business schools, OR generally falls within the domain of operations management or management science. Most mathematics departments also offer introductory OR courses at the junior or senior undergraduate level.

The use of OR expanded beyond the military to include other government organizations and private companies. The petroleum and chemical industries were early users of OR. They improved the performance of plants, developed natural resources and planned strategy. In the 1990s, OR models were critical enablers for multinational companies to become integrated global planners of facility operations and resource management.

In the last decade the field of operations research has gained added traction under the buzzword analytics. Numerous organizations have created analytics groups. A simple internet search yields: Google Analytics, Twitter Analytics, Pinterest Analytics, IBM Analytics, etc. Each spring INFORMS offers a Business Analytics Conference as well as special analytics conferences in areas such as healthcare.

Today, operations research and business analytics play important roles in industry and government as in:

- Airline, hospitality and entertainment industry – scheduling planes cruise ships, managing the capital investment, pricing tickets, taking reservations
- Pharmaceutical industry – managing research and development and designing sales territories;
- Delivery services – planning routes and developing pricing strategies
- Financial services – credit scoring, marketing, and internal operations
- Internet and marketing – managing the traffic to websites around the globe, tracking customers to target marketing programs
- Healthcare – hospital and healthcare clinic management, control of epidemics, effectiveness of procedures and causes of
- Commodities and lumber industry – managing mining, growing forests and cutting timber
- Local government – deploying emergency services
- Policy studies and regulation – environmental pollution, air traffic safety, AIDS, and criminal justice policy.

CHAPTER 1:

Make Hard Decisions—Multi-Criteria Decision Making (MCDM)

Section 1.0: Introduction to Making Hard Decisions

We all face decisions in our jobs, in our communities, and in our personal lives. For example,
- Where should a new airport, manufacturing plant, power plant, or health care clinic be located?
- Which college should I attend, or which job should I accept?
- Which car, house, computer, stereo, or health insurance plan should I buy?
- Which supplier or building contractor should I hire?

Decisions such as these involve comparing alternatives that have strengths or weaknesses with regard to multiple objectives of interest to the decision. For example, your criteria in buying health insurance might be to minimize cost *and* maximize protection. Sometimes these multiple criteria get in each other's way.

Multi-criteria decision making (MCDM) is used when one needs to make a hard decision with many criteria. In this chapter, you will see one form of multi-criteria decision making. The method introduced in this chapter is a structured methodology designed to handle the tradeoffs among multiple criteria.

A Little History

One of the first applications of this method of MCDM involved the study of possible locations for a new airport in Mexico City in the early 1970s. The criteria considered included cost, capacity, access time to the airport, safety, social disruption, and noise pollution.

The problems in this chapter use the steps of multi-criteria decision making to make hard decisions. MCDM is a systematic approach to quantify an individual's preferences. Measures of interest are rescaled to numerical values on a 0–1 scale, with 0 representing the worst value of the measure and 1 representing the best. This allows the direct comparison of many diverse measures. In other words, with the right tool, it really is possible to compare apples to oranges! The result of this process is an evaluation of the alternatives in a rank order that reflects the decision makers' preferences.

For example, individuals, college sports teams, Master's degree programs, or even hospitals can be ranked in terms of their performance on many diverse measures. Another example is the Bowl Championship Series (BCS) in college football that attempts to identify the two best college football teams in the United States to play in a national championship bowl game. This process has reduced, but not eliminated, the annual end-of-year arguments as to which college should be crowned national champion.

Section 1.1: Choosing a Wireless Plan

Choosing a wireless plan is an important decision for many people. In fact, most teenagers own smart phones. When choosing a wireless plan, there are many factors to consider.

1. What factors would you consider if you were choosing a wireless plan?

In this chapter, you will develop a process for making important decisions, such as choosing a wireless plan, with many competing features. Before doing so, you will complete an opening activity.

1.1.1 Opening Activity

In this activity, you will make a decision about what wireless plan you would choose if you were considering a new plan. To do so, complete the following steps:

1. Make a list of possible wireless plans that you would consider using.

2. Collect data on each of these plans that you would find useful in making a decision.

3. Choose one of the plans based on your data.

4. Explain why you chose this plan over the others.

2. What possible issues do you foresee with using these steps to choose a wireless plan?

In the following sections, the steps of the MCDM process will be explained in the context of a high school student and her friend helping her parents to choose a wireless plan. Isabelle Nueva needs to help her mother and father decide on the best wireless plan to buy for their family. She and her friend, Angelo Franco, will use the MCDM process they learned in their math class to help her parents make this decision. Follow along with Isabelle as she and Angelo use the MCDM process to make this decision.

1.1.2 Identify Criteria and Measures

The first thing they do is identify the **criteria** of a wireless plan that were important to Isabelle's family. From discussions she had with her mother and father, Isabelle knew that the criteria that were important to them were cost, contract features, and phone service.

3. If you were choosing a wireless plan, what criteria would be important to you?

Isabelle and Angelo know that they need to find at least one way to **measure** each of the criteria. They decide to measure the Cost criterion using Monthly Charge, Monthly Access Fee, and Overage Fee. They decide to measure the Contract Features criterion using Number of GB of Data per Month, Rollover Data, and Contract Length. The measure of Phone Service is defined as Quality of Service. Each criterion and its measures are provided in Table 1.1.1.

Criteria	Measures
Cost	Monthly Charge
	Access Fee per Line
	Overage Fee ($/GB)
Contract Features	Data Plan
	Rollover Data
	Contract Length
Phone Service	Quality of Service

Table 1.1.1: Criteria and measures for choosing a wireless plan

4. How would you measure each of your criteria?

The value of three measures—the Monthly Charge, Overage Fee, and Access Fee per Line—could be any numerical amount within a reasonable range. These are examples of **continuous measures**. That is, these measures can take on any numerical value within a range.

Isabelle and Angelo decide that the data they collected for the other three measures can be grouped into a finite number of categories. All of the plans they looked at before focusing on just three plans, had values that were multiples of 2.5 GB. They ranged from a low of 7.5 GB to a high of 15 GB. Thus, they decided to treat this as a **categorical measure** with only four possible values for Data Plan: 7.5, 10, 12.5, and 15 GB.

To obtain data on Quality of Service they decide to use ratings from a consumer magazine. The magazine considered dropped or disconnected calls, static and interference, and voice distortion to rate the quality of service. Isabelle and Angelo decide to only consider plans the magazine rated "Good", "Very Good", or "Excellent". Therefore, this measure has three categories.

Another categorical measure is Contract Length—the shortest time a customer must remain with a particular plan to avoid paying a fee to cancel the service. Isabelle's parents were concerned about being locked into a plan for a long period of time. The plans under consideration have only three different Contract Lengths (0, 1 year, and 2 years). All plans they investigate seem to use one of these. Thus, the Contract Length measure has three possible values. The categorical measures and their possible values are provided in Table 1.1.2.

5. Of the measures you listed in question 4, which are continuous and which should be treated as categorical?

6. Create a table similar to Table 1.1.2 for your categorical measures identified in the previous question. In order to do this, you will need to research possible wireless plans. What sort of research would you need to do?

Categorical Measure	Categorical Values (from best to worst)
Data Plan	15 GB 12.5 GB 10 GB 7.5 GB
Rollover Data	Yes No
Quality of Service	Excellent Very Good Good
Contract Length	0 1 year 2 years

Table 1.1.2: Categorical variables with categories and numeric values

1.1.3 Collect Data

Isabelle's parents were considering three wireless plans: Trot, UST&T, and Horizon. Isabelle and Angelo collected the data they need to help her parents make their decision. The first data they collected were the basic monthly fees that appear in Table 1.1.3.

Angelo and Isabelle discussed the impact of the monthly access fee on the family's cost. The Nueva family planned to initially sign up for four lines, one each for the parents and their two older teenagers. Angelo suggested that instead of two measures, these data should be combined into one measure, Total Monthly Charge. This is calculated by multiplying the per line fee by the number of lines and adding it to the base monthly fee. With this calculation, the monthly fee would be $180 for Trot. The monthly fee for UST&T would be $250. Lastly, Horizon would cost $220 per month. However, Isabelle raised the possibility that her youngest brother who is in middle school might be given a fifth line. However, after some thought, they both agreed the cost of a fifth line should not be included in the decision analysis for now.

	Plan		
	Trot	UST&T	Horizon
Base Monthly Charge ($)	100	130	20
Monthly Access Fee ($/line)	20	30	50
Total Monthly Charge for Four Lines ($)	180	250	220

Table 1.1.3: Wireless plans monthly cost

The other data they collected about the various plans are included in Table 1.1.4 alongside the monthly cost of four lines.

Plan	Trot	UST&T	Horizon
Total Monthly Charge ($)	180	250	220
Overage Fee ($/GB)	50	15	10
Contract Length	2 years	1 years	None
Data Plan (GB/month)	10	15	7.5
Rollover Data	No	Yes	No
Quality of Service	Excellent	Very Good	Good

Table 1.1.4: Isabelle and Angelo's wireless plan data

7. Create a table similar to Table 1.1.4 for your wireless plan data.

1.1.4 Find the Range of Each Measure

Next, Isabelle and Angelo specify a range for each measure. They first specify the range for the two continuous measures (Total Monthly Charge and Overage Fee per GB). For each of these measures, they decide to use the range of the actual data they collected. That is, for Total Monthly Charge, the range was $180 to $250. The range Overage Fee was $10 to $50. For each of the categorical measures, Isabelle and Angelo simply list the two extreme values for each category. The **scale ranges** for each of Isabelle and Angelo's measures are given in Table 1.1.5.

Measure	Scale range
Total Monthly Charge	$180 to $250
Overage Fee per GB	$10 to $50
Contract Length	0 to 2 years
Data Plan	7.5 GB to 15 GB
Rollover Data	Yes or No
Quality of Service	Good to Excellent

Table 1.1.5: Ranges of each measure

8. Specify the ranges for each of your measures, and create a table similar to Table 1.1.5.

1.1.5 Rescale Data on All Continuous Measures to a Common Unit

It would be difficult to compare the three plans using these raw data. For example, how would one compare a $10 difference in the monthly service charge to a one-year difference in minimum contract length? In order to avoid such problems, operations researchers rescale the raw data of each measure to common unit values between zero and one. This creates a **common unit** that varies from zero to one for each measure. Zero always represents the worst value and one the best value for each measure.

For both of the continuous measures, Isabelle and Angelo use a **proportional scale** to assign a score to intermediate values. For example, the range for the Total Monthly Charge measure is $180 to $250. The smallest possible value here is the best option. Since the value one represents the best option, $180 is converted to a common unit value of one. Similarly, the largest possible value of the monthly service charge is the worst option. Thus, $250 converted to zero. That is,

$180 \rightarrow 1$
$250 \rightarrow 0$

Next, Isabelle and Angelo convert the price of Horizon's plan to a common unit value. They must decide what $220 should be converted to when it is compared to the best and worst values for Total Monthly Charge. The graph in Figure 1.1.1 illustrates this.

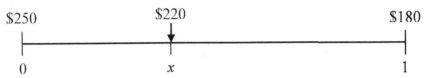

Figure 1.1.1: Determining the common unit values for the Total Monthly Charge measure

9. What do you think $220 should be converted to?

10. Is $220 closer to the best or the worst option?

11. How far is $220 from the best option? How far from the worst?

Isabelle and Angelo solve a proportion to arrive at the common unit value for the Total Monthly Charge of $220. To find the common unit value for $220 using proportions, Isabelle and Angelo write two equivalent fractions of the form $\frac{part}{whole}$. Figure 1.1.2 illustrates this.

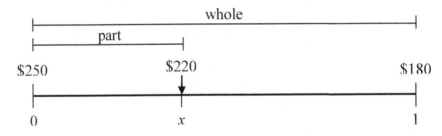

Figure 1.1.2: Determining the proportion to find the common unit values

In the first fraction, the "part" refers to the distance between $250 and $220, and the "whole" refers to the distance between $250 and $180. In the second fraction, the "part" refers to the distance between 0 and x, and the "whole" refers to the distance between 0 and 1. As can be seen in Figure 1.1.2, these two fractions are equivalent.

Isabelle and Angelo solve for the unknown in the equivalent fractions, using absolute value to find the distance between two values.

$$\frac{|220-250|}{|180-250|} = \frac{|x-0|}{|1-0|}$$

$$\frac{30}{70} = \frac{x}{1}$$

$$0.42 = x$$

Therefore, the raw value $220 is converted to the common unit value 0.42.

Notice, each time these equivalent fractions are developed, the fraction on the right will always be:

$$\frac{|x-0|}{|1-0|} = \frac{x}{1} = x$$

Therefore, there is no need to write the entire fraction. Simply x can be used instead.

12. What other ways could Isabelle and Angelo use to calculate the common unit value for $220?

13. Find the common unit values for the Overage Fee per GB measure.

1.1.6 Rescale Each Categorical Measure to a Common Unit

For the four categorical measures, Isabelle and Angelo assign a common unit value of zero to the worst option and one to the best option. For the Rollover Data measure, the only possible values are "yes" and "no". Yes was assigned a one, because is preferable; and no was assigned zero, because it is the worse value. When there was something between the best and worst values, Isabelle and Angelo discussed what to assign the intermediate values. With regard to Quality of Service, they simply assigned the one intermediate value, "very good," a score of 0.5. They used analogous reasoning for the two intermediate values of the Data Plan measure. They assigned common units proportionately: 10 GB was 0.33 and 12.5 was 0.67. However, they knew that Isabelle's parents really liked the idea of not being tied into a contract. A one-year contract was not much better than a two-year contract. They therefore assigned just 0.25 to a one-year contract. These conversions are summarized in Table 1.1.6.

Categorical Measure	Categorical Values	Common Units
Contract Length	0	1
	1 year	0.25
	2 years	0
Data Plan	15 GB	1
	12.5 GB	0.67
	10 GB	0.33
	7.5 GB	0
Rollover Data	Yes	1
	No	0
Quality of Service	Excellent	1
	Very Good	0.5
	Good	0

Table 1.1.6: Common unit values for the categorical measures

Isabelle and Angelo use the relationships developed above to convert the data for each plan into values between zero and one. The results of this conversion are presented in Table 1.1.7.

Plan	Trot	UST&T	Horizon
Total Monthly Charge ($)	1	0	0.43
Overage Fee ($/GB)	0	0.88	1
Contract Length	0	0.25	1
Data Plan (GB/month)	0.33	1	0
Rollover Data	0	1	0
Quality of Service	1	0.5	0
Total Points	**2.33**	**3.63**	**2.43**
Average Points	**0.39**	**0.61**	**0.41**

Table 1.1.7: Wireless plan data converted to a common unit

When Isabelle and Angelo looked at these results, they noticed that each plan received the top common unit value of one on two of the measures. They also noticed that each plan received at least one common unit value of zero. Therefore, it is not obvious to them which plan they should choose.

14. Based on the common unit values, which plan do you think Isabelle should recommend to her parents?

Angelo thinks they should use the total of all of the common units to get a total score for each plan. The totals are also listed in Table 1.1.7. Isabelle thinks it will be more meaningful to compute the average common unit scores for each plan. To do so, she divided the total score for each plan by six (the total number of measures and therefore the highest possible score). The averages she obtained are given in the bottom row of Table 1.1.7.

15. Do you think it makes more sense to use the sum or the average to make a decision?

16. Based on the total and average scores, which plan do you think Isabelle should recommend to her parents? Why?

17. What are some reasons why Isabelle may not recommend Horizon to her parents?

18. What are some reasons why Isabelle may think Trot would be a better choice for her parents?

19. What are some reasons why Isabelle may think UST&T would be a better choice for her parents?

20. Calculate the total scores and the average scores for each of your wireless plans.
    a. Based on these values, which plan would you choose?
    b. What are some reasons why these plans may not be the best choice for you?
    c. Was this plan what you expected to choose based on the opening activity? Why or why not?

Whether they use the sum or the average, Isabelle and Angelo realize that each plan has something in its favor. They wonder how to reach a decision. Then Isabelle remembers that her parents were really worried about the Total Monthly Charge, and not as worried about Contract Length. They decide that they need a system that does not treat all of the measures as equally important, as the sum and average do. They need a system that weights each measure according to how important it is to Isabelle's parents.

1.1.6 Conduct an Interview to Calculate Weights

In order to learn how important each measure is to her parents, Isabelle and Angelo decide to interview them. They want to learn which measure Isabelle's parents believe is most important to them. To do so, the parents will need to look closely at the most preferred value and least preferred value for each measure. Angelo and Isabelle explore with her parents how Isabelle's parents would rank order the six measure ranges. Mr. and Mrs. Nueva decide that the difference between the highest and lowest monthly payments was most important to them. The difference between lowest and highest is $70 per month; this is substantial. Therefore, they rank the Total Monthly Charge measure number one.

They knew their teenagers wanted to use their smart phones to download large files. They, therefore, rank Data Plan as the second most important measure. The Nueva's rated the Quality of Service as the third most important measure. They might have ranked it higher if the scale included poor service. However, since the minimum was "good," they were comfortable ranking it third most important. They really liked that Horizon offered a plan with no contract and therefore listed Contract Length as fourth. They were confident their children would strive to live within the monthly GB of data budget. However, they feared every once and a while they would lose track. In that case they could be shocked with a huge overage fee; they ranked Overage Fee fifth. They assumed their children would rarely have GBs of data to rollover into the next month. This measure was ranked last.

Table 1.1.8 shows their rank-ordering of the measures. For example, Total Monthly Charge is the most important measure to Isabelle's parents and Rollover Data is the least important. This table also includes the least and the most preferred values for each measure.

Measure	Least Preferred Value	Most Preferred Value	Rank
Total Monthly Charge ($)	250	180	1
Overage Fee ($/GB)	50	10	5
Contract Length	2 years	0	4
Data Plan (GB/month)	7.5	15	2
Rollover Data	No	Yes	6
Quality of Service	Good	Excellent	3

Table 1.1.8: Rank-order of the measures according to Isabelle's parents

21. Rank-order each of your measures.

Next, Isabelle and Angelo ask her parents to assign points to each measure to better capture the magnitude of the differences between two rankings. To make their decision-making model even more useful, they want a sense of how much more important one measure is than another. For example, if one measure is twice as important as another, then the assigned points should be twice as much for the higher ranked measure.

Isabelle and Angelo ask Mr. and Mrs. Nueva to assign 100 points to Total Monthly Charge, the measure they ranked number one. Then, they ask them to assign a number of points less than 100 to the second-ranked measure, Data Plan. In doing so, they ask Isabelle's parents to pick a number that reflects how important Data Plan is compared to the Total Monthly Charge.

Mr. and Mrs. Nueva decide to assign 90 points to Family Data, because they know their children like to download large files. It is almost as important as the Total Monthly Charge. Quality of service was also important to them and only slightly less important than Data Plan. This was given 80 points. Although they liked not having a contract, it really was far less important than the first three measures. They assigned it 40 points, or half the weight of Quality of Service. The high overage fee was a risk they thought they could manage and gave it only 20 points. They did not think there was much value to their family of Rollover Data. They assigned it only 10 points. The Nueva's preferences are summarized in Table 1.1.9.

Measure	Least Preferred Value	Most Preferred Value	Rank	Points
Total Monthly Charge ($)	250	180	1	100
Overage Fee ($/GB)	50	10	5	20
Contract Length	2 years	0	4	40
Data Plan (GB/month)	7.5	15	2	90
Rollover Data	No	Yes	6	10
Quality of Service	Good	Excellent	3	80

Table 1.1.9: Points assigned to each of the measures

22. Assign points to each of your measures, and create a table similar to Table 1.1.9.

Now, Isabelle and Angelo total all of the assigned points and obtain 340. Then, they divide the point assignment for each measure by that total. This number is the **weight** of that measure. For example, monthly charge was assigned 100 points. Thus, the weight of this measure is:

$$\frac{100}{340} = 0.29$$

One way of interpreting the weight of 0.29 for Total Monthly Charge is that 29% of the final decision will be based on this measure. The results of Isabelle and Angelo's interview of her parents are summarized in Table 1.1.10.

Measure	Least Preferred Value	Most Preferred Value	Rank	Points	Weight
Total Monthly Charge ($)	250	180	1	100	0.29
Overage Fee ($/GB)	50	10	5	20	0.06

Contract Length	2 years	0	4	40	0.12
Family Data (GB/month)	7.5	15	2	90	0.26
Rollover Data	No	Yes	6	10	0.03
Quality of Service	Good	Excellent	3	80	0.24

Table 1.1.10: Calculated weight for each measure

23. What measure has the largest weight? Which has the smallest?

24. What is the ratio of the largest weight to the smallest weight?

25. What should this ratio mean in the context of the decision?

26. Assign points to each of your measures, and create a table similar to Table 1.1.10.

1.1.7 Calculate Total Scores

Now, Isabelle and Angelo calculate a **total score** for each plan. The total score is an example of a **weighted average**. They multiply each common unit value from Table 1.1.7 by the corresponding weight from Table 1.1.10. Then for each plan, they sum those six products together to get the total score. The data from these two tables are placed side-by-side in Table 1.1.11. The results of these computations are given in Table 1.1.12. Notice that this weighted average captures how important the various measures are to Isabelle's parents.

Measure	Weight	Trot	UST&T	Horizon
Total Monthly Charge ($)	0.29	1	0	0.43
Overage Fee ($/GB)	0.06	0	0.88	1
Contract Length	0.12	0	0.25	1
Family Data (GB/month)	0.26	0.33	1	0
Rollover Data	0.03	0	1	0
Quality of Service	0.24	1	0.5	0

Table 1.1.11: Measure weights and wireless plan scores

Measure	Weight	Trot	UST&T	Horizon
Total Monthly Charge ($)	0.29	$1 \times 0.29 = 0.29$	$0 \times 0.29 = 0$	$0.43 \times 0.29 = 0.13$
Overage Fee ($/GB)	0.06	$0 \times 0.06 = 0$	$0.88 \times 0.06 = 0.05$	$1 \times 0.06 = 0.06$
Contract Length	0.12	$0 \times 0.12 = 0$	$0.25 \times 0.12 = 0.03$	$1 \times 0.12 = 0.12$
Family Data (GB/month)	0.26	$0.33 \times 0.26 = 0.09$	$1 \times 0.26 = 0.26$	$0 \times 0.26 = 0$
Rollover Data	0.03	$0 \times 0.03 = 0$	$1 \times 0.03 = 0.03$	$0 \times 0.03 = 0$
Quality of Service	0.24	$1 \times 0.24 = 0.24$	$0.5 \times 0.24 = 0.12$	$0 \times 0.24 = 0$
Wireless Plan's Total Score		**0.62**	**0.49**	**0.31**

Table 1.1.12: A weighted total score is computed for each plan.

27. Multiply the common unit values by the corresponding weights for each of your plans, and create a table similar to Table 1.1.12.

28. Would everyone's score results lead to the same preferred choice? Explain.

1.1.8 Determine Strengths/Weaknesses and Make Final Decision

Trot is clearly the preferred plan. UST&T is a distant second. Isabelle and Angelo decide to closely examine the results. They clearly do not produce the same results as the sum or average methods.

29. For which measures does Trot have a higher weighted score than UST&T? For which does UST&T outscore Trot?

When Isabelle and Angelo compare Trot with UST&T, they see that Trot had higher weighted scores for the first- and third-ranked measures, Total Monthly Cost and Quality of Service. UST&T scored higher on the other four measures. However, the magnitude of the difference for measures ranked four, five, and six was always small. In each case the difference was only 0.03 or less. These could not overcome the advantage Trot had on Total Monthly Cost, the highest ranked measure. Their weighting system did what it was supposed to do; it took into account Mr. and Mrs. Nueva's preferences. They decide to recommend the Trot plan to Isabelle's parents.

1.1.9 Alternative Trot Plan

Trot recently announced an alternative that comes with a larger Data Plan. This plan comes with 12.5 GB of data per month. It also costs $15 a month more. The two plans are compared in Table 1.1.13.

Plan	Old Trot	New Trot
Total Monthly Charge ($)	180	195
Overage Fee ($/GB)	50	50
Contract Length	2 years	2 years
Data Plan (GB/month)	10	12.5
Rollover Data	No	No
Quality of Service	Excellent	Excellent

Table 1.1.13: Alternative Trot Plan

To compare the two plans, Isabelle and Angelo must first convert the new values to common units between zero and one. Then they will need to multiply the values by their corresponding weights.

30. What is the common unit value for the monthly charge of $195?

31. What is the common unit value for the Data Plan, 12.5 GB per month?

32. Should Isabelle and Angelo recommend to Mr. and Mrs. Nueva that they adopt the new Trot plan?

1.1.10 Summary

In this problem, Isabelle and Angelo wanted to help Isabelle's parents choose a wireless plan. They completed the following steps:

1. Identify criteria and measures
2. Collect data
3. Find the range of each measure
4. Rescale each measure to a common unit

After completing these steps, Isabelle and Angelo found the total score and the average score for each wireless plan. However, they noticed that these values treated all measures under consideration as being equally important. This was not a reasonable way to make a decision. They needed a way to weight some measures more than others, because Isabelle's parents were more concerned about the cost of the plan than anything else.

In order to take Mr. and Mrs. Nueva's preferences regarding a wireless plan into account, Isabelle and Angelo completed four additional steps:

5. Conduct an interview to rank order measures and assign points

6. Calculate the weight of each measure
7. Calculate a total score for each alternative
8. Interpret results

This eight-step process will be applied in the next two sections and in the homework problems to make slightly more complicated decisions. This process is also a life skill, because you may find it useful to help you make important decisions in your future.

Section 1.2: Enrique Ramirez Chooses a College

Enrique Ramirez lives in Brooklyn, a borough of New York City. He's a senior in high school and has been accepted at four colleges: Canisius College in Buffalo, NY; Clark University in Worcester, MA; Drexel University in Philadelphia, PA; and Suffolk University in Boston, MA. Now he must decide which one to attend.

Enrique asks his friend Anna for help. Enrique and Anna realize that there are many different issues to consider when making this decision. They also realize that the issues of interest to Enrique and their relative importance are not the same as those for Anna.

To make this decision, Enrique, with the help of Anna, follows the steps of MCDM that were presented in the previous section. These steps are given in Table 1.2.1.

General Steps	Descriptions for this Particular Decision
1. Identify Criteria and Measures	First, generate a list containing general criteria that are important when choosing a college. These criteria will be broad in nature and will be based on objective and subjective goals. Next, specify at least one measure for each criterion.
2. Collect Data	For each college, collect the data for each measure.
3. Find the Range of Each Measure	Specify a reasonable scale for each measure.
4. Rescale Each Measure to a Common Unit	Rescale each measure to common units from 0 to 1, with 0 being the worst alternative and 1 being the best alternative.
5. Conduct an Interview to Calculate Weights	With the help of an interviewer, rank-order the measures, assign points from 0 to 100 to each measure, and calculate a proportional weight between 0 and 1 for each measure.
6. Calculate Total Scores	Calculate a total weighted score for each college. These weights will yield a ranking of the colleges, allowing you to identify the best option based on your preferences.
7. Interpret Results	Review the results to understand the strengths and weaknesses of your top alternatives before finalizing your decision.

Table 1.2.1: Steps of MCDM

1.2.1 Identify Criteria and Measures

With Anna's help, Enrique decides that academics, cost, location, and social life are the criteria most critical in his choice of a school. Next, Enrique and Anna take his list of four criteria and specify two or three measures for each criterion. His criteria and measures are given in Table 1.2.2.

Criteria	Measures
Academics	Average SAT Score (based on last year's freshman class)
	U.S. News & World Report Ranking
Cost	Room & Board (annual)
	Tuition (annual)
Location	Average Daily High Temperature
	Nearness to Home
Social Life	Athletics
	Reputation
	Size

Table 1.2.2: Enrique's criteria and measures

1.2.2 Collect Data

For each measure, Enrique and Anna collect data, which is listed in Table 1.2.3. Some of the measures are naturally categorical. For example, *U.S. News & World Report* ranks schools into four categories:
   1. Nationally ranked
   2. Regionally ranked
   3. Regionally tier 3
   4. Regionally tier 4

Enrique and Anna divide Athletics into three categories:
   1. Division 1
   2. Division 2
   3. Division 3

Similarly, they divide Reputation into three categories:
   1. Seriously academic
   2. Balanced academics and social life
   3. Party school

The data for the remaining measures are numerical values. Enrique and Anna are able to find the average values for SAT score and daily high temperature and the exact values for room and board cost, tuition cost, nearness to home, and size.

Measure	Canisius	Clark	Drexel	Suffolk
Average SAT Score	1590	1750	1700	1480
U.S. News & World Report Ranking	22nd (regional)	91st (national)	109th (national)	Tier 3 (regional)
Room & Board	$10,150	$8,850	$12,135	$11,960
Tuition	$28,157	$33,900	$30,470	$25,850
Average Daily High Temperature	56°	56°	64°	59°
Nearness to Home	297 mi	157 mi	81 mi	191 mi
Athletics	Division 1	Division 3	Division 1	Division 3
Reputation	Balanced	Seriously academic	Seriously academic	Balanced
Size	3,300 students	2,175 students	12,348 students	4,985 students

Table 1.2.3: Raw data for Enrique's four schools

1.2.3 Find the Range of Each Measure

Next, Enrique and Anna choose an appropriate scale for each of the nine measures. Some of the measures are continuous (e.g., SAT score), while others are categorical (e.g., athletics).

For the Nearness to Home measure, Enrique believes that exact mileage is not important, but rather broad ranges of mileage better represent his concerns. Therefore, Enrique and Anna convert this measure from continuous to categorical.

1. Looking at Table 1.2.4, what other measure was converted from continuous to categorical?

Enrique and Anna also realize that the range of each scale is important. For example, the theoretical range of the average combined SAT score is 600–2400, but in actuality, the range of the average combined SAT score at the colleges Enrique is considering is 1480–1750, which is a much narrower range. Enrique and Anna decide that it is much more realistic to use a range that is close to the actual range.

2. In the previous section, the ranges for the continuous measures were simply the ranges of the data collected. In this section, the ranges are expanded slightly. For example, instead of the SAT range staying as 1480-1750, Enrique and Anna choose the range 1400-1800. Why might one prefer to use the ranges of the data collected? Why might one prefer to round the ranges?

3. Looking at Table 1.2.4, for what other measures do Enrique and Anna create realistic ranges? Do you agree with their ranges? Why or why not?

The type and scale range of each measure are given in Table 1.2.4.

Measure	Type	Scale range
Average SAT Score	Continuous	1400–1800
U.S. News & World Report Ranking	Categorical	1: Nationally ranked 2: Regionally ranked 3: Regionally tier 3 4: Regionally tier 4
Room & Board	Continuous	$8,000–$14,000
Tuition	Continuous	$25,000–$35,000
Average Daily High Temperature	Continuous	50°–70°F
Nearness to Home	Categorical	1: 1 hr to 2 hr drive (50-100 mi) 2: Within 4 hr drive (101–200 mi) 3: Within a day's drive (201–300 mi)
Athletics	Categorical	1: Division 1 2: Division 2 3: Division 3
Reputation	Categorical	1: Seriously academic 2: Balanced academics and social life 3: Party school
Size	Categorical	1: Under 3,000 students 2: 3,001–6,000 students 3: 6,001–12,000 students 4: Over 12,000 students

Table 1.2.4: Types and ranges of measures

Before continuing, Enrique and Anna convert the values of the categorical measures into the numerical values based on the ranges of each measure. The converted data for the categorical measures are given in Table 1.2.5.

Measure	Canisius	Clark	Drexel	Suffolk
Average SAT Score	1590	1750	1700	1480
U.S. News & World Report Ranking	2	1	1	3
Room & Board	$10,150	$8,850	$12,135	$11,960
Tuition	$28,157	$33,900	$30,470	$25,850
Average Daily High Temperature	56°	56°	64°	59°
Nearness to Home	3	2	1	2
Athletics	1	3	1	3
Reputation	2	1	1	2
Size	2	1	4	2

Table 1.2.5: Converted categorical data for Enrique's four schools

1.2.4 Rescale Each Measure to a Common Unit

Once Enrique and Anna choose appropriate scales for each of the measures, Anna reminds Enrique that if they compared the data in its current form, it would be like comparing apples to oranges. They decide to convert the data to common units. To do so, they assign 1 to the best value and 0 to the worst value in the range of each measure. Recall from the previous section, the method for determining intermediate values differs for continuous and categorical measures.

Converting Continuous Measures

For the continuous measures, Enrique and Anna use a proportional scale. For example, the Average SAT Score at Canisius is 1590. The range for this measure is 1400–1800, so 1400 (the least desirable score) should be converted to 0 and 1800 (the most desirable score) to 1. But what should the proportional value for Canisius be?

$$1800 \rightarrow 1$$
$$1590 \rightarrow x$$
$$1400 \rightarrow 0$$

Figure 1.2.1 illustrates this example.

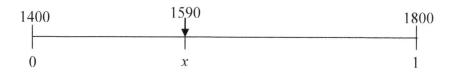

Figure 1.2.1: Determining the common unit values for the SAT scores measure

Using the same method as in the previous section, Enrique and Anna solve a proportion to find x.

$$\frac{|1590-1400|}{|1800-1400|} = \frac{|x-0|}{|1-0|}$$

$$\frac{190}{400} = \frac{x}{1}$$

$$x = 0.475$$

Therefore, an average SAT score of 1590 is converted to a common unit value of 0.475. Enrique and Anna decide to use proportional common units for each of the measures that have a continuous scale.

Converting Categorical Measures: Proportional or Non-proportional Scale

For the categorical measures, Enrique and Anna begin by assigning the best value a 1 and the worst value a 0. Then, Enrique and Anna decide how to apportion the common units. In some cases, apportionment is proportional, while in other cases it is not. They decide to use proportional common units for Nearness to Home, Athletics, and Reputation.

On the other hand, Enrique feels that some categorical measures should not be apportioned proportionately. For example, Enrique and Anna decided that there is a big difference between being ranked nationally and regionally on the *U.S. News & World Report* Ranking measure. Thus, they decide to have the following common units:

Nationally Ranked → 1
Regionally Ranked → 0.5
Tier 3 → 0.25
Tier 4 → 0

Enrique prefers a smaller school. Therefore, he assigns the following common units:

Under 3,000 students → 1
3,000-6,000 students → 0.75
6,001-12,000 students → 0.25
Over 12,000 students → 0

Table 1.2.6 contains the results of Enrique and Anna's rescaling of each measure to common units.

Measure	Canisius	Clark	Drexel	Suffolk
Average SAT Score	0.475	0.875	0.750	0.200
U.S. News & World Report Ranking	0.50	1	1	0.25
Room & Board	0.642	0.858	0.311	0.340
Tuition	0.684	0.110	0.453	0.915
Average Daily High Temperature	0.30	0.30	0.70	0.45
Nearness to Home	1	0.5	0	0.5
Athletics	1	0	1	0
Reputation	1	0.5	0.5	1
Size	0.75	1	0	0.75

Table 1.2.6: Each measure rescaled to common units

4. From Table 1.2.3, Clark University has the highest average combined SAT score and the highest tuition. Why does it make sense in Table 1.2.6 that Clark has the highest common unit value on one of those measures, but the lowest common unit value on the other?

5. Looking at the Nearness to Home measure in Table 1.2.6, what was the most desirable distance to Enrique? What was least desirable?

6. Looking at the Athletics measure in Table 1.2.6, what was the most desirable division to Enrique? What was least desirable?

7. Looking at the Reputation measure in Table 1.2.6, what was the most desirable reputation to Enrique? What was least desirable?

To review, there are essentially three steps to rescale data to common units.

Step 1: Assign 1 to the best value in the range.
Assign 0 to the worst value in the range.

Step 2: For continuous data, assign intermediate scores proportionally:

$$\text{Scaled Score} = \frac{|\text{score} - \text{least preferred score}|}{|\text{most preferred score} - \text{least preferred score}|}$$

Step 3: For categorical data, assign intermediate scores proportionally or based on your own opinions and values.

1.2.5 Conduct an Interview to Calculate Weights

Next, Enrique and Anna assign weights to each of the measures to reflect the relative importance Enrique attaches to each of them. They decide Anna will interview Enrique. She makes observations to ensure that Enrique understands the measures he chose and the effects of the weights he assigns to each of them. As a reference tool during the interview, they create Table 1.2.7.

Anna: We have some measures and their ranges for making a decision about your college preference. Focus first on the column of least preferred values. Which one of the measures would you most want to increase from the least preferred value to its most preferred value? For example, is it more important to you to move the SAT score from 1400 to 1800 or to reduce tuition from $35,000 to $25,000?

Enrique: Lower the tuition!

Anna: Are you sure that lowering the tuition to $25,000 is the most important improvement in the whole list?

Enrique: Yes, so I think we should rank tuition number one.

Anna: Enrique, what would be the next most important measure to move from least preferred to most preferred?

Enrique: *U.S. News & World Report* ranking is important, so let's rank that second, and SAT score third.

They continue like this until each measure has been ranked, as shown in Table 1.2.7.

The next task is to subjectively assign points from 0 to 100 for each measure based on the rank order. The points assigned reflected the relative importance Enrique places on each measure. They continue the interview to assign these points. During this interview process, Anna encourages Enrique to think about the relative importance of moving between the best and worst values of the two measures being considered. This is seen in the next part of the interview.

Criterion	Measure	Least preferred	Most preferred	Rank order	Points (0–100)
Academics	Average SAT Score	1400	1800	3	85
	U.S. News & World Report Ranking	Tier 4	Nat'l. Rank	2	90
Cost	Room & Board	$14,000	$8,000	4	80
	Tuition	$35,000	$25,000	1	100
Location	Average Daily High Temperature	50° F	70° F	9	20
	Nearness to Home	Within 1 hr.	Within 1 day	6	60
Social life	Athletics	Div. 3	Div. 1	8	30
	Reputation	Party	Balanced	7	50
	Size	> 12,000	< 3,000	5	70
				Sum:	585

Table 1.2.7: Enrique's ranking and point assignment

Anna: Let's start by assigning 100 points to the tuition range, which you've ranked first. Now, you've ranked *U. S. News & World Report* rating second. How important is this rating, from worst to best, compared to reducing the cost of tuition from $35,000 to $25,000? If it's close, you should use a number close to 100.

Enrique: I think it's about 90% as important, so let's use 90 points for that one, and SAT scores are almost as important, so we'll use 85 points for that range.

Table 1.2.7 contains the rest of the points Enrique assigns to each of his measures.

The interview continues:

Anna: Enrique, what did you get for the total number of points for all your measures? Once you have the point total, you'll need to divide the points for each measure by this total to get the weight.

Enrique: I got 585 total points. Now I can calculate the individual measure weights.

The weights Enrique calculates appear in Table 1.2.8. These were calculated by dividing the points for a particular measure by the total points. For example, Average SAT Score has a point value of 85. So, the weight for this measure is:

$$\frac{85}{585} = 0.145.$$

Criterion	Measure	Least preferred	Most preferred	Rank order	Points (0–100)	Weight (Points/Sum)
Academics	Average SAT Score	1400	1800	3	85	0.145
	U.S. News & World Report Ranking	Tier 4	Nat'l. Rank	2	90	0.154
Cost	Room & Board	$14,000	$8,000	4	80	0.137
	Tuition	$35,000	$25,000	1	100	0.171
Location	Average Daily High Temperature	50° F	70° F	9	20	0.034
	Nearness to Home	Within 1 hr.	Within 1 day	6	60	0.103
Social life	Athletics	Div. 3	Div. 1	8	30	0.051
	Reputation	Party	Balanced	7	50	0.085
	Size	> 12,000	< 3,000	5	70	0.120
				Sum:	585	1.000

Table 1.2.8: Enrique's assignment of weights to each measure

Next, Anna wants to ensure that Enrique has assigned an appropriate weight to each *criterion*. The interview continues.

Anna: Enrique, what is the total weight for each criterion?

Enrique: I get a total of 0.299 for academics, 0.308 for cost, 0.137 for location, and 0.256 for social life.

Anna: Which criterion has the greatest weight assigned to it?

Enrique: It looks like cost, with 0.308.

Anna: Are there criteria with similar weights?

Enrique: It looks like academics and cost are almost the same.

Anna: Are these the criteria you feel are the most important criteria for choosing a college, and do you think they're about the same in importance?

Enrique: I didn't realize I placed so much importance on academics.

Anna: What did you expect to happen?

Enrique: I thought social life would be at the top of the list!

Anna: Well, you gave athletics only 30 points, reputation 50 points, and size 70 points. Do you want to change anything?

Enrique: No, I really think academics and cost are most important.

1.2.6 Calculate Total Scores

Finally, Enrique and Anna calculate a total score for each school. They use the data from Table 1.2.6, where common units were computed, and the weights calculated in the last column of Table 1.2.8 to calculate a score for each school on each measure. In Table 1.2.9 below, Enrique has calculated the product of the weight, W, and the corresponding common unit, CU:

$$\text{Score} = W \cdot CU$$

For example, the common unit score for the average SAT score at Canisius College is 0.475, and the weight Enrique has assigned to average SAT score is 0.145. Multiplying these two numbers yields 0.069. This value appears opposite SAT score and below Canisius in Table 1.2.9. It is 10% of the total score for Canisius. The rest of the values in Table 1.2.9 are computed in the same way. Then, totaling the scores for each measure for each college yields the total scores that appear in the last row in Table 1.2.9 Now Enrique can see which of his college choices best suits his preferences.

Measure	Weight	Canisius	Clark	Drexel	Suffolk
Average SAT Score	0.145	0.145 · 0.475 = 0.069	0.127	0.109	0.029
U.S. News & World Report Ranking	0.154	0.077	0.154	0.154	0.038
Room & Board	0.137	0.088	0.117	0.043	0.046
Tuition	0.171	0.117	0.019	0.077	0.156
Average Daily High Temperature	0.034	0.010	0.010	0.024	0.015
Nearness to Home	0.103	0.103	0.051	0	0.051
Athletics	0.051	0.051	0	0.051	0
Reputation	0.085	0.085	0.043	0.043	0.085
Size	0.120	0.090	0.120	0	0.090
Total Score:	**1.000**	**0.690**	**0.641**	**0.501**	**0.512**

Table 1.2.9: Calculating the measure score and total scores of Enrique's schools

1.2.7 Interpreting the Results

Enrique reviews these results carefully. He notices that Drexel and Suffolk have scored much lower than his top-ranked choice, so he excludes them from further study. However, he decides to take a closer look at the relative strengths and weaknesses of Canisius, ranked first, and Clark, ranked second. There is only a 0.049 difference between the two, and he is not sure that it is enough evidence to make this critical life decision.

8. What are some reasons why Enrique may not choose Canisius, even though it was ranked first?

9. On many of the measures, Clark received better scores than Canisius. Why did Canisius end up having the higher total score?

10. Suppose Enrique was offered a scholarship at Clark for $5,000. How do you think this would affect Enrique's decision?

Section 1.3: Judy Purchases a Used Car

Judy is trying to decide which used car to purchase from among four possibilities: a 2006 Honda Civic Hybrid, a 2006 Toyota Prius, and a 2007 Nissan Versa that she has found at dealerships, as well as a 2005 Ford Focus that Judy's uncle Roger is trying to sell by himself. Judy asks her friend Dave to help her structure her thoughts in a consistent manner and to use the steps in the process of multi-criteria decision making (see Section 1.2 for a list of the steps).

With Dave's help, Judy decides that the criteria most important for her choice of a used car are minimizing total cost and maximizing condition, accessories, and aesthetics. They identify two measures for each criterion, as shown in Table 1.3.1.

Criterion	Measures
Total cost	Purchase price
	Miles per gallon, based on the EPA rating when new
Condition	Odometer reading
	Body condition
Accessories	Functional air conditioner and heater
	Sound system
Aesthetics	Color
	Body design

Table 1.3.1: Judy's list of criteria and measures

Judy and Dave collect data on each of the cars being considered. Their data appear in Table 1.3.2.

Measure	Honda Civic Hybrid	Toyota Prius	Ford Focus	Nissan Versa
Purchase Price	$15,000	$15,500	$7,700	$11,000
Miles per Gallon	43	46	25	33
Odometer Reading	85,000	80,000	95,000	65,000
Body Condition	Good	Good	Good	Excellent
Functional Air Conditioner and Heater	Both Work	Both Work	Both Work	Both Work
Sound System	Radio and CD Players	Radio and CD Players	Radio, CD, and MP3 Players	Radio and CD Players
Color	Red	Silver	Blue	White
Body Design	Sedan	Sedan	Wagon	Hatch-back

Table 1.3.2: Judy's data on four used cars

After collecting data and determining the scale range for each measure, Judy and Dave create Table 1.3.3. At this point, for each of the categorical measures, she assigned an integer value. For example, the five possible colors were given values from 1 for blue, the least preferred to 5 for the most preferred color. She created four categories for the sound system and numbered them from 1 to 4.

Measure	Scale range		Type
Purchase Price	$6,000–$16,000		Continuous
Miles per Gallon	20–50 mpg		Continuous
Odometer Reading	50,000–100,000 miles		Continuous
Body Condition	1	Fair	Categorical
	2	Good	
	3	Excellent	
Functional Air Conditioner and Heater	1	Neither works	Categorical
	2	Only one works	
	3	Both work	
Sound System	1	None	Categorical
	2	Radio only	
	3	Radio and CD player	
	4	Radio, CD, and MP3	
Color	1	Blue	Categorical
	2	Red	
	3	Silver	
	4	White	
	5	Black	
Body Design	1	Wagon	Categorical
	2	Hatchback	
	3	Sedan	

Table 1.3.3: The type and range for each of Judy's measures

Next, Judy and Dave convert the data for each car into the numerical values given in Table 1.3.4.

Measure	Honda Civic Hybrid	Toyota Prius	Ford Focus	Nissan Versa
Purchase Price	$15,000	$15,500	$7,700	$11,000
Miles per Gallon	43	46	25	33
Odometer Reading	85,000	80,000	95,000	65,000
Body Condition	2	2	2	3
Functional Air Conditioner and Heater	3	3	3	3
Sound System	3	3	4	3
Color	2	3	1	4
Body Design	3	3	1	2

Table 1.3.4: Converted categorical data for Judy's four cars

However, to determine a total score, each value must be rescaled to values between 0 and 1. Judy and Dave convert each continuous measure proportionally (on a scale from 0 to 1) and each categorical measure subjectively, based on Judy's preferences. Table 1.3.5 shows each common unit value. Use this table to answer the questions below.

Measure	Honda Civic Hybrid	Toyota Prius	Ford Focus	Nissan Versa
Purchase price	0.1	0.05	0.83	0.5
Miles per gallon	0.767	0.867	0.167	0.433
Odometer reading	0.3	0.4	0.1	0.7
Body condition	0.5	0.5	0.5	1
A/C and heater	1	1	1	1
Sound system	0.75	0.75	1	0.75
Color	0.25	0.5	0	0.75
Body design	1	1	0	0.5

Table 1.3.5: Each measure rescaled to common units

1. Consider the continuous measures. Determine how Judy and Dave calculated the common unit values for each continuous measure.
   a. Purchase Price
   b. Miles per Gallon
   c. Odometer Reading

2. Now consider the categorical measures. Determine Judy's preferences for the following measures based on the information in Table 1.3.5:
   a. Body Condition
   b. Functional Air Conditioner and Heater
   c. Sound System
   d. Color
   e. Body Design

While the information given in Table 1.3.5 is informative, it does not take Judy's preferences into consideration because each measure is weighted equally.

Therefore, Dave interviews Judy to determine how they would weight each measure. The rank order, points, and weights that came out of this interview can be seen in Table 1.3.6. Use this table to answer the questions below.

Criterion	Measure	Least preferred	Most preferred	Rank order	Points (0–100)	Weight (Points/Sum)
Total cost	Purchase Price	$16,000	$6,000	1	100	0.2
	Miles per Gallon	20 mpg	50 mpg	2	95	0.19
Condition	Odometer Reading	100,000 mi	50,000 mi	5	60	0.12
	Body Condition	1 (fair)	3 (excellent)	3	75	0.15
Accessories	Functional Air Conditioner and Heater	1 (neither works)	3 (both work)	7	35	0.07
	Sound System	1 (none)	4 (radio, CD, MP3)	6	50	0.1
Aesthetics	Color	1 (blue)	5 (black)	8	10	0.02
	Body Design	1 (wagon)	3 (sedan)	3	75	0.15
				Sum =	500	1

Table 1.3.6: Judy's rank ordering, point assignment, and weight calculation for her measures

3. Which measure is most important to Judy? How do you know?

4. Which measure is least important to Judy? How do you know?

5. Which two measures have equal importance to Judy? How do you know?

6. How would you describe Judy's feelings towards Purchase Price versus her feelings towards Miles per Gallon?

7. How would you describe Judy's feelings towards Miles per Gallon versus her feelings towards Body Condition?

8. How were the weights calculated?

9. What criterion is most important to Judy? How do you know?

Finally, Judy and Dave calculate the total scores for each car, as shown in Tables 1.3.7 and 1.3.8. Use these tables to answer the questions below.

Measure	Weight	Honda Civic Hybrid	Toyota Prius	Ford Focus	Nissan Versa
Purchase price	0.2	0.1	0.05	0.83	0.5
Miles per gallon	0.19	0.767	0.867	0.167	0.433
Odometer reading	0.12	0.3	0.4	0.1	0.7
Body condition	0.15	0.5	0.5	0.5	1
A/C and heater	0.07	1	1	1	1
Sound system	0.1	0.75	0.75	1	0.75
Color	0.02	0.25	.5	0	0.75
Body design	0.15	1	1	0	0.5

Table 1.3.7: Judy's weights and common units for each measure

Measure	Honda Civic Hybrid	Toyota Prius	Ford Focus	Nissan Versa
Purchase price	0.02	0.01	0.166	0.1
Miles per gallon	0.1457	0.1647	0.0317	0.0823
Odometer reading	0.036	0.048	0.012	0.084
Body condition	0.075	0.075	0.075	0.15
A/C and heater	0.07	0.07	0.07	0.07
Sound system	0.075	0.075	0.1	0.075
Color	0.005	0.010	0.000	0.015
Body design	0.15	0.15	0	0.075
Total Score	**0.582**	**0.613**	**0.470**	**0.636**

Table 1.3.8: Judy's calculation of the measure subtotal score and total score for each used car

10. How were the scores for each measure calculated?

11. Looking at the total scores, which car would you recommend to Judy? Is this choice obvious?

12. What significant advantages does the Toyota Prius have over the Nissan Versa?

13. What significant advantages does the Nissan Versa have over the Toyota Prius?

When the numeric values are this close, the ultimate answer may be that the decision maker will be equally satisfied with either choice.

14. If you were choosing among these cars, which car would you choose? Was your choice impacted by the total scores calculated in this problem?

1.3.1 Use Excel to Calculate Scores

The spreadsheet format of Excel offers an ideal tool to calculate scores. The data in table 1.3.5 are the common unit scores for each car on each measure. These were input into EXCEL as seen in Figure 1.3.2. To determine the total score, for example, for the Honda, Judy will need to multiply the weights in column B by the common units in column C. Judy can carry out this computation in either of two ways. The simpler method involves using an EXCEL function named SUMPRODUCT. This command involves specifying the two sets of numbers that are to be multiplied and then summed. In this example we want to multiply the values in cells B3 through B10 by the corresponding values in Cells C3 to C10. In EXCEL you specify a range with a semicolon as in B3:B10 and as in C3:C10. The two ranges are separated by a comma in the command. The SUMPRODUCT command multiplies the value in B3 and by the value in C3, multiplies the value in B4 and multiplies by C4, etc. and then sums the value to obtain the total score.

=SUMPRODUCT(B3:B10,C3:C10)

	A	B	C	D	E	F
1			Data Rescaled to Common Units (Between 0 and 1)			
2			Honda	Toyota	Ford	Nissan
3	Measure	Weights	Civic Hybrid	Prius	Focus	Versa
4	Purchase price	0.2	0.100	0.050	0.830	0.500
5	Miles per gallon	0.19	0.767	0.867	0.167	0.433
6	Odometer reading	0.12	0.300	0.400	0.100	0.700
7	Body condition	0.15	0.500	0.500	0.500	1.000
8	Air conditioner and heater	0.07	1.000	1.000	1.000	1.000
9	Sound system	0.1	0.750	0.750	1.000	0.750
10	Color	0.02	0.250	0.500	0.000	0.750
11	Body design	0.15	1.000	1.000	0.000	0.500
12	Total Score	1	0.577	0.603	0.455	0.651

Figure 1.3.2: Use SUMPRODUCT to calculate total score in one step

To obtain Toyota's total score, Judy will use column B but this time multiply it by column D.

=SUMPRODUCT(B3:B10,D3:D10)

Judy recalls that EXCEL allows one to copy and paste functions from one cell into another. She wants to copy the SUMPRODUCT formula from cell C11 to cells D11, E11, and F11. However she notices that for each of the cars, the first range will always refer to B3 to B10. To ensure that column B always appears in the function, she places a $ sign before each B letter. The $ before the B ensures that when the cell is copied into another cell the B column is unchanged. Here is how she wrote the function.

=SUMPRODUCT($B3:$B10,C3:C10)

When Judy copied C11 into D11 the result was

=SUMPRODUCT($B3:$B10,D3:D10).

She repeated this for cells E11 and F11.

Multi-step Method with more Details

The above method determines the total score but does not show the individual measure components of each total score. Thus, Judy is unable to tell how much the purchase price contributes to the total score of each car. She decided to use EXCEL's capabilities to replicate Table 1.3.8 which has the detailed information. (See also Table 1.1.12 and Table 1.2.10.) She set up a new area in the spreadsheet in rows 16 through 26 to calculate the individual subtotals. This is displayed in Figure 1.3.4. To obtain the value in cell C18, she multiplied C3 by $B3. She again placed a $ symbol before the B because she was going to use the copy and paste function to complete the table.

	A	B	C	D	E	F
15			Weight * Common Unit			
16			Honda	Toyota	Ford	Nissan
17	Measure		Civic Hybrid	Prius	Focus	Versa
18	Purchase price		0.020	0.010	0.166	0.100
19	Miles per gallon		0.146	0.165	0.032	0.082
20	Odometer reading		0.036	0.048	0.012	0.084
21	Body condition		0.075	0.075	0.075	0.150
22	Air conditioner and heater		0.070	0.070	0.070	0.070
23	Sound system		0.075	0.075	0.100	0.075
24	Color		0.005	0.010	0.000	0.015
25	Body design		0.150	0.150	0.000	0.075
26	**Total Score**		**0.577**	**0.603**	**0.455**	**0.651**

Figure 1.3.3: Calculate subtotals: multiply weight by common unit and sum the subtotals

Judy then copied cell C18 into cells C19 through C25. Judy then used the SUM function to calculate the total score. In cell C26 she wrote

=SUM(C18:C25)

She then copied the entire column of values C18 through C26 to columns D, E, and F. She now can see the impact of the subtotals on each car's total score. She noticed that the purchase price contributes only 0.02 to Honda's total score of 0.582. In contrast, the purchase price contributes 0.10 to Nissan's total score of 0.636.

Chapter 1 (MCDM) Homework Questions

1. Olivia wants to pursue a career in medicine, but she is not sure which profession would be best for her. After some preliminary research, she narrows her choices to physician, nurse, and pharmacist. Olivia decides to consider four criteria to help structure her decision: professional preparation, personal fulfillment, financial compensation, and lifestyle. The table below shows these criteria and the measures she has decided to use for each.

Criterion	Measure	Type of Scale	Type of Data
Professional Preparation	Schooling		
	Internship		
	Difficulty		
Personal Fulfillment	Job satisfaction		
	Personal interest		
Financial Compensation	Initial salary		
	Median salary		
Lifestyle	Likely schedule		
	Maternity leave		
	Prestige		

   a. Decide which type of scale would be appropriate for each measure, either *continuous* or *categorical*.

   b. Determine which of the data will have to be collected through *research* (objective) and what will be based on personal *opinion* (subjective).

   c. The table below shows some of the data Olivia has collected for the professional preparation criterion. Based on the scale ranges, determine what you would consider most preferred and least preferred for each measure.

Criterion	Measure	Scale Range	Physician (M.D.)	Nurse (R.N.)	Pharmacist (Pharm.D.)
Professional Preparation	Schooling (years)	2-8	8	4	6
	Internship (years)	0-4	3	0	1
	Difficulty (rank)	1-3	1	3	2

   d. What else must be done before obtaining common unit values?

   e. Fill in the following table with scores scaled to common units.

Criterion	Measure	Physician (M.D.)	Nurse (R.N.)	Pharmacist (Pharm.D.)
Professional Preparation	Schooling			
	Internship			
	Difficulty			

f. Suppose Olivia weights Schooling at 0.109, Internship at 0.091, and Difficulty at 0.073. Complete the following table with the weighted subtotal score for each measure for each alternative..

Criterion	Measure	Physician (M.D.)	Nurse (R.N.)	Pharmacist (Pharm.D.)
Professional Preparation	Schooling			
	Internship			
	Difficulty			

2. Rana is trying to decide what part time job to take during the school year.

   a. Identify 3 or more criteria she could use to determine her preferred job.

   b. For each criterion, specify at least two measures.

   c. Specify the type of scale for each measure

   d. Assume now the decision involves taking a full-time job in the summer. Identify at least one measure to be eliminated from your list. Identify at least one criterion and measure to be added to the evaluation.

Criterion	Measure

3. Give an example of a measure that could be a continuous scale but you would choose to create a categorical scale instead. Explain your answer.

4. Give an example of a measure that uses a categorical scale, but might not be converted to common units proportionally. Explain your answer.

Chapter 1   Make Hard Decisions—Multi-Criteria Decision Making (MCDM)

5. In problem 1.1 of the chapter, Isabelle Nueva is helping her mother and father decide on the best wireless plan for her family.

   a. What additional measures do you think should be considered?

   b. Add and describe a categorical measure for the problem and create 3 categories for this new measure.

   c. Add and describe a numerical measure for the problem.

6. In problem 1.2 of the chapter, Enrique Ramirez is selecting a college to attend.

   a. What additional measures do you think should be considered?

   b. Add and describe a categorical measure for the problem and create 3 categories for this new measure.

   c. Add and describe a numerical measure for the problem.

7. In problem 1.3 of the chapter, Judy is choosing which used car to purchase from among four possibilities.

   a. What additional measures do you think should be considered?

   b. Add and describe a categorical measure for the problem and create 3 categories for this new measure.

   c. Add and describe a numerical measure for the problem.

8. A high school student wants to buy a digital camera. Checking the experts' recommendations, she creates a list of important features and ranks them as follows. She ranks Price as the most important measure and, therefore, assigns 100 points it. Brand name is slightly less important than price. It is ranked 2nd and she assigns 90 points to it. She thinks that having an Anti-shake system is much less important than brand name and assigns 60 points to it. Size of view screen is ranked below anti-shake system and has a little bit less importance, thus she assigned it 55 points. Finally, ease of use is the least important factor with 40 points. Calculate the weight assigned to each measure.

Measure	Rank	Points	Weight
Size of view screen	4	55	
Price	1	100	
Brand name	2	90	
Anti-shake system	3	60	
Easy to use	5	40	
		Total	

Lead Authors: Thomas Edwards and Kenneth Chelst

9. Suppose you are looking to buy a digital camera for yourself.

   a. Suggest and add a relevant categorical measure in the table below. Describe the new measure.

   b. Suggest and add a relevant numerical measure in the table below. Describe the new measure.

   c. Use your personal preferences and rank the measures. Then, assign points to each measure and calculate the weight of each measure.

		Assign	Calculate
Measure	**Rank**	**Points**	**Weight**
Size of view screen			
Price			
Brand name			
Anti-shake system			
Easy to use			
New categorical measure:			
New numerical measure:			
	Total		

10. Kim is interested in purchasing a desktop computer for her office. After reviewing the specification of different models, she ended up with the following measures. Classify each measure as numerical or categorical.

Measure	Type: Numerical or Categorical
Computational power	
Monitor size	
Years of warranty	
Operating system	
Price	

11. A high school has selected one of its students to be the chair of a committee planning a class trip. One of her first responsibilities is to pick a co-chair for planning the trip. Suggest two measures for each criterion that could be used to help select a co-chair for planning the trip. In specifying measures be sure they are relevant to this co-chair selection. Specify the type of each measure.

Criterion	Measure	Type: Numerical or Categorical
Knowledge		
Reliability		
Personality		

12. Sam and his wife were just married and are looking for an apartment in a safe area close to Sam's school. After discussing their preferences, they came up with the following measures that are very important to them.

Measure	Description
Spaciousness	Size and design
Price	Monthly rental
Condition	Freshly painted, floors, age of appliances
Apartment building rating	Based on previous tenants' rating in www.apartmentrating.com

a. After searching in a 10-mile radius around his school, they ended up with the following three apartments they like. Sam summarized the data as follows:

Measure	Ap1	Ap2	Ap3
Spaciousness	Good	Medium	Poor
Price ($/month)	700	650	550
Condition (0-1)	0.6	0.9	0.7
Apartment building rating (between 1 and 5)	4	4.5	3.8

b. Specify the range for each measure and then determine the common unit for each of them. Insert the common units in the following table.

Measure	Ap1	Ap2	Ap3
Spaciousness			
Price			
Condition(0-1)			
Apartment building rating			

c. After considering the measures, Sam and his wife ranked the measures as in the following table. Use the assigned points to calculate the weights.

Measure	Rank	Points	Weight
Spaciousness	3	70	
Price	1	100	
Condition(0-1)	2	90	
Apartment Building Rating	4	50	

d. For each alternative, calculate the product of the weight and the corresponding common unit for each measure. Determine the total score for each alternative.

Measure	Ap1	Ap2	Ap3
Spaciousness			
Price			
Condition			
Apartment Building Rating			
Total Score			

e. Which alternative is ranked 1st and what measures contribute the most to it being ranked 1st?

13. James and George are seeking a team member for their final project in their senior year that will involve a lot of data analysis. It is a very demanding project that requires a wide range of skills. To help evaluate potential teammates, they created the following list of measures.

Measure
Writing Skills
GPA of Math courses
Total GPA
Reliability and commitment
Communication skills

a. After considering all their classmates who were not yet assigned to any project, they ended up with following three people. They summarized the data for these three as follows:

Measure	Ed	Ken	Thad
Writing skills	Excellent	Acceptable	Good
GPA in math courses	3.5	3.9	3.0
Total GPA	3.6	3.8	3.3
Reliability and commitment	Acceptable	Good	Good
Communication skills	Good	Acceptable	Excellent

b. Specify the range for each measure and then determine the common unit for each of them. Insert the common units in the following table. (Assume proportionality.)

Measure	Ed	Ken	Thad
Writing skills			
GPA in math courses			
Total GPA			
Reliability and commitment			
Communication skills			

c. They are not sure how to rank the measures. Based on your personal preferences, rank the measures and fill out the rest of table.

Measure	Rank	Points	Weight
Writing skills			
GPA in math courses			
Total GPA			
Reliability and commitment			
Communication skills			
Total			

d. For each alternative, calculate the product of the weight and the corresponding common unit for each measure. Determine total score for each alternative.

Measure	Ed	Ken	Thad
Writing skills			
GPA in math courses			
Total GPA			
Reliability and commitment			
Communication skills			
Total Score			

e. Which alternative is ranked 1st and what measures contribute the most to him being ranked 1st?

14. Neil is trying to find a location in Michigan to open a convenience store. Location is very important for convenience stores. Thus, he wants to be very precise in this process. After talking to some consultants and other store managers, he plans to use the following measures.

Measure	Description
Traffic through intersection	Daily number of the cars passing the intersection
Population within 2 mile	Total population over the age of 15
Distance to the nearest competitor	Miles to nearest convenience store
Cost of the property	Purchase price of property

a. After considering all available properties in the area, he ends up with the following three locations. The data for these three locations is summarized below.

Measure	L1	L2	L3
Traffic through intersection (vehicles)	16,000	15,000	19,000
Population within 2 miles	50,000	45,000	55,000
Distance to the nearest competitor (miles)	1.5	2	0.5
Cost of the property ($)	210,000	180,000	250,000

b. Specify the range for each measure and then determine the common unit for each of them. Insert the common units in the following table.

Measure	L1	L2	L3
Traffic through intersection			
Population within 2 mile			
Distance to the nearest competitor			
Cost of the property			

c. After considering the measures, he ranks the measures as in the following table. Use assigned points to calculate the weights.

Measure	Rank	Point	Weight
Traffic through intersection	2	85	
Population within 2 mile	3	80	
Distance to the nearest competitor	4	70	
Cost of the property	1	100	
	Total		

d. For each alternative, calculate the product of the weight and the corresponding common unit for each measure. Determine total score for each alternative.

Measure	L1	L2	L3
Traffic through intersection			
Population within 2 mile			
Distance to the nearest competitor			
Cost of the property			
Total Score			

e. Which alternative is ranked 1st and what measures contribute the most to it being ranked 1st?

15. Gerald and his friends are trying to decide where to go for spring break.

    a. Identify 3 or more criteria he and his friends could use to determine the preferred spring break location.

    b. For each criterion, specify at least two measures.

    c. Specify the type of scale for each measure

16. Identify a multi-criterion decision context that you or anyone in your family is facing within the next year. Explain how this is a multi-criterion decision.

Chapter 1 Summary

What have we learned?

We have learned that the multi-criteria decision making process provides a framework for making a subjective decision when considering several alternatives, each of which has advantages and disadvantages. As the person making the decision, you must structure the decision. What criteria or objectives will be considered? What measures of your criteria will be included? How will you rank and weight these measures to help make a decision that is best for your values and priorities?

This process allows for direct comparison and evaluation of complex alternatives. The steps are as follows:

1. Identify Criteria and Measures
2. Collect Data
3. Find the Range of Each Measure
4. Rescale Each Measure to a Common Unit
5. Conduct an Interview to Calculate Weights
6. Calculate Total Scores
7. Interpret Results

Terms

Categorical Measure A measure whose scores are classifications

Common Unit A value that varies from 0 to 1, where 0 always represents the worst value, 1 the best value, and intermediate values are found using a proportional or non-proportional scale

Continuous Measure A measure whose scores are numeric values that can take on any value in a certain range

Categorical Measure A measure that is divided into distinct categories. These categories could be natural such as the color of a car or they can be created by grouping numerical values into ranges.

Criteria Objectives or aspects of the alternatives that you wish to either maximize or minimize

Measure A trait that will quantify an aspect of a criterion

Proportional Scale The rescaled score for intermediate values of continuous measures (calculated by dividing the difference between the particular score and the least preferred score by the scale range). A categorical measure can also use a proportional to assign a common unit to an intermediate category.

Non-Proportional Scale For Categorical Measure The rescaled score for intermediate values of categorical measures need not be proportional to where the category falls on the list. For example, with 3 categories, the intermediate category need not be assigned a common unit of 0.5. (Non-proportional scales can also be used for continuous variables as well. This concept is beyond the scope of this course.)

Scale Range The range of possible values for a measure.

Total score For each alternative, multiply the rescaled score by the weight for each measure. The sum of all these weighted, rescaled scores is the total score.

Weighted Subtotal Scores The rescaled score for each measure, weighted according to its importance (calculated by multiplying each scaled score by the corresponding weight of the measure)

Chapter 1 (MCDM) Objectives

You should be able to:

- List the sequence of steps in the multi-criteria decision making process
- Explain the purpose of each step in the process
- Identify criteria you will use to choose between several alternatives
- Select measure(s) for each criterion
- Distinguish between categorical and continuous measures
- Determine scale types and ranges for measures
- Scale scores
- Rescale scores to common units
- Weight scores for each measure
- Calculate a total score for each alternative
- Evaluate the results of the multi-criteria decision making process by comparing the strengths and weaknesses of the top two alternatives

Chapter 1 Study Guide

1. Explain why the Multi-Criteria Decision Making (MCDM) process is useful.

2. Discuss the differences between a *criterion* and a *measure*.

3. When choosing between the same alternatives, why might you and a classmate, both using MCDM, come to a different decision?

4. Compare and contrast *continuous* and *categorical* measures.

5. Give an example of a scale range in which one end is most preferable for you, but the other end may be preferable to a classmate. Explain.

6. Why do we scale all scores between zero and one?

7. Describe how scores are scaled differently for continuous and categorical measures.

8. Describe how scaled scores are rescaled to common units differently for continuous and categorical measures.

9. Identify which steps in MCDM involve you inserting your own preferences and priorities into the process and describe how this occurs?

10. What role do the weights of the measures play in determining which alternative is the best?

11. Describe the process that occurs from collecting raw data for measures to obtaining a total score for an alternative.

12. Should you always choose the alternative with the highest total score?

References

Chelst, K. and Canbolat, Y. B. (2012). *Value Added Decision Making for Managers*. CRC Press a division of Taylor & Francis Group.

Lenhart, A. (2010). *Teens, cell phones and texting*. Pew Internet & American Life Project: Pew Research Center.

CHAPTER 2:

Optimize Product Mix: Profit Maximization with Linear Programming

Section 2.0: Mathematical Programming

The next five chapters in the text focus on mathematical programming. The father of mathematical programming is George Dantzig. Between 1947 and 1949, Dantzig developed the basic concepts used for framing and solving linear programming problems. During WWII, he worked on developing various plans which the military called "programs." After the war he was challenged to find an efficient way to develop and solve these programs.

Dantzig recognized that these programs could be formulated as a system of linear inequalities. Next, he introduced the concept of a goal. At that time, goals usually meant rules of thumb for carrying out a goal. For example, a navy admiral might have said, "Our goal is to win the war, and we can do that by building more battleships." Dantzig was the first to express the selection of a plan to reach a goal as a mathematical function. Today it is called the *objective function*.

All of this work would not have had much practical value without a way to solve the problem. Dantzig found an efficient method called the simplex method. This mathematical technique finds the optimal solution to a set of linear inequalities that maximizes (profit) or minimizes (cost) an objective function.

Economists were excited by these developments. Several attended an early conference on linear programming and the simplex method called "Activity Analysis of Production and Allocation." Some of them later won Nobel prizes in economics for their work. They were able to model fundamental economic principles using linear programming.

The first problem Dantzig solved was a minimum cost diet problem. The problem involved the solution of nine inequalities (nutrition requirements) with seventy-seven decision variables (sources of nutrition). The National Bureau of Standards supervised the solution process. It took the equivalent of one man working 120 days using a hand-operated desk calculator to solve the problem. Nowadays, a standard personal computer could solve this problem in less than one second. Excel spreadsheet software includes a standard add-in called "Solver," a tool for solving linear programming problems.

Mainframe computers became available in the 1950s and grew more and more powerful. This allowed many industries, such as the petroleum and chemical industries, to use the simplex method to solve practical problems. The field of linear programming grew very fast. This led to the development of non-linear programming, in which inequalities and/or the objective function are not linear functions. Another extension is called integer programming, in which the variables can only have integer values. Together, linear, non-linear and integer programming are called **mathematical programming**.

2.1 An Introductory Problem: Lego Furniture

In order to get a feel for mathematical programming, this chapter begins with a problem that has a concrete model. This model can be built from Lego bricks. When a mathematical model of a real world situation is constructed in symbolic form, it is often helpful to construct a physical or visual model at the same time. The role of the latter model is to help the model builder to understand the real-world situation as well as its mathematical model.

The Problem

A certain furniture company makes only two products: tables and chairs. The manufacturing of tables and chairs can be modeled using Lego bricks. To make a table requires two large and two small bricks, and a chair requires one large and two small bricks. Figure 2.1.1 shows a table and a chair made from Legos.

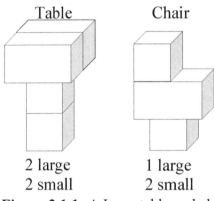

Figure 2.1.1: A Lego table and chair

If the resources needed to build tables and chairs were unlimited, the company would just manufacture as many of each as it thought it could sell. In the real world, however, resources are not unlimited. Suppose that the company can only obtain six large and eight small bricks per day. Figure 2.1.2 shows these limited resources.

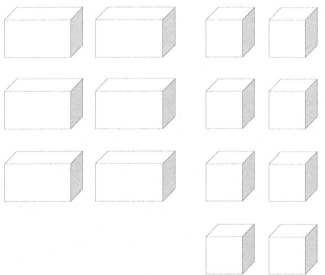

Figure 2.1.2: The furniture company's limited resources

The profit from each table is $16, and the profit from each chair is $10. The production manager wants to find the rate of production of tables and chairs per day that earns the most profit. **Production rate** refers to the number of tables and chairs this company can produce per day.

1. What do you think the production rates should be in order to generate the most profit?

2. Does the number of table and chairs produced each day have to be an integer value?

3. Using only eight small and six large Lego bricks, build a physical model of this problem. If Legos are unavailable, draw pictures to explore some possibilities. Create several combinations of tables and chairs this company could make using your model.

Solving the Problem

There are many possible product mixes this company could make. A **product mix** is a combination of each product being manufactured. The various product mixes could be explored using the Lego model.

First, the company could begin by making as many tables as possible since the profit from a table is much greater than the profit from a chair. Each table requires two large bricks and two small bricks. There are only six large and eight small bricks available. Therefore, only three tables can be built. This generates 3($16) = $48 profit. There are two small bricks left over, but nothing can be built from them. Thus, $48 is the total profit if three tables (and no chairs) are built.

Three tables and zero chairs was one possible product mix. There could be other production rates that generate more profit.

No more than three tables could be made due to the limited resources available, and making three tables yielded a profit of $48. Now, suppose two tables are made. Manufacturing two tables uses four large and four small bricks. Now there are two large and four small bricks left over. These are just enough resources to build two chairs. The profit on two tables and two chairs would be 2($16) + 2($10) = $52. This is more profit than building three tables. However, the production manager wonders, "Is $52 the greatest profit possible? Is there another product mix that could generate more profit?"

4. In a Table 2.1.1, record other combinations of tables and chairs the company could produce. For each combination, write the production rate of tables, the production rate of chairs, and the profit for each possibility.

Production Rate of Tables (tables per day)	Production Rate of Chairs (chairs per day)	Total Profit ($ per day)

Table 2.1.1: Exploring the total profit for each combination of tables and chairs

5. Which production rates generate the most profit?

6. Did any product mix yield a profit greater than $52?

It is impossible to find the total profit for every product mix because there are infinitely many possibilities. However, most likely no one in the class found a profit greater than $52. In the next section, you will learn how to know for certain you found the product mix with the greatest profit.

Notice that in Table 2.1.1 you used a set of similar equations to compute the profit for each possibility. These equations are the basis for the **objective function**. The two production rates *varied* across each possible product mix, and exploring these variations allows a *decision* about production to be made. Therefore, the production rates for tables and chairs are known as the **decision variables** for this problem.

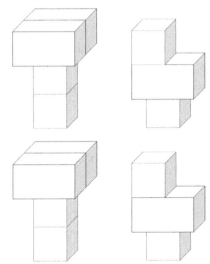

Figure 2.1.3: Two tables and two chairs yield the most profit

Because the profit has been optimized, the solution in Figure 2.1.3 is called the **optimal solution**. Besides the optimal solution, there are many other possible solutions. Although they are not optimal, each possible solution is still a **feasible** solution. Building four tables is an example of an **infeasible** solution.

7. Why is choosing to build four tables an example of an infeasible solution?

8. Give another example of an infeasible solution.

Stepping Beyond the Current Optimal Solution

Operations researchers understand that there is more to their work than merely finding solutions to problems. Once a solution is found, it must be interpreted. One sort of interpretation is called **sensitivity analysis**. Sensitivity analysis involves exploring how sensitive the solution is to changes in the parameters of the problem. For example, in the Lego problem above, one of the parameters of the problem is the availability of large bricks.

9. Would it make a difference if *seven* large bricks were available instead of six (there are still eight small bricks)? If so, what is the new optimal solution, and how much profit does it generate?

10. Would it make a difference if *nine* small bricks were available instead of eight (there are still six large bricks)? If so, what is the new optimal solution, and how much profit does it generate?

11. Would it make a difference if *seven* large bricks and *nine* small bricks were available? If so, what is the new optimal solution, and how much profit does it generate?

Growing the Problem

Suppose now that the furniture company has decided to dramatically expand production. Now it is able to obtain 27 small and 18 large Lego bricks per day. The profit on tables and chairs remains the same.

12. Use Table 2.1.2 to complete the following:
    a. Make a suggestion for the number of tables and chairs to produce in this expanded problem.
    b. Calculate the combined total number of small Lego bricks and large Lego bricks that are used in this suggestion. Be sure that both of the totals do not exceed the available number of Lego bricks.
    c. Calculate the total profit for the suggested production rate.
    d. Repeat these steps four more times, for a total of five suggestions. With each new suggestion, try to produce more profit than the suggestion before.

Suggested Production Plan	Production Rate of Tables (tables per day)	Production Rate of Chairs (chairs per day)	Small Lego Bricks Used (# per day)	Large Lego Bricks Used (# per day)	Total Profit ($ per day)
1					
2					
3					
4					
5					

Table 2.1.2: Exploring the total profit for different combinations of tables and chairs

13. What difficulties did you experience when developing the feasible production plans in question 12?

Section 2.2: Computer Flips, a Junior Achievement Company

Junior Achievement (JA) is an educational program available worldwide. JA uses hands-on experiences to help young people understand the economics of life. In partnership with businesses and educators, JA brings the real world to students. The JA Company Program provides basic economic education for high school students by using support and guidance of volunteer consultants from the local business community. By organizing and operating an actual business, students learn how businesses function. They also learn about the structure of the free enterprise system and the benefits it provides.

Gates Williams is the production manager for Computer Flips, a Junior Achievement company. Computer Flips purchases a basic computer at wholesale prices and then adds a display, extra memory cards, extra USB ports, or a CD-ROM or DVD-ROM drive. The company also purchases these extra components at wholesale prices. The computers, with the added features, are then resold at retail prices.

Computer Flips produces two models: Simplex and Omniplex. The profit on each Simplex is $200, and on each Omniplex, the profit is $300. The Simplex model has fewer add-ons, so it requires only 60 minutes of installation time. The Omniplex has more add-ons and requires 120 minutes of installation time. Five JA students do all of the installation work. Each of them works 8 hours per week. Gates Williams must decide the rate of production per week of each computer model in order to maximize the company's weekly profit.

To make decisions such as the one Gates Williams faces, operations researchers use a technique known as **linear programming**. Answering the following questions will help you understand this technique.

2.2.1 Exploring the Problem

One way to approach the problem is to make some guesses and test the profit generated by each guess. For example, suppose Gates Williams decides the company should make 20 of each model.

1. How much profit would be generated?

2. Is there enough installation time available to make that number of each model?

3. Answer the same two questions if Gates Williams decides to make:
   a. 10 Simplex computers and 30 Omniplex computers
   b. 30 Simplex and 10 Omniplex

4. Can you find a product mix for which there is enough installation time?

5. How much profit do the production rates you found generate for the company?

2.2.2 Generalizing the Problem

Sometimes it is helpful to visualize things. The numbers, variables, and their relationships in a problem can be represented by a graph. Before graphing the Computer Flips problem, you must translate the information in the problem into mathematical statements—equations or inequalities.

First, let
x_1 represent the weekly production rate of Simplex computers and
x_2 represent the weekly production rate of Omniplex computers.

The variables x_1 and x_2 are called **decision variables** because Gates Williams uses them to help make his decision. Mr. Williams's goal is to make as much money as possible. He does this by selling as many computers as he is able. Therefore, Mr. Williams can calculate his weekly profit (z) as a function of x_1 and x_2. Because the objective is to maximize profit, the profit function is called the **objective function**.

6. Write an equation for the profit (z) the company would earn in a week. [Hint: Look back at Section 2.1.1 and see how you calculated profit for 20 Simplex and 20 Omniplex computers.]

7. Write a mathematical statement in terms of x_1 and x_2 that describes the relationship between the installation time required and the installation time available. [Hint: Look back at Section 2.1.1 and see how you determined if there was enough installation time to produce 20 Simplex and 20 Omniplex computers.]

8. Can Computer Flips produce a negative number of either model?

9. Write two mathematical inequalities that represent the answer to the previous question.

The mathematical statements created in this section will be used to find the optimal solution in the following sections.

2.2.3 A Visual Approach

At this point, it should be clear that Gates Williams cannot decide to make any number of each model he chooses, because there is only a certain amount of installation time available each week. That is, the available installation time *constrains* the number of Simplex and Omniplex computers that can be made each week. The inequality that captures this relationship (from question 7) is called a **constraint**. The other two inequalities (from question 9) express the fact that the decision variables in this problem cannot be negative. Thus, they are called **non-negativity constraints**.

These constraints can be graphed on a coordinate place. This graph gives a visual representation of the possible production rates for each computer model.

10. On the same coordinate axes, graph each of the three inequalities you wrote in the previous section (one from question 7 and one from question 9). For uniformity, place x_1 on the horizontal axis and x_2 on the vertical axis.

11. Give one point that satisfies all three inequalities.

12. Where are all of the points that satisfy all three inequalities?

13. What is the connection between the points identified in the previous question and the Simplex and Omniplex computers?

The points that satisfy each of the constraint inequalities represent a mix of Simplex and Omniplex computers that could be produced each week. Recall that this region of the coordinate system is called the **feasible region**, because those points represent feasible production mixes.

14. Choose any point in the feasible region, and compute the weekly profit that would be generated by producing that mix of Simplex and Omniplex computers.

15. Choose a second point in the feasible region that generates the <u>same</u> weekly profit as the first point.
    a. Draw a line through the two points.
    b. Write the equation for this line in terms of x_1 and x_2.

Every point on the line you have drawn generates the same weekly profit. For this reason, such a line is called a **line of constant profit**.

16. Suppose Computer Flips generates $6,000 of profit each week.
    a. Write an equation to represent this situation.
    b. Graph this equation.

Note that the points (0, 20), (15, 10), and (30, 0) are on the line you drew, and each coordinate pair generates a profit of $6,000 when substituted into the objective function.

17. For each of the following profits, write an equation and then graph that equation (on the same coordinate plane).
    a. $0 profit
    b. $3600 profit
    c. $4800 profit
    d. $7200 profit

18. What do you notice about the three lines you have drawn?

19. Which of the lines generates the largest weekly profit?

20. If you were to continue drawing lines in this way, where does the line that generates the largest weekly profit intersect the feasible region? What is the profit at that point?

2.2.4 Solving the Problem

Hopefully, the previous line of investigation has suggested that the point or points representing the largest possible weekly profit are close to the boundary of the feasible region. That is, in order to maximize profits, Computer Flips' production rates should be as large as possible, while still keeping within the available installation time.

21. Choose a point on the boundary of the feasible region, but not at a corner (vertex), and evaluate the profit there.

22. Continue to choose points on the boundary, but try to increase the amount of profit each time.

23. Finally, evaluate the profit at each of the corner points of the feasible region.

24. What is the relationship between the corner points and the feasible region?

25. What combination of products generates the greatest profit?

Notice that as the amount of constant profit increases, the lines are higher and further right in the first quadrant. Try to visualize a single line moving upward or to the right while its slope remains constant. The last point(s) in the feasible region that such a moving line touches will be optimal, because the profit is the greatest of any feasible points.

26. There is not always only one optimal solution.
    Draw an example of a feasible region that could have more than one optimal solution.

2.2.5 Complicating the Problem

After several weeks of operation, one of the students in the sales department of Computer Flips does some market research. Based on this research, she decides that the company cannot sell more than 20 Simplex computers in any given week.

27. Write an inequality that expresses this market constraint.

28. Graph the new system of constraint inequalities.

29. What do you notice about the optimal solution you found earlier?

30. What is the optimal solution after adding the market constraint?

Now the students in the sales department of Computer Flips decide to extend the market research to the Omniplex model. On the basis of their research, they decide that Computer Flips cannot sell more than 16 Omniplex computers in any given week.

31. Write an inequality for this new market constraint.

Graph the new feasible region.

32. Does the previous optimal solution lie in the new feasible region?

33. What is the optimal solution after the addition of the second market constraint?

The students at Computer Flips notice that they are getting a lot of returns. Every computer that was returned had a problem with one of the add-ons. They realize that they need to test their finished products before shipping them. They decide to assign the task of testing the computers to only one of the student installers. To accommodate this change, the other four student installers agree to work 10 hours per week, so that the total available installation time remains 40 hours per week. The student who will do the testing also works 10 hours per week. It takes her 20 minutes to test a Simplex and 24 minutes to test an Omniplex.

34. Write an inequality for the testing constraint based on the information in the previous paragraph.

35. Graph the new feasible region.

36. Using the new feasible region, what is the optimal solution?

37. Why is it possible to have a non-integer solution?

2.2.6 Success Breeds—An Even More Complicated Problem

Computer Flips has some initial success, so the students are considering producing two additional models: Multiplex and Megaplex. Multiplex will have more add-ons than Simplex, but not as many as Omniplex. Each Multiplex will generate $250 profit. Megaplex, as the name implies, will have more add-ons than any of the other models. Each Megaplex will generate $400 profit.

38. What are the decision variables in the new problem? What do they represent?

39. Write an equation for the profit (z) the company would earn in a week.

The installation and testing times for each computer appear in Table 2.1.1. In addition, market research indicates that the *combined* sales of Simplex and Multiplex cannot exceed 20 computers per week, and the *combined* sales of Omniplex and Megaplex cannot exceed 16 computers per week.

	Simplex	Omniplex	Multiplex	Megaplex
Installation Time	60 min.	120 min.	90 min.	150 min.
Testing Time	20 min.	24 min.	24 min.	30 min.

Table 2.2.1: Installation and Testing times for all four computer models

40. Using the information above, formulate the constraints after the Multiplex and Megaplex models have been added to the product mix.

41. Is it possible to solve this problem by graphing? Why or why not?

In the next section, you will see another way to solve linear programming problems. In particular, the following section explores solving problems without graphing. You may wonder why this graphing approach cannot be used to solve every linear programming problem. If a problem contains three decision variables, it would be difficult for many people to visualize the graph. If a problem contains four or more decision variables, a graph is not even possible.

Section 2.3: SK8MAN, Inc.

SK8MAN, Inc. manufactures and sells skateboards. A skateboard is made of a deck, two trucks that hold the wheels (see Figure 2.3.1), four wheels, and a piece of grip tape. SK8MAN, Inc. manufactures the decks of skateboards in its own factory and purchases the rest of the components.

Figure 2.3.1: A skateboard truck

To produce a skateboard deck, the wood must be glued and pressed, then shaped. After a deck has been produced, the trucks and wheels are added to the deck to complete a skateboard. Skateboard decks are made of either North American maple or Chinese maple. A large piece of maple wood is peeled into very thin layers called veneers. A total of seven veneers are glued at a gluing machine and then placed in a hydraulic press for a period of time (see Figure 2.3.2). After the glued veneers are removed from the press, eight holes are drilled for the truck mounts. Then the new deck goes into a series of shaping, sanding, and painting processes. Figure 2.3.3 shows a deck during the shaping process.

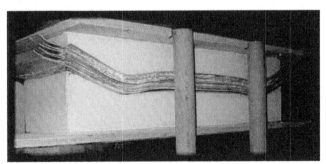

Figure 2.3.2: Maple veneers in a hydraulic press

Figure 2.3.3: Shaping a deck

Currently, SK8MAN, Inc. manufactures two types of skateboards: Sporty (Figure 2.3.4, top) and Fancy (Figure 2.3.4, bottom).

Figure 2.3.4: Sporty and Fancy skateboards

G. F. Hurley, the production manager at SK8MAN, Inc., needs to decide the production rate for each type of skateboard in order to make the most profit. Each Sporty board earns $15 profit, and each Fancy board earns $35 profit. However, Mr. Hurley might not be able to produce as many boards of either style as he would like, because some of the necessary resources, such as the North American maple and Chinese maple, are limited. That is, the production rates are *constrained* by the availability of the resources.

The Sporty board is a less expensive product, because its quality is not as good as the Fancy board. Chinese maple is used in the manufacture of Sporty decks. North American maple is used for Fancy decks. Because Chinese maple is soft, it is easier to shape. On average, it takes a worker 5 minutes to shape a Sporty board. However, a Fancy board requires 15 minutes to shape. G. F. Hurley needs to determine the production rates of Sporty and Fancy boards that will yield the maximum profit.

1. Develop a table to organize the information about Sporty and Fancy boards.

2.3.1 Problem Formulation

To find how to maximize profit, G. F. Hurley uses linear programming. The first step in the formulation of a linear programming problem is to define the decision variables in the problem.

Let:
x_1 represent the weekly production rate of Sporty boards and
x_2 represent the weekly production rate of Fancy boards.

The decision variables are then used to define the objective function. This function captures the goal in the problem, which, in this case, is to *maximize* the company's profits per week. Therefore, the objective function should represent the weekly profit from the sale of the two different styles of skateboards. The variable z is used to represent the amount of profit SK8MAN, Inc. earns per week.

Now, since the profit for each style of skateboard is known ($15 and $35, respectively), G. F. Hurley writes the objective function by expressing the profit (z) in terms of the decision variables (x_1 and x_2):

Maximize: $z = 15x_1 + 35x_2$.

The last step in the formulation of the problem is to represent any constraints in terms of the decision variables. G. F. Hurley cannot just decide to make as many boards as he wants, because the number made is *constrained* by the available shaping time. Therefore, shaping time will be a constraint.

Suppose SK8MAN, Inc. is open for 8 hours a day, 5 days a week, which is a 40-hour workweek. However, since the information about shaping time is expressed in minutes, 40 hours is converted to 2,400 minutes. If SK8MAN, Inc. makes x_1 Sporty boards and x_2 Fancy boards per week, they use $5x_1 + 15x_2$ minutes of shaping time.

For example, making 100 Sporty boards and 150 Fancy boards would take $5(100) + 15(150) = 2{,}750$ minutes. Note that since 2,750 minutes is greater than 2,400 minutes, this production mix is not feasible.

Thus, the shaping time constraint is:
$$5x_1 + 15x_2 \leq 2400$$

There are also two not-so-obvious but completely logical constraints. G. F. Hurley knows the production rate cannot be a negative number for either type of skateboard, so he writes the non-negativity constraints: $x_1 \geq 0$ and $x_2 \geq 0$.

The complete linear programming formulation looks like this:

Decision Variables
　　Let:　x_1 = the weekly production rate of Sporty boards
　　　　　x_2 = the weekly production rate of Fancy boards

Objective Function
　　Maximize:　$z = 15x_1 + 35x_2$,
　　where z = the amount of profit SK8MAN, Inc. earns per week.

Constraints
　　Subject to:
　　　　Shaping Time (min):　$5x_1 + 15x_2 \leq 2400$
　　　　Non-Negativity:　$x_1 \geq 0$ and $x_2 \geq 0$

This formulated linear programming problem can now be solved graphically. To do so, G. F. Hurley sets up a coordinate plane with x_1 as the horizontal axis and x_2 as the vertical axis. Then, he graphs the constraints, as shown in Figure 2.3.5.

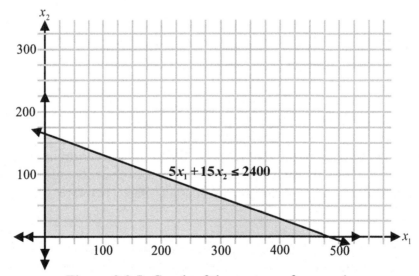

Figure 2.3.5: Graph of the system of constraints

G. F. Hurley recalls that every point in the shaded region satisfies all three constraints and is thus called the feasible region. Each ordered pair in the feasible region represents a combination of Fancy and Sporty boards that SK8MAN, Inc. could produce without violating any of the constraints. There are an infinite number of points in the feasible region, and the solution to the problem of maximizing profit is the one point that generates the most profit.

Rather than try to test an infinite number of points in the objective function, the optimal solution can be found by testing only a few points. This is due to the Corner Point Principle.

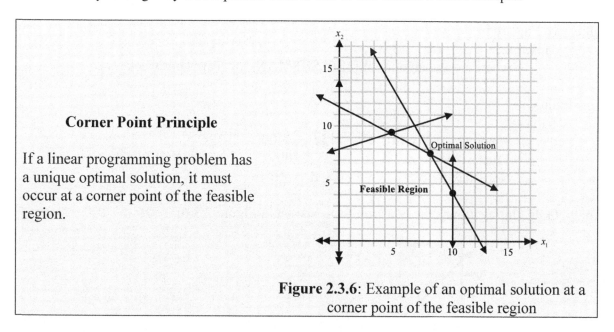

Corner Point Principle

If a linear programming problem has a unique optimal solution, it must occur at a corner point of the feasible region.

Figure 2.3.6: Example of an optimal solution at a corner point of the feasible region

The Corner Point Principle allows us to simply evaluate the objective function at each corner point of the feasible region. Instead of there being an infinite number of possibilities for the optimal solution, there are only as many possibilities as there are corners of the feasible region.

Therefore, G.F. Hurley tests only the corner points of the feasible region in the objective function, as shown in Table 2.3.1.

Point	Profit
(0, 0)	$15(0) + $35(0) = $0
(480, 0)	$15(480) + $35(0) = $7,200
(0, 160)	$15(0) + $35(160) = $5,600

Table 2.3.1: Corner points and their profits

Based on this information, SK8MAN, Inc. should produce 480 Sporty boards and 0 Fancy boards each week. This product mix will generate a weekly profit of $7,200.

2.3.2 Add a New Constraint

G. F. Hurley just found out that the company that supplies the trucks for SK8MAN Inc.'s boards can provide at most 2,800 trucks per month. To make the problem easier, G. F. Hurley considers a month to be four weeks, and therefore there are 700 trucks available per week. Since each skateboard needs two trucks, this new information represents another constraint.

The new complete linear programming formulation is as follows:

Decision Variables
Let: x_1 = the weekly production rate of Sporty boards
x_2 = the weekly production rate of Fancy boards

Objective Function
Maximize: $z = 15x_1 + 35x_2$,
where z = the amount of profit SK8MAN, Inc. earns per week.

Constraints
Subject to:
Shaping Time (min): $5x_1 + 15x_2 \leq 2400$
Trucks (#): $2x_1 + 2x_2 \leq 700$
Non-Negativity: $x_1 \geq 0$ and $x_2 \geq 0$

Again, G. F. Hurley graphs the constraints, as shown in Figure 2.3.6.

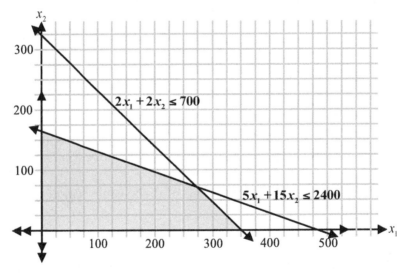

Figure 2.3.6: The feasible region after adding the truck constraint

The new constraint changes the feasible region. The previous optimal solution is no longer included.

To find the new optimal solution, G. F. Hurley evaluates all the new corner points in the objective function, as seen in Table 2.3.2.

Point	Profit
(0, 0)	$15(0) + $35(0) = $0
(350, 0)	$15(350) + $35(0) = $5,250
(285, 65)	$15(285) + $35(65) = $6,550
(0, 160)	$15(0) + $35(160) = $5,600

Table 2.3.2: New corner points and their profits

Now G. F. Hurley can easily see that the maximum weekly profit SK8MAN, Inc. can earn is $6,550, and the company does so by manufacturing 285 Sporty skateboards and 65 Fancy skateboards each week.

2.3.3 Add a Third Constraint

The U.S. Congress recently enacted legislation regulating the consumption of North American maple by U.S. manufacturers. As a consequence, SK8MAN, Inc.'s supplier told the company that it can provide no more than 840 veneers per week. The law leads to a new constraint. Recall that to make a skateboard, seven veneers are glued together and then placed in a hydraulic press (see Figure 2.3.2). Also recall that North American maple is used only for Fancy decks (Sporty decks are made from Chinese maple).

G. F. Hurley develops the new complete linear programming formulation:

Decision Variables
    Let: x_1 = the weekly production rate of Sporty boards
             x_2 = the weekly production rate of Fancy boards

Objective Function
    Maximize: $z = 15x_1 + 35x_2$,
        where z = the amount of profit SK8MAN, Inc. earns per week.

Constraints
    Subject to:
        Shaping Time (min): $5x_1 + 15x_2 \leq 2400$
        Trucks (#): $2x_1 + 2x_2 \leq 700$
        North American Maple (#): $7x_2 \leq 840$
        Non-Negativity: $x_1 \geq 0$ and $x_2 \geq 0$

Again, G. F. Hurley graphs the constraints, as shown in Figure 2.3.7.

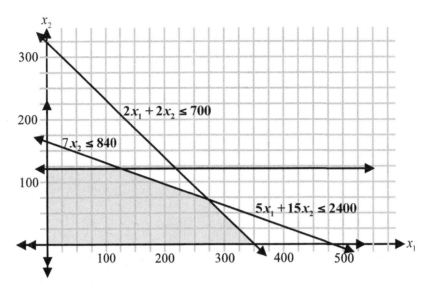

Figure 2.3.7: The feasible region after adding the North American maple constraint

When the new constraint is graphed, the feasible region changes, but the previous optimal solution (285 Sporty boards and 65 Fancy boards) is still included. Applying the corner point principle confirms that the maximum profit is unchanged because the optimal solution without the North American maple constraint remains in the feasible region after the North American maple constraint is added to the formulation. Table 2.3.3 shows the corner point calculations with the new constraint added to the formulation.

Point	Profit
(0, 0)	$15(0) + $35(0) = $0
(350, 0)	$15(350) + $35(0) = $5,250
(285, 65)	$15(285) + $35(65) = $6,550
(120, 120)	$15(120) + $35(120) = $6,000
(0, 120)	$15(0) + $35(160) = $5,600

Table 2.3.3: Evaluating the objective function at each corner point of the new feasible region

Therefore, the optimal solution remains at 285 Sporty boards and 65 Fancy boards. Since the North American maple constraint has no effect on the optimal product mix, it is called a **non-binding** constraint. The optimal product mix uses only 65(7) = 455 of the 840 available North American maple veneers (because the Sporty boards do not use North American maple, and the Fancy boards use seven North American maple veneers per board). Not all of the available resource is expended in producing the optimal solution; thus there is a **slack** of 840 – 455 = 385 North American maple veneers. The ideas of non-binding constraints and slack will be explored throughout the chapter.

2.3.4 Add a Fourth Constraint

Finally, SK8MAN, Inc.'s Chinese maple supplier has decided to limit its exports and will deliver a maximum of 1,470 veneers per week. Now G. F. Hurley needs to determine the new mix of products that will maximize weekly profit. As before, this information leads to a new constraint, but the decision variables, objective function, and previous constraints remain the same. G. F. Hurley develops the new complete linear programming formulation:

Decision Variables
Let: x_1 = the weekly production rate of Sporty boards
x_2 = the weekly production rate of Fancy boards

Objective Function
Maximize: $z = 15x_1 + 35x_2$,
where z = the amount of profit SK8MAN, Inc. earns per week.

Constraints
Subject to:
Shaping Time (min): $5x_1 + 15x_2 \leq 2400$
Trucks (#): $2x_1 + 2x_2 \leq 700$
North American Maple (#): $7x_2 \leq 840$
Chinese Maple (#): $7x_1 \leq 1470$
Non-Negativity: $x_1 \geq 0$ and $x_2 \geq 0$

Figure 2.3.8 shows the new graph of the constraints.

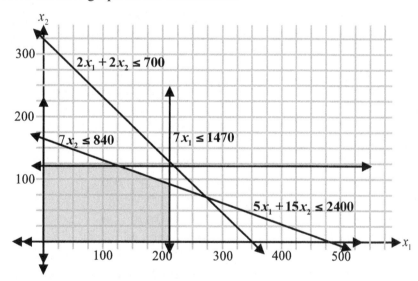

Figure 2.3.8: The feasible region after adding the Chinese maple constraint

Point	Profit
(0, 0)	$15(0) + $35(0) = $0
(210, 0)	$15(210) + $35(0) = $3,150
(210, 90)	$15(210) + $35(90) = $6,300
(120, 120)	$15(120) + $35(120) = $6,000
(0, 120)	$15(0) + $35(160) = $5,600

Table 2.3.4: Evaluating the objective function after the last constraint is added

As the graph in Figure 2.3.8 shows, the feasible region changes again. The previous optimal solution, (285, 65), is no longer feasible, so each corner point must be tested. Table 2.3.4 presents the profit associated with each corner point feasible solution. Now SK8MAN, Inc.'s maximum profit is $6,300 per week. The product mix that achieves that profit is 210 Sporty skateboards and 90 Fancy skateboards. Notice the tendency for maximum profit to decrease as the number of constraints increases.

2.3.5 Add a Third Decision Variable

SK8MAN Inc. is introducing a new product—the Pool-Runner skateboard—which is made from Chinese maple. It is wider and shorter than the Sporty board so that it will be easy to use in a pool. It takes four minutes to shape a Pool-Runner board, and SK8MAN, Inc. earns $20 for each one sold. G. F. Hurley needs to determine the new constraints and the optimal product mix. He begins by developing the new complete linear programming formulation:

2. Develop a table to organize the information about Sporty, Fancy, and Pool-Runner boards.

Decision Variables
Let: x_1 = the weekly production rate of Sporty boards
x_2 = the weekly production rate of Fancy boards
x_3 = the weekly production rate of Pool-Runner boards

Objective Function
Maximize: $z = 15x_1 + 35x_2 + 20x_3$, where z = SK8MAN, Inc.'s weekly profit.

Constraints
Subject to:
Shaping Time (min): $5x_1 + 15x_2 + 4x_3 \leq 2400$
Trucks (#): $2x_1 + 2x_2 + 2x_3 \leq 700$
North American Maple (#): $7x_2 \leq 840$
Chinese Maple (#): $7x_1 + 7x_3 \leq 1470$
Non-Negativity: $x_1 \geq 0, x_2 \geq 0$, and $x_3 \geq 0$

Adding the third decision variable makes solving this problem graphically very difficult. Graphing the feasible region with three decision variables would require three dimensions. While it is possible to do so, visualizing such a feasible region is very difficult for most people. There are two other possible ways to solve linear programming problems involving three or more decision variables. The first way is to apply a paper-and-pencil technique called the Simplex Method. This method will not be described here. Instead, the use of a spreadsheet Solver will be explored. A spreadsheet Solver applies a computer procedure to solve linear programming problems. The following directions will walk you through the steps needed to use the Solver function in Microsoft Excel to solve this problem.

2.3.6 Excel Solver and SK8MAN

Here we describe how to use Solver, an add-in to Excel, to solve linear programming problems. The steps involved are outlined below. It begins with making sure that Solver is an active add-in to Excel. There are then two major phases. The first phase comprises the steps for setting up the spreadsheet to represent all of the information in the linear programming formulation. The second phase encompasses the steps for setting up Solver to execute an algorithm to optimize the linear programming problem.

0. Install Excel Add-in "Solver"

Phase I. Set up spreadsheet
1. Input the components of the linear programming formulation
2. Write the equation for the objective function
3. Write the equations for the constraints

Phase II. Set up Solver
4. Open Solver window and specify location of objective function
5. Specify problem is maximization
6. Specify location of values of decision variables
7. Add in the regular constraints
8. Specify all decision variables are non-negative
9. Specify solution method
10. Select the appropriate options
11. Solve and Print Reports

Step 0: Add-in Solver
Open Excel and click on the "Data" tab. The Solver option should be at the top right of the menu (see Figure 2.3.9). If it is not available, it needs to be added. To add Solver, go to "Options" under the "File" menu (see Figure 2.3.10) and click on "Add-Ins" (see Figure 2.3.11). Next, choose the "Solver Add-in" and click "Go." Finally, check "Solver Add-in" and click "OK" (See Figure 2.3.12). It may take some time to install.

Figure 2.3.9: Location of Solver in Microsoft Excel 2010

Chapter 2 — Optimize Product Mix – Profit Maximization with Linear Programming

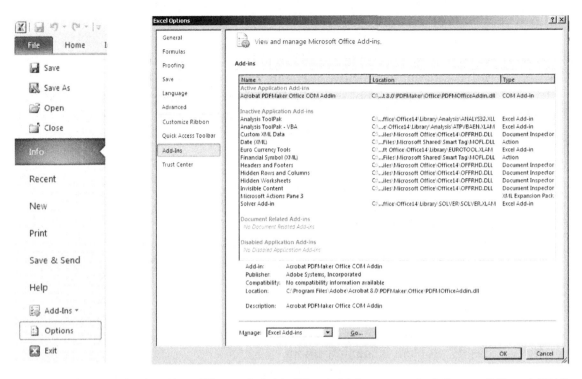

Figure 2.3.10: Choose "Options" under the "File" menu

Figure 2.3.11: The Add-Ins menu in Microsoft Excel 2010

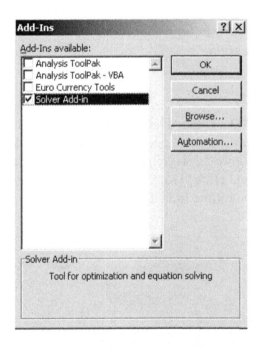

Figure 2.3.12: Choose "Solver Add-in" and click "OK"

Chapter 2 — Optimize Product Mix – Profit Maximization with Linear Programming

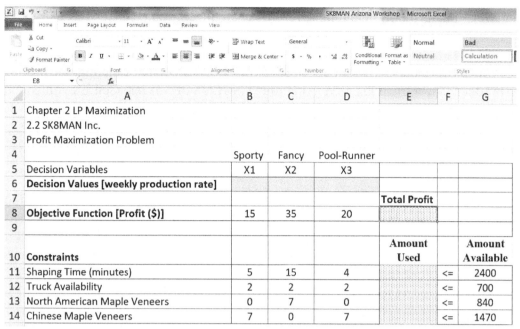

Figure 2.3.13: Setting up the problem formulation in an Excel spreadsheet

Step 1: Set Up a Spreadsheet

To set up the spreadsheet, keep in mind the complete linear programming formulation of the SK8MAN problem with three decision variables. Begin by putting a title in column A (see Figure 2.3.13). This description will make things easier when coming back to this spreadsheet at some later date.

In planning the input of the elements of the linear programming model, it is important to understand that there are 3 components of each equation: coefficients, decision variables and equation. For example, the objective function for the three decision Snowboard problem is

$$z = 15x_1 + 35 + 20x_3,$$

This equation has 3 elements.
1. A coefficient for each decision variable that represents its profit margin
2. The individual values of the decision variable represented by x_i
3. The mathematical equation which is the sum total of the product of each coefficient and each decision variable.

In Excel, the coefficients for the objective function are placed in a separate row from the values of the decision variables. Then a formula is written into a cell to perform the calculation. The same three elements apply to each constraint; the coefficients represent the rate at which a resource is consumed by one unit of that decision variable.

In the setup presented in Figure 2.3.13, Row 4, columns B, C, D contain the original names of the decision variables. Row 6 lists the corresponding representations, X_1, X_2, and X_3, used in the linear programming formulation. The information in Rows 4 and 5 is simply for clarification. Row 6 will contain the actual values of the decision variables that Solver will optimize. The cells are initially empty and treated as zeroes. Because the values in these cells are continually changing, Solver calls them "changing cells" or "changing variable cells." To identify the rows of the spreadsheet, we type in "Decision Variables" in A5 and "Decision Values" in A6.

Row 8 columns B, C, and D contain the coefficients of the objective function, the profit margin for each product. The term "Objective Function" is written in cell A8. Cell E8 will contain the formula for the total profit with a label "Total Profit" placed in the cell above, E7.

Next, we use column A to label each of the for constraints that are to be recorded in rows 11 through14.

Now, the coefficients for each of the constraints are entered in rows 11 through 14, columns B through D. The direction of each constraint inequality, all ≤ in this case, is entered in cells F11-F14. This reminds us of the nature of each individual constraint but is not actually used in any computations. Finally, the resources available that appear on right-hand side of each constraint are entered in cells G11-G14. We insert "Constraints" and "Amount Available" in cells A10 and G10 to identify the information that appears below. This is all displayed in Figure 2.3.13. Lastly, we write "Amount Used" in cell E10, to let us know that cells E11-E14 will contain the formula that calculate the amount of each resource used by the set of decision variables. This is discussed in step 3.

Step 2: Develop a Formula for the Objective Function

The objective function must be defined mathematically, so that the Solver software can compute its value. This is accomplished by entering a formula into cell E8. Notice that the label "Total Profit" is directly above this cell.

First, recall that the objective function is $z = 15x_1 + 35 + 20x_3$. To compute its value, 15 is multiplied by the weekly production rate of Sport skate boards, 35 is multiplied by the weekly production rate of Fancy skateboards, and 20 is multiplied by the weekly production rate of Pool-Runner skateboards. The objective function coefficients are stored in row 8 and the values of the decision variables in row 6. The total profits for each product is computed and then the profits are summed by typing the formula "=B6*B8+C6*C8+D6*D8 in cell E8 (see Figure 2.3.14, and notice that the formula is displayed next to *fx* in the bar above the horizontal ruler).

Chapter 2   Optimize Product Mix – Profit Maximization with Linear Programming

	E8		fx	=B6*B8+C6*C8+D6*D8		

	A	B	C	D	E	F	G
1	Chapter 2 LP Maximization						
2	2.2 SK8MAN Inc.						
3	Profit Maximization Problem						
4		Sporty	Fancy	Pool-Runner			
5	Decision Variables	X1	X2	X3			
6	Decision Values [weekly production rate]						
7					Total Profit		
8	Objective Function [Profit ($)]	15	35	20	$0.00		
9							
10	Constraints				Amount Used		Amount Available
11	Shaping Time (minutes)	5	15	4		<=	2400
12	Truck Availability	2	2	2		<=	700
13	North American Maple Veneers	0	7	0		<=	840
14	Chinese Maple Veneers	7	0	7		<=	1470

Figure 2.3.14: The formula for the objective function

Step 3: Develop Formulas for the Left-Hand Side of Each Constraint
The formulas for the left-hand sides of the constraints are written the same way as the formula for the objective function. Again, the coefficients are multiplied by the cells containing the weekly production rate of each skateboard, namely B6, C6, and D6, and then added together. These formulas are typed in cells E11-E14. In Figure 2.3.15, we have highlighted cell E11 which contains the formula for the shaping constraint "=B6*B11+C6*C11+D6*D11. Notice B6, C6, and D6 appear again since they correspond to the values of the decision variables. (see Figure 2.3.15). Note that these expressions compute the sum total of each resource that is consumed by a particular set of values of the decision variables. Currently, cells E11-E14 report a value of 0 because all of the decision variables are still 0.

	E11		fx	=B6*B11+C6*C11+D6*D11		

	A	B	C	D	E	F	G
1	Chapter 2 LP Maximization						
2	2.2 SK8MAN Inc.						
3	Profit Maximization Problem						
4		Sporty	Fancy	Pool-Runner			
5	Decision Variables	X1	X2	X3			
6	Decision Values [weekly production rate]						
7					Total Profit		
8	Objective Function [Profit ($)]	15	35	20	$0.00		
9							
10	Constraints				Amount Used		Amount Available
11	Shaping Time (minutes)	5	15	4	0.0	<=	2400
12	Truck Availability	2	2	2	0.0	<=	700
13	North American Maple Veneers	0	7	0	0.0	<=	840
14	Chinese Maple Veneers	7	0	7	0.0	<=	1470

Figure 2.3.15: Formulas for the constraint left-hand sides are entered

At this point, G. F. Hurley decided to explore the spreadsheet and the formulas in the spreadsheet. He considered a production plan of 40 Sporty, 90 Fancy and 60 Pool-Runner skateboards (See Figure 2.3.16). This solution is feasible because each of the resources used is less than the amount available. For example, 1790 minutes of shaping time is used which is less than the 2400 minutes that are available. The total profit for this solution is $4,950.

Mr. Hurley then explored the impact of doubling production of the highly profitable Fancy skateboards to 180 per week (See Figure 2.3.17). The profit for this plan would be $8,100. However, when he checked the resource constraints, he found this plan was infeasible. This plan required 3,140 minutes of shaping time which is more than the available shaping time minutes. It also needed 1,260 North American Maple Veneers but there were only 840 available each week.

	A	B	C	D	E	F	G
1	Chapter 2 LP Maximization						
2	2.2 SK8MAN Inc.						
3	Profit Maximization Problem						
4		Sporty	Fancy	Pool-Runner			
5	Decision Variables	X1	X2	X3			
6	Decision Values [weekly production rate]	40	90	60			
7					Total Profit		
8	Objective Function [Profit ($)]	15	35	20	$4,950.00		
9							
10	Constraints				Amount Used		Amount Available
11	Shaping Time (minutes)	5	15	4	1,790.0	<=	2400
12	Truck Availability	2	2	2	380.0	<=	700
13	North American Maple Veneers	0	7	0	630.0	<=	840
14	Chinese Maple Veneers	7	0	7	700.0	<=	1470

Figure 2.3.16: The spreadsheet showing a feasible solution

	A	B	C	D	E	F	G
1	Chapter 2 LP Maximization						
2	2.2 SK8MAN Inc.						
3	Profit Maximization Problem						
4		Sporty	Fancy	Pool-Runner			
5	Decision Variables	X1	X2	X3			
6	Decision Values [weekly production rate]	40	180	60			
7					Total Profit		
8	Objective Function [Profit ($)]	15	35	20	$8,100.00		
9							
10	Constraints				Amount Used		Amount Available
11	Shaping Time (minutes)	5	15	4	3,140.0	<=	2400
12	Truck Availability	2	2	2	560.0	<=	700
13	North American Maple Veneers	0	7	0	1,260.0	<=	840
14	Chinese Maple Veneers	7	0	7	700.0	<=	1470

Figure 2.3.17: The spreadsheet showing an infeasible solution

Step 4: Open Solver

Before the problem can be solved, parameters need to be set up in the Solver program. These parameters include all of the parts of the problem formulation: the decision variables, the objective function, and the constraints. Solver needs to be told where in the spreadsheet each of these parameters is located. First, click on the cell containing the objective function (cell E8). Second, go to the Data menu and choose Solver (see Figure 2.3.18). A Solver Parameters window should come up showing $E$8 as the target cell (see Figure 2.3.19). Notice that the target cell is the cell in which the objective function is defined. The dollar signs merely indicate that specific cell. When the Solver Parameters window is opened, if cell E8 is not already selected, you can type E8 with or without the $ signs.

	A	B	C	D	E	F	G
1	Chapter 2 LP Maximization						
2	2.2 SK8MAN Inc.						
3	Profit Maximization Problem						
4		Sporty	Fancy	Pool-Runner			
5	Decision Variables	X1	X2	X3			
6	Decision Values [weekly production rate]						
7					Total Profit		
8	Objective Function [Profit ($)]	15	35	20	$0.00		
9							
10	Constraints				Amount Used		Amount Available
11	Shaping Time (minutes)	5	15	4	0.0	<=	2400
12	Truck Availability	2	2	2	0.0	<=	700
13	North American Maple Veneers	0	7	0	0.0	<=	840
14	Chinese Maple Veneers	7	0	7	0.0	<=	1470

E18 =B6*B8+C6*C8+D6*D8

Figure 2.3.18: Cell E8 highlighted when Solver is opened

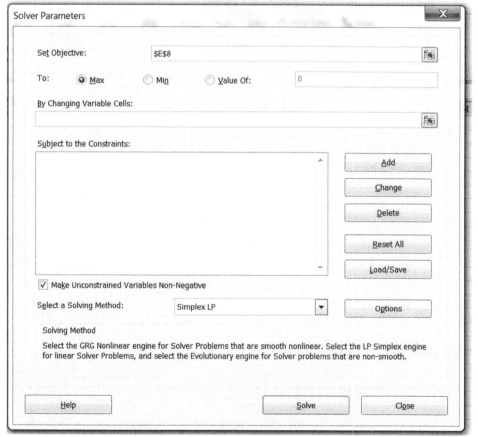

Figure 2.3.19: Solver Parameters dialogue window when Cell E8 is highlighted in the spreadsheet

Step 5: Choose the Type of Linear Programing Problem
Recall that the objective for this problem is to *maximize* profit. Therefore, in the Solver Parameters window, make sure the "Max" circle is filled in (see Figure 2.3.20).

Step 6: Identify the Decision Variable Cells – "By Changing Variable Cells"
Next, look at the "By Changing Variable Cells" title. Solver needs to be told that the decision variable values are in cells B6, C6, and D6. To do so, you could type in B6, C6, D6. However, it is preferable to use the shortcut B6:E6. In Excel a semicolon ";" between two cells, means use all cells in the range from B6 to E6. Instead of typing in this range, you can also click on cell B6 and drag the pointer across to E6. Solver will add dollar signs in the cell names, and you can leave them as they are (see Figure 2.3.20).

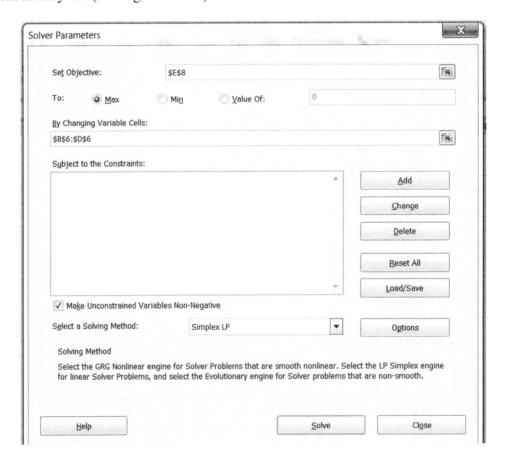

Figure 2.3.20: Identifying the decision variable cells – "By Changing Variable Cells"

Step 7: Set Up the Constraints
Click inside the box labeled "Subject to the Constraints." The constraints will be added, one at a time. Click "Add." As shown in Figure 2.3.21, a window will appear titled "Add Constraint."

For the shaping constraint in row 11, the value of the formula in cell E11 should be less than or equal to the value in cell G11. The amount of shaping time used needs to be less than or equal to 2400 minutes. To do this, type F11 (or click on F11) into "Cell Reference" and type G11 into "Constraint" (or click on F11). Use the drop-down menu to change the inequality symbol, if necessary. The default is ≤ which applies to each of the constraints in the SK8MAN example. Click "Add" and then continue to the next constraint. After adding the last constraint, click OK. Figure 2.3.22 shows the Solver Parameters window after all four constraints have been added.

Figure 2.3.21: Drop-down menu for adding a constraint

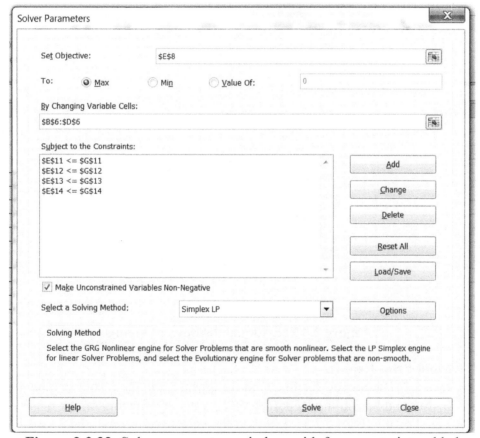

Figure 2.3.22: Solver parameters window with four constraints added

Step 8: Verify Non-Negativity Constraints

Notice that the non-negativity constraints were not included in the spreadsheet setup because Solver has a shortcut for doing so. Under the "Subject to the Constraints" box, make sure "Make Unconstrained Variables Non-Negative" is checked (see Figure 2.3.22).

Step 9: Select Solving Method

Next, a method of solution must be selected. If the box next to "Select a Solving Method" does *not* say "Simplex LP", use the drop-down menu to select "Simplex LP" (See Figure 2.3.22).

Step 10: Set Up the Solver Options

Click on the "Options" button in the Solver Parameters window (see Figure 2.3.22). In the "All Methods" tab, check the box that says "Use Automatic Scaling." Also check the box that says "Ignore Integer Constraints." Solution precision can be controlled to some extent using the "Constraint Precision" option. For most examples, the default setting (0.000001) will suffice. Finally, you can set "Solving Limits." Setting "Max Time (Seconds)" to 100 and "Iterations" to 100 works well for most problems. Figure 2.3.23 shows these settings. Click "OK" to finish.

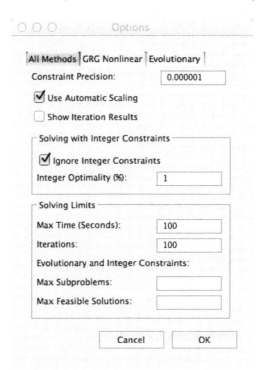

Figure 2.3.23: Adding the Solver Options

Step 11: Solve and Print Reports

Finally, click "Solve." The Solver Results window appears, and the results can be seen in the spreadsheet. Answer and Sensitivity Reports are generated by clicking on the name of the report in the window, as shown in Figure 2.3.24. Solver then creates a new tab that corresponds to a worksheet for each report.

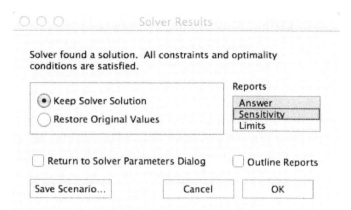

Figure 2.3.24: Generating Answer and Sensitivity Reports in Solver

To review, the steps for solving a maximization linear programming problem using Excel Solver are given in Table 2.3.5.

Step	Description
0	Add in Solver (skip this step once Solver has been added).
1	Set up a spreadsheet using the linear programming formulation.
2	Develop a formula for the objective function.
3	Develop formulas for left-hand sides of the constraints, label the type of constraint ($\leq$, $=$, $\geq$), and type in the values for the right-hand sides of the constraints.
4	Open Solver from the Data menu.
5	Choose the type of linear programming problem.
6	Identify the decision variable cells.
7	Set up the constraints.
8	Verify non-negativity constraints.
9	Select solving method from the drop-down menu.
10	Set up the Solver options.
11	"Solve" and generate Answer and Sensitivity Reports.

Table 2.3.5: Steps for solving a linear programming problem using Excel Solver

You might also look at the YouTube videos we developed. There are two file in the playlist: -excel basics and -excel solver tutorial

http://www.youtube.com/playlist?list=PLBcLt4kL3jChKdrE2ex5RV0OWymb70gAk

2.3.7 Three Excel Shortcuts

There are three shortcuts that speed up the process of creating the formula and setting up Solver. The shortcuts also reduce the likelihood of making an error.

Shortcut 1. The copy and paste feature of the spreadsheet can be used to speed up the process of adding the constraint formulas. When all of those formulas have the same structure as the formula for the objective function, the objective function formula can simply be copied and pasted into each of the cells containing the left-hand side of a constraint. Before copying the formula, however, the row numbers for the decision variables must be locked, so that Solver will always refer to the specific row containing the values of the decision variables. In our example, row 6 is locked by inserting a $ to the left of the 6 in each term of the objective function formula. When finished, the formula looks like this: =B8*B$6+C8*C$6+D8*D$6. Now, when this formula is pasted into cells E11-E14, the row number of the first factor in each term of the formula will be changed to the current row number. However, the row number of the second factor in each term is locked at row 6 where the decision variables are. For example, when pasted into row 11, the formula will read: =B11*B$6+C11*C$6+D11*D$6.

Shortcut 2. The second shortcut uses a built-in spreadsheet function, SUMPRODUCT. The SUMPRODUCT function allows you to multiply each value in one array by a corresponding value in an equal sized array and sum all of the products. To write the formula for the objective function simply type in cell F8.

=SUMPRODUCT(B6:D6,B8:D8)

The colon between B6 and D6 tells Excel to use every value in the range of cells between B6 and D6. After placing a comma, type in the second range B8:D8. Be sure to have parentheses before and after the ranges.

	E8	▼	*f_x*	=SUMPRODUCT(B$6:D$6,B8:D8)		
	A	B	C	D	E	F	G
1	Chapter 2 LP Maximization						
2	2.2 SK8MAN Inc.						
3	Profit Maximization Problem						
4		Sporty	Fancy	Pool-Runner			
5	Decision Variables	X1	X2	X3			
6	Decision Values [weekly production rate]						
7					Total Profit		
8	Objective Function [Profit ($)]	15	35	20	$0.00		
9							
10	Constraints				Amount Used		Amount Available
11	Shaping Time (minutes)	5	15	4	0.0	<=	2400
12	Truck Availability	2	2	2	0.0	<=	700
13	North American Maple Veneers	0	7	0	0.0	<=	840
14	Chinese Maple Veneers	7	0	7	0.0	<=	1470

Figure 2.3.25: SUMPRODUCT Function

We are going to copy and paste cell E8 into cells E11-E14. We need to insert the $ sign to establish row 6, the decision variable row, as unchanging.

=SUMPRODUCT(B$6:D$6,B8:D8)

Now the formula in cell E8 can be copied and pasted into cells E11-E14 for the constraint left-hand side values. Cell F11 should now contain

= SUMPRODUCT(B$6:D$6,B11:D11)

Shortcut 3. In the Solver Parameters window, it is possible to add a group of constraints of the same type in one step. In this example all of the constraints are of the type <=. To use this shortcut, after clicking "Add" in the Solver Parameters window, type in a range of cells in each box. In the "Cell Reference" box record D11:D14 and in the "Constraint" box record G11:G14 boxes (see Figure 2.3.26). Solver interprets these ranges as meaning E11 must be less than or equal to G11, E12 must be less than or equal to G12, etc. Figure 2.3.27 shows how this changes the way constraints are listed in the Solver Parameter window.

Figure 2.3.26: Entering four constraints at one time.

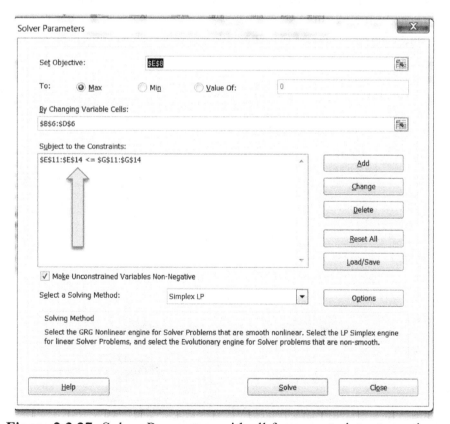

Figure 2.3.27: Solver Parameters with all four constraints at one time.

Section 2.4: The Pallas Sport Shoe Company

The Pallas Sport Shoe Company manufactures six different lines of sport shoes: High Rise, Max-Riser, Stuff It, Zoom, Sprint, and Rocket. Table 2.4.1 displays the amount of profit generated by each pair of shoes for each of these six lines. The production manager of the company would like to determine the daily production rates for each line of shoes that will maximize profit.

Product	High Rise	Max-Riser	Stuff It	Zoom	Sprint	Rocket
Profit	$18	$23	$22	$20	$18	$19

Table 2.4.1: Profit per pair for six lines of sport shoes

There are six main steps in the production of a pair of sport shoes at Pallas. Some of these steps can be seen in Figure 2.4.1.

1. **Stamping**: The parts that go together to form the upper portion of the shoe are cut using patterns on a large stamping machine. This process resembles cutting dough with a cookie cutter.
2. **Upper Finishing**: These parts are stitched or cemented together to form an upper, and holes for the laces are punched.
3. **Insole Stitching**: An insole is stitched to the sides of the upper.
4. **Molding**: The completed upper is then placed on a plastic mold, called a *last*, to form the final shape of the shoe.
5. **Sole-to-Upper Joining**: After the upper has been molded, it is cemented to the bottom sole using heat and pressure.
6. **Inspecting**: Finally, the shoe is inspected, and any excess cement is removed.

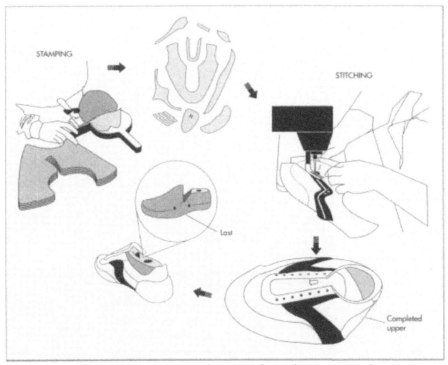

Figure 2.4.1: The steps in manufacturing a sport shoe

The time it takes to complete each of these steps differs across the six lines, and the total time available for each process constrains the daily production rates. Table 2.4.2 shows the time, in minutes, required for each of the six production steps for each of the six lines of sport shoe produced by Pallas as well as the total number of minutes available per day for each step.

Task	High Rise	Max-Riser	Stuff It	Zoom	Sprint	Rocket	Total Time Available
Stamping	1.25	2	1.5	1.75	1	1.25	420
Upper Finishing	3.5	3.75	5	3	4	4.25	1,260
Insole Stitching	2	3.25	2.75	2.25	3	2.5	840
Molding	5.5	6	7	6.5	8	5	2,100
Sole-to-Upper Joining	7.5	7.25	6	7	6.75	6.5	2,100
Inspecting	2	3	2	3	2	3	840

Table 2.4.2: Time, in minutes, per production task for each line of shoes and total time available

1. Based on the information in Tables 2.4.1 and 2.4.2, predict what the optimal solution will be for this problem. Explain your reasoning.

2.4.1 Problem Formulation

The linear programming formulation of the Pallas Sport Shoe Company problem appears below.

<u>Decision Variables</u>
Let:
 x_1 = the daily production rate of High Rise
 x_2 = the daily production rate of Max-Riser
 x_3 = the daily production rate of Stuff It
 x_4 = the daily production rate of Zoom
 x_5 = the daily production rate of Sprint
 x_6 = the daily production rate of Rocket

<u>Objective Function</u>
Maximize: $z = 18x_1 + 23x_2 + 22x_3 + 20x_4 + 18x_5 + 19x_6$,
where z = the amount of profit Pallas Sport Shoe Company earns per day

<u>Constraints</u>
Subject to:
Stamping Time: $1.25x_1 + 2x_2 + 1.5x_3 + 1.75x_4 + x_5 + 1.25x_6 \leq 420$
Upper Finishing Time: $3.5x_1 + 3.75x_2 + 5x_3 + 3x_4 + 4x_5 + 4.25x_6 \leq 1,260$
Insole Stitching Time: $2x_1 + 3.25x_2 + 2.75x_3 + 2.25x_4 + 3x_5 + 2.5x_6 \leq 840$
Molding Time: $5.5x_1 + 6x_2 + 7x_3 + 6.5x_4 + 8x_5 + 5x_6 \leq 2,100$
Sole-to-Upper Joining Time: $7.5x_1 + 7.25x_2 + 6x_3 + 7x_4 + 6.75x_5 + 6.5x_6 \leq 2,100$
Inspecting Time: $2x_1 + 3x_2 + 2x_3 + 3x_4 + 2x_5 + 3x_6 \leq 840$
Non-Negativity: $x_1 \geq 0, x_2 \geq 0, x_3 \geq 0, x_4 \geq 0, x_5 \geq 0, x_6 \geq 0$

Figure 2.4.2 contains this formulation in an Excel spreadsheet format for use with Solver.

	A	B	C	D	E	F	G	H	I	J
1	Chapter 2: LP Maximization									
2	2.3 Pallas Sport Show Company									
3	Profit Maximization									
4										
5	Decision Variable	High Rise (x_1)	Max-Riser (x_2)	Stuff It (x_3)	Zoom (x_4)	Sprint (x_5)	Rocket (x_6)			
6	Decision Values [daily production rate]									
7										Total Profit
8	Objective Function [Profit ($)]	18	23	22	20	18	19			$0.00
9										
10	Constraints							Used		Available
11	Cutting Time (minutes)	1.25	2	1.5	1.75	1	1.25	0	≤	420
12	Upper Finishing Time (minutes)	3.5	3.75	5	3	4	4.25	0	≤	1260
13	Insole Stitching Time (minutes)	2	3.25	2.75	2.25	3	2.5	0	≤	840
14	Molding Time (minutes)	5.5	6	7	6.5	8	5	0	≤	2100
15	Sole-to-Upper Joining Time (minutes)	7.5	7.25	6	7	6.75	6.5	0	≤	2100
16	Inspecting Time (minutes)	2	3	2	3	2	3	0	≤	840

Figure 2.4.2: An Excel spreadsheet formulation of the Pallas Shoe problem

2.4.2 Problem Solution

After solving this linear programming problem in Excel, an Answer Report can be generated, as shown in Figure 2.4.3. Figure 2.4.4 shows this Answer Report.

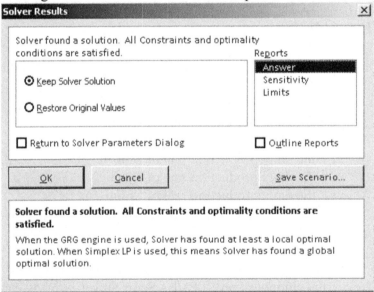

Figure 2.4.3: Generating an Answer Report in Excel Solver

Objective Cell (Max)

Cell	Name	Original Value	Final Value
$J$8	Objective Function [Profit ($)] Total Profit	$0.00	$6,132.57

Variable Cells

Cell	Name	Original Value	Final Value	Integer
$B$6	Decision Values [daily production rate] High Rise (x1)	0	0	Contin
$C$6	Decision Values [daily production rate] Max-Riser (x2)	0	4.282944345	Contin
$D$6	Decision Values [daily production rate] Stuff It (x3)	0	45.12172352	Contin
$E$6	Decision Values [daily production rate] Zoom (x4)	0	72.32746858	Contin
$F$6	Decision Values [daily production rate] Sprint (x5)	0	104.9019749	Contin
$G$6	Decision Values [daily production rate] Rocket (x6)	0	89.82118492	Contin

Constraints

Cell	Name	Cell Value	Formula	Status	Slack
$H$11	Cutting Time (minutes) Used	420	$H$11<=$J$11	Binding	0
$H$12	Upper Finishing Time (minutes) Used	1260	$H$12<=$J$12	Binding	0
$H$13	Insole Stitching Time (minutes) Used	840	$H$13<=$J$13	Binding	0
$H$14	Molding Time (minutes) Used	2100	$H$14<=$J$14	Binding	0
$H$15	Sole-to-Upper Joining Time (minutes) Used	2100	$H$15<=$J$15	Binding	0
$H$16	Inspecting Time (minutes) Used	799.3421903	$H$16<=$J$16	Not Binding	40.65780969

Figure 2.4.4: Answer Report for Pallas Sport Shoe Company

Notice that the Answer Report is split into three sections: (1) Objective Cell (Max), (2) Variable Cells, and (3) Constraints.

2. Use the "Objective Cell (Max)" section of the Answer Report in Figure 2.4.4 to complete the following.
   a. Why is "$J$8" listed under "Cell?"
   b. Why do you think "Objective Function [Profit ($)] Total Profit" is listed under "Name?"
   c. Why do you think the "Original Value" is 0?
   d. What is the "Final Value" referring to?

3. Use the "Variable Cells" section of the Answer Report in Figure 2.4.4 to complete the following.
   a. What are the "Original Value" and "Final Value" columns referring to?
   b. Interpret the information given in the "Final Value" column in terms of the problem context.
   c. How do you think the production manager should handle the decimal values that appear in the "Final Value" column?

4. Use the "Constraints" section of the Answer Report in Figure 2.4.4 to complete the following.
   a. Interpret the information given in the "Cell Value" column in terms of the problem context.
   b. The first five constraints are binding and the sixth one is not. What does this mean in terms of the problem context?
   c. How are the values in the "Slack" column calculated?

d. If you were only given the slack value for a constraint, how could you determine whether that constraint is binding?

5. Which of the six sport shoe lines should be produced, and at what daily rates, in order to maximize profit? (Approximate to two decimal places.)

Suppose Pallas Sport Shoe Company considers adding another five minutes of cutting time each day. Therefore, the cutting time constraint is changed from 420 to 425. The Answer Report for this new scenario is shown in Figure 2.4.5.

Objective Cell (Max)

Cell	Name	Original Value	Final Value
$J$8	Objective Function [Profit ($)] Total Profit	$0.00	$6,160.47

Variable Cells

Cell	Name	Original Value	Final Value	Integer
$B$6	Decision Values [daily production rate] High Rise (x1)	0	0	Contin
$C$6	Decision Values [daily production rate] Max-Riser (x2)	0	6.700179533	Contin
$D$6	Decision Values [daily production rate] Stuff It (x3)	0	48.89766607	Contin
$E$6	Decision Values [daily production rate] Zoom (x4)	0	74.55655296	Contin
$F$6	Decision Values [daily production rate] Sprint (x5)	0	100.4452424	Contin
$G$6	Decision Values [daily production rate] Rocket (x6)	0	85.86714542	Contin

Constraints

Cell	Name	Cell Value	Formula	Status	Slack
$H$11	Cutting Time (minutes) Used	425	$H$11<=$J$11	Binding	0
$H$12	Upper Finishing Time (minutes) Used	1260	$H$12<=$J$12	Binding	0
$H$13	Insole Stitching Time (minutes) Used	840	$H$13<=$J$13	Binding	0
$H$14	Molding Time (minutes) Used	2100	$H$14<=$J$14	Binding	0
$H$15	Sole-to-Upper Joining Time (minutes) Used	2100	$H$15<=$J$15	Binding	0
$H$16	Inspecting Time (minutes) Used	800.0574506	$H$16<=$J$16	Not Binding	39.94254937

Figure 2.4.5: Answer Report for Pallas Sport Shoe Company with 425-minute Cutting Time constraint

6. How does the Answer Report in Figure 2.4.5 differ from the one in Figure 2.4.4?

7. Is the first constraint still binding? Do you think Pallas Sport Shoe Company should add this extra five minutes of cutting time each day? Explain your reasoning.

Next, suppose Pallas Sport Shoe Company considers subtracting (rather than adding) five minutes of cutting time each day. Therefore, the cutting time constraint is changed from 420 to 415. The Answer Report for this new scenario is shown in Figure 2.4.6.

Objective Cell (Max)

Cell	Name	Original Value	Final Value
$J$8	Objective Function [Profit ($)] Total Profit	$0.00	$6,104.67

Variable Cells

Cell	Name	Original Value	Final Value	Integer
$B$6	Decision Values [daily production rate] High Rise (x1)	0	0	Contin
$C$6	Decision Values [daily production rate] Max-Riser (x2)	0	1.865709156	Contin
$D$6	Decision Values [daily production rate] Stuff It (x3)	0	41.34578097	Contin
$E$6	Decision Values [daily production rate] Zoom (x4)	0	70.0983842	Contin
$F$6	Decision Values [daily production rate] Sprint (x5)	0	109.3587074	Contin
$G$6	Decision Values [daily production rate] Rocket (x6)	0	93.77522442	Contin

Constraints

Cell	Name	Cell Value	Formula	Status	Slack
$H$11	Cutting Time (minutes) Used	415	$H$11<=$J$11	Binding	0
$H$12	Upper Finishing Time (minutes) Used	1260	$H$12<=$J$12	Binding	0
$H$13	Insole Stitching Time (minutes) Used	840	$H$13<=$J$13	Binding	0
$H$14	Molding Time (minutes) Used	2100	$H$14<=$J$14	Binding	0
$H$15	Sole-to-Upper Joining Time (minutes) Used	2100	$H$15<=$J$15	Binding	0
$H$16	Inspecting Time (minutes) Used	798.62693	$H$16<=$J$16	Not Binding	41.37307002

Figure 2.4.6: Answer Report for 415-minute Cutting Time constraint

8. How does the Answer Report in Figure 2.4.6 differ from the one in Figure 2.4.4?

9. Is the first constraint still binding? Do you think Pallas Sport Shoe Company should subtract this five minutes of cutting time each day? Explain your reasoning.

Now, suppose Pallas Sport Shoe Company considers adding another 130 minutes of cutting time each day. Therefore, the cutting time constraint is changed from 420 to 550. The Answer Report for this new scenario is shown in Figure 2.4.7.

Objective Cell (Max)

Cell	Name	Original Value	Final Value
$J$8	Objective Function [Profit ($)] Total Profit	$0.00	$6,781.57

Variable Cells

Cell	Name	Original Value	Final Value	Integer
$B$6	Decision Values [daily production rate] High Rise (x1)	0	0	Contin
$C$6	Decision Values [daily production rate] Max-Riser (x2)	0	61.57068063	Contin
$D$6	Decision Values [daily production rate] Stuff It (x3)	0	131.9371728	Contin
$E$6	Decision Values [daily production rate] Zoom (x4)	0	123.1413613	Contin
$F$6	Decision Values [daily production rate] Sprint (x5)	0	0	Contin
$G$6	Decision Values [daily production rate] Rocket (x6)	0	0	Contin

Constraints

Cell	Name	Cell Value	Formula	Status	Slack
$H$11	Cutting Time (minutes) Used	536.5445026	$H$11<=$J$11	Not Binding	13.45549738
$H$12	Upper Finishing Time (minutes) Used	1260	$H$12<=$J$12	Binding	0
$H$13	Insole Stitching Time (minutes) Used	840	$H$13<=$J$13	Binding	0
$H$14	Molding Time (minutes) Used	2093.403141	$H$14<=$J$14	Not Binding	6.596858639
$H$15	Sole-to-Upper Joining Time (minutes) Used	2100	$H$15<=$J$15	Binding	0
$H$16	Inspecting Time (minutes) Used	818.0104712	$H$16<=$J$16	Not Binding	21.9895288

Figure 2.4.7: Answer Report for 550-minute Cutting Time constraint

10. How does the Answer Report in Figure 2.4.7 differ from the one in Figure 2.4.4?

11. Do you think Pallas Sport Shoe Company should add this extra 130 minutes of cutting time each day? Explain your reasoning.

Recall that the decision variable x_1 is not in the optimal solution. A logical question to ask is whether increasing the profitability of x_1 could allow it to enter the optimal solution and, if so, how much of an increase would be necessary. Return again to the spreadsheet in Figure 2.4.2, reset the cutting time constraint to its original value of 420 minutes, and change the value of the objective function coefficient of x_1 from 18 to 19. The Answer Report for this new scenario is shown in Figure 2.4.8.

12. In terms of the problem, what does this change represent?

Objective Cell (Max)

Cell	Name	Original Value	Final Value
$J$8	Objective Function [Profit ($)] Total Profit	$0.00	$6,216.90

Variable Cells

Cell	Name	Original Value	Final Value	Integer
$B$6	Decision Values [daily production rate] High Rise (x1)	0	170.6951872	Contin
$C$6	Decision Values [daily production rate] Max-Riser (x2)	0	8.983957219	Contin
$D$6	Decision Values [daily production rate] Stuff It (x3)	0	125.7754011	Contin
$E$6	Decision Values [daily production rate] Zoom (x4)	0	0	Contin
$F$6	Decision Values [daily production rate] Sprint (x5)	0	0	Contin
$G$6	Decision Values [daily production rate] Rocket (x6)	0	0	Contin

Constraints

Cell	Name	Cell Value	Formula	Status	Slack
$H$11	Cutting Time (minutes) Used	420	$H$11<=$J$11	Binding	0
$H$12	Upper Finishing Time (minutes) Used	1260	$H$12<=$J$12	Binding	0
$H$13	Insole Stitching Time (minutes) Used	716.4705882	$H$13<=$J$13	Not Binding	123.5294118
$H$14	Molding Time (minutes) Used	1873.15508	$H$14<=$J$14	Not Binding	226.8449198
$H$15	Sole-to-Upper Joining Time (minutes) Used	2100	$H$15<=$J$15	Binding	0
$H$16	Inspecting Time (minutes) Used	619.8930481	$H$16<=$J$16	Not Binding	220.1069519

Figure 2.4.8: Answer Report for $19 High Rise shoe profit

13. How does the Answer Report in Figure 2.4.8 differ from the one in Figure 2.4.4?

14. Why may Pallas Sport Shoe Company not want this new optimal solution?

15. Explain, in your own words, the usefulness of Answer Reports.

Chapter 2 (LP Maximization) Homework Questions

1. Anderson Cell Phone Company has started cell phone production. It produces smart phones and standard phones. Initially, they hired 10 workers for the assembly line. The workers are paid for eight hours per day. However, they spend only seven hours assembling the phones because of a 30-minute lunch break and two 15-minute breaks. A smart phone takes 2.5 minutes to assemble and a standard phone takes 1.5 minutes to assemble. The company receives a delivery of 2000 LCD screens per day from its supplier. Profit margins for a smart phone and a standard phone are $40 and $30, respectively. Anderson Company is interested in determining the product mix that gives them the highest daily profit.

    a. Define the decision variables for the problem.

    b. Use the decision variables to define the objective function.

    c. Use decision variables to define the assembly constraint.

    d. Use decision variables to define the screen constraint.

    e. Graph the feasible region.

    f. Pick two points in the feasible region and calculate the profit for each point.

    g. What are the corner points of the feasible region?

    h. Calculate the profit at each corner point.

    i. Which point gives the highest profit? What is the optimal product mix?

    j. Find the optimal production plan with Excel Solver to verify the optimal solution.

2. GA Sports makes professional and regular soccer balls. Typically, soccer balls are made up of four elements: the cover, the stitching, the lining, and the bladder. GA Sports uses a synthetic leather cover for professional balls and puts four layers of cotton linings underneath. It uses a rubber cover for regular balls and puts three layers of cotton linings underneath. The cover and the linings are created by cutting large pieces of each material into 32 panels for a professional soccer ball and 18 panels for a regular soccer ball. It takes 14 minutes to carefully cut the casing for a professional ball. It takes only seven minutes to cut the material for a regular ball. GA Sports has one experienced cutter who works seven hours per day. The 32 panels of a professional soccer ball are stitched together. An experienced worker can stitch four balls in a workday. The company has four experienced stitchers. The regular balls are thermally molded on a special machine in twelve minutes. The company has one machine that is run by a machinist for six hours per day. The company makes $15 profit on professional balls and $10 on regular balls.

    a. Define the decision variables for the problem.

b. Use the decision variables to define the objective function.

c. Use decision variables to define the cutting time constraint.

d. Use decision variables to define the stitching time constraint.

e. Formulate the problem.

f. Find the feasible region graphically.

g. Find the optimal production plan using MS Excel Solver.

3. Family Cow is a small dairy. The owner is the only person who works in the dairy; he works eight hours per day. He makes and sells cream and butter and donates the skim milk to a charity nearby. He is able to make 1 quart of cream from 10 quarts of milk or 1 pound of butter from 20 quarts of milk. It takes 20 minutes to get a quart of cream and 30 minutes to make a pound of butter. The dairy has 300 quarts of milk daily. Family Cow's prices are higher because the products are organic: a quart of organic cream is sold for $7 and a pound of organic butter for $12. He can only sell eight pounds of butter per day to a local store.

    a. Define the decision variables for the problem.

    b. Use the decision variables to define the objective function.

    c. Formulate the problem.

    d. Determine the optimal solution graphically.

    e. Find the optimal production plan using MS Excel Solver.

4. The owner of The Family Cow dairy has been using the optimal production plan for several months. He is considering adding plain yogurt and cheese to the products he sells. It takes 10 quarts of milk to produce seven pounds of yogurt or three pounds of cheese. Prices of yogurt and cheese are $3.5/lb. and $5.5/lb., respectively. It takes 10 minutes to make a pound of yogurt and 15 minutes for a pound of cheese.

    a. Define the additional decision variables.

    b. Formulate the problem.

    c. Find the new optimal product mix and the daily profit using MS Excel Solver. How many different products does he make? How much more money will he make by offering two more products? Would you recommend he change his production plan?

5. Katia has won $200,000 from the lottery. She considers investing the money in a bond fund and a domestic stock fund. The projected annual return for the bond fund is 8% and for the stock fund is 15%. Her friend is experienced with investments. She suggests she invest at most $75,000 in the stock fund. However, the stock and bond funds could go down in value. The brokerage firm told Katia that the most she could lose in the next year is 5% of her original investment in the bond fund and 20% of her original investment in the stock fund. She wants to limit her total potential losses to no more than $20,000. How should she invest her winnings?

   a. Define the decision variables for the problem.

   b. Formulate the problem.

   c. Determine the optimal investment plan using MS Excel Solver.

6. In addition to the bond fund and the domestic stock fund, Katia also considers investing in an international stock fund. The projected annual return for that stock fund is 22%. The maximum investment in any stocks is still $75,000. The most she could lose in the next year in this international stock is 25%.

   a. What is her optimal investment strategy?

   b. Will she make more money than before?

7. John Farmer is studying operations research in school. He is curious about applying what he learns in class to actual problems. John's father owns a farm in Missouri and has 640 acres under cultivation. John would like to help his father determine the mix of corn and soybeans to plant that would maximize his father's profit. Table 2.4.1 contains some data that John collected about corn and soybean production in Missouri.

	Corn	Soybeans
Price ($/bushel)	2.90-3.50	7.85-8.85
Yield (bushels/acre)	155.8	41.4
Seed Cost ($/acre)	45.50	34.00
Fertilizer ($/acre)	88.10	37.80
Fuel ($/acre)	24.40	10.00
Worker ($/acre)	13.00	9.50

Table 1: 2007-08 USDA corn and soybean estimates for Missouri

   a. In order to formulate the problem, John must know the price per bushel at which corn and soybeans can be sold. However, the USDA data contains a range of values for both of these crops. What value do you think John should use for each? Why?

   b. What is the largest gross revenue before considering expenses that Mr. Farmer can make? What crop mix produces this gross revenue?

c. What is the largest net revenue after expenses that Mr. Farmer can make? What crop mix produces this net revenue?

d. Suppose Mr. Farmer has budgeted only $60,000 to cover all of the expenses of his crop production. What is the largest net revenue he can earn under this constraint?

8. Corn and soybeans are used in the production of biofuels. Biofuel consumption is important for the environment, because greenhouse gas emissions are reduced 12% by ethanol combustion and 41% by biodiesel combustion. The total corn and soybean production in the United States can meet only 12% of the demand for gasoline and 6% of the demand for diesel fuel. In order to encourage corn production, the USDA pays a subsidy to increase the price per bushel to $3.50. However, to get the subsidy, the farmer must produce at least 40,000 bushels of corn. Assuming that the price per bushel that Mr. Farmer can sell his corn for without the subsidy is $2.90, should Mr. Farmer accept the constraint of producing at least 40,000 bushels of corn in order to earn the subsidy?

9. After doing some research, John learned that soybean followed by corn a year after, increases corn yield by 7.5% and saves 25% of the soybean residue nitrogen, which will reduce the fertilizer use in corn production by $2.50 per acre. The reduction in fertilizer use will also reduce its harmful effect to the environment. In the news, he heard that nitrate leaching causes surface and ground water to degrade. This will harm the living things that use water for drinking or swimming. For all of these reasons, John wants to convince his father, who planted 200 acres of land with soybean last year, to begin to rotate corn and soybeans. Add a constraint that the maximum amount of corn is equal to these 200 acres. Should John's father begin to rotate his corn and soybean crops by increasing his corn planting?

10. John gathered some data related to wheat production (Table 2.4.2). Go back to the original formulation stated in question 7. Is it profitable to produce some wheat? Use the midpoint of price range ($3.95) to determine net revenue per acre

	Corn	Soybean	Wheat
Price ($/bushel)	2.90-3.50	7.85-8.85	3.75-4.15
Yield (bushels/acre)	155.8	41.4	60
Seed Cost ($/acre)	45.50	34	24
Fertilizer ($/acre)	88.10	37.80	69.25
Fuel ($/acre)	24.40	10	40
Worker ($/acre)	13	9.50	9

Table 2: 2007-2008 Corn, soybean, and wheat estimates for Missouri

11. At the end of Problem 2.1, you were asked to formulate the Computer Flips problem after the addition of two new products. Your formulation should have used four decision variables and should have included eight constraints, including four non-negativity constraints. In an Excel spreadsheet, set up this formulation, and use Solver to obtain a solution.

   a. What is the optimal solution?

b. What does this optimal product mix imply about the planned addition of new products?

c. What research do you think the Sales Department should conduct before implementing the optimal product mix?

12. Elegant Fragrances, Ltd. decides to produce two new perfumes, *L'Arbre d'Amour* and *Evening Rose*. The factory management asked the industrial engineering department to develop a mathematical model for maximizing the profit obtained from producing these products. There are also some limitations in resources, budget and the capacity of the factory that should be considered in the model. At the beginning, the industrial engineering team studied the problem and gathered information from which a model can be developed. The team collected the following information:
   - Each perfume is made of two main components: a fragrant perfume oil and a solvent. The solvent, such as a combination of ethanol and water, is necessary to reduce the allergic reactions of skin to the perfume oil. The solvent is a large percentage of the overall final product.
   - A fragrant oil for *L'Arbre d'Amour* is obtained from Mango Pulp, Tea Leaves, and Juniper Berry, and a fragrant oil for *Evening Rose* is obtained from Mango Pulp, Tea Leaves, and White Rose.
   - There are two main processes in the production of these perfumes: extraction and blending. In the extraction stage, physical and chemical processes change the raw materials and the perfume oil is extracted. In the blending stage, the perfume oil is blended with the solvent. However, Elegant Fragrances does not work directly with the fragrance raw materials, but instead purchases fragrance essences from suppliers and only blends them.
   -

			L'Arbre d'Amou	*Evening Rose*
Income ($ per pound)			370	215
Percentage of raw materials	**Fragrance oil (pure)**	Mango	3%	5%
		Tea leaves	5%	7%
		Juniper berry	7%	0
		White rose	0	8%
	Solvent	Ethanol and water	85%	80%
Process costs ($ per pound)		Blending	12	12

Table 3: Some information about the perfumes

Component	Availability (pounds per year)	Cost ($ per pound)
Mango	1,400	14
Tea leaves	1,600	10
Juniper berry	1,000	384
White rose	1,500	56
Ethanol and water	Unlimited	7

Table 4: Ingredient information

The total cost for making each perfume is the $12 processing cost plus the cost of ingredients. The cost for ingredients to make ***L'Arbre d'Amou*** is determined by multiplying the percentage of each ingredient by the corresponding costs.

Ingredient cost = .03(14) + .05(10) + .07(384) + 0(56) + .85(7) = $33.75
Net Profit = 370.00 − 12 − 33.75 = $324.25

    a. Determine the net profit for ***Evening Rose***.

    b. Define the decision variables as the amount of annual production of each perfume and help the industrial engineering team formulate this problem for maximizing the profit within the given constraints.

    c. Use Solver to obtain the optimal solution for the Elegant Fragrances problem.

13. The management at Elegant Fragrances is considering producing two new perfumes, *Evergreen* and *Embrasser du Soir*. The income and costs, as well as the key ingredients and their proportions for the new perfumes are given in Table 2.4.5 below. If the availability of resources is unchanged, what are the optimal production rates for each of the four perfumes so as to maximize profit?

			Evergreen	*Embrasser du Soir*
	Income ($ per pound)		490	235
Percentage of raw materials	Fragrance Oil (Pure)	Mango	8%	7%
		Tea leaves	7%	5%
		Juniper berry	9%	0
		White rose	0	10%
	Solvent	Ethanol and water	76%	78%
Process costs ($ per pound)		Blending	$12	$12

Table 5: New Perfumes Information

Chapter 2 Summary

What have we learned?

Linear programming is a process of taking a real world situation, modeling it with inequalities, and finding the best or optimal solution.

- Modeling the situation
    - We start by finding the decision variables – what things can you choose? Typically this is how much to make of a particular product or how much to invest in a particular option. We assign a variable for each of these choices.
    - Next we write an objective function that captures the goal of the problem. – What will determine when you have found the optimal solution? This is the equation that we want to maximize.
    - Finally we define the constraints - those things that limit our choices. These are typically the amount of money, time, people, or resources available.

- Once we have defined our problem, we use a spreadsheet program such as Microsoft Excel to find the optimal solution. After entering the inequalities we set up the Solver parameters and run Solver. This gives us an answer and sensitivity report.

- The answer report will show:
    - The objective function's final value
    - The value for each of the decision variables
    - The amount of each constraint that is used

Terms

Answer Report	A report that details the optimal solution, lists whether constraints are binding or non-binding, and gives the slack for each constraint
Binding Constraint	A constraint that is satisfied as a strict equality in the optimal solution; all of the available constraint is used
Constraint	A condition that must be satisfied, represented by equations or inequalities
Decision Variable	A quantity that the decision-maker controls
Feasible Solution	A solution that satisfies all the constraints
Final Value	A column in the Answer and Sensitivity Reports that refers to the number in the decision variable cells *after* you use Solver (i.e., the decision variable and objective function values for the optimal solution)
Line of Constant Profit	A line representing the objective function, where every point on the line generates the same profit
Linear Programming	A mathematical technique for finding the optimal value of a linear objective function subject to linear constraints when the decision variables can take on fractional values.
Mathematical Programming	A mathematical approach to allocating limited resources among options in an optimal manner (includes linear programming, integer programming, and binary programming)
Non-Binding Constraint	A constraint for which all available resources are not used
Objective Function	The function that is to be optimized
Optimal Solution	The feasible solution with the best value for the objective function
Original Value	A column in the Answer Report that refers to the number in the decision variable cells *before* you use Solver
Product Mix	The composition of all goods and/or services being produced

Production Rate   The number of products made in a given period of time

Slack   The amount of a resource that is not used in the optimal solution

Chapter 2 (LP Maximization) Objectives

You should be able to:

- Identify the decision variables

- Define the objective function by finding the goal to be solved for the situation

- Identify the constraints and write inequalities to model them.

- Enter each of these into Microsoft Excel

- Use Solver to find the Optimal Solution and generate Answer and Sensitivity Reports

- Analyze the Answer Report

Troubleshooting

What can go wrong?

Troubleshooting is a valuable skill when using Excel and Solver. Quite often a problem is developed and solved and the answer will not make sense. As you work through this book try to remember the mistakes you make so you can avoid them in the future.

- One thing that students may do when working through the problems with the text is to look ahead. They realize that the answers to the questions are on the next page and will copy them onto their spreadsheet. This can lead to problems running Solver. When setting up the constraints, the right side must be a value – how much of the resource is available. The left side must be a formula that multiplies the decision variables by the amount used for each one. This formula does not appear on the screen, only the value appears. Typing values printed in the book rather than the formula will cause a problem.
- Another common mistake is to forget to change the direction of the inequality in Solver. The default in Solver is a less than or equal (<=) constraint. For example, if you have up to 40 hours to produce something, you use the <= constraint. Some situations require something to be at least a certain value. For example if you make tables and chairs you have to make at least 4 chairs for each table. This requires you to use the greater than or equal (>=) constraint.
- When a model changes, don't forget to make the changes in Solver. Recall that when we develop a spreadsheet, we start by entering our decision variables, then write our objective function, and finally a row for each constraint. We then start Solver and enter the decision variables, objective function and constraints. If you revise a problem, either by adding new decision variables or adding new constraints, you follow the same process. First change the decision variables and/or constraints in Excel but then you must make the same changes in Solver. Otherwise the computer will not recognize that the situation has changed.

Chapter 2 Study Guide

1. What are decision variables? Where do they come from in the word problem?

2. What is the objective function? Where does it come from in the word problem?

3. What are constraints? Where do they come from in the word problem?

4. What information have we used from the Answer Reports?

5. Write a definition for each in your own words.
   a. Binding constraint-

   b. Non-binding constraint-

6. What is the "Final Value" on an Answer Report?

7. What is slack? What does the cell value tell us?

References

Kolecki, C. (2010). How running shoe is made - material, manufacture, used, parts, components, machine, Raw Materials, Design, The Manufacturing Process of running shoe, Quality Control. Retrieved November 8, 2010, from http://www.madehow.com/Volume-1/Running-Shoe.html

Pendergraft, N. (1997). "Lego of my simplex." *OR/MS Today* 24(1):128

Skateboard - Wikipedia, the free encyclopedia. (2011, April 28). Retrieved April 28, 2011, from https://secure.wikimedia.org/wikipedia/en/wiki/Skateboard

Sk8 Factory. (2010). Retrieved April 28, 2011, from http://www.sk8factory.com/

Appendix A: Using Excel Solver in Microsoft Office 2003 and 2007

The majority of the steps for using Excel Solver are the same in all versions of Microsoft Office. However, there a few differences.

Microsoft Office 2003

To add in Solver in Microsoft Office 2003, go to Add-Ins under the Tools menu and click on it. The Add-Ins window will appear. Then, check the Solver Add-In box and then click OK.

Next, set up the spreadsheet in the same way as detailed in Steps 1-4 of Section 2.2.6.

Once the spreadsheet is properly set up, click on the cell containing the objective function and choose Solver from the Tools menu. A Solver Parameters window will open, as shown in Figure 2.A.1.

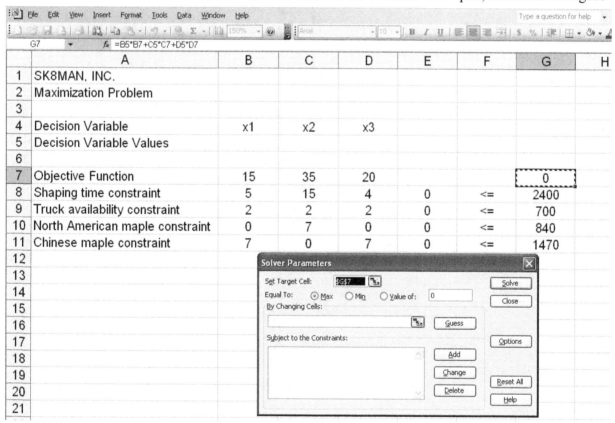

Figure 2.A.1: Solver Parameters window in Microsoft Office 2003

Verify that the "Max" circle is filled in. Then fill in the "By Changing Cells" and "Subject to the Constraints" windows as described in Steps 7-8 in Section 2.2.6

To include the non-negativity constraint, choose Options and check the box that says "Assume Non-Negative." Also, check the box that says "Assume Linear Model." See Figure 2.A.2 for an illustration of this.

Chapter 2 — Optimize Product Mix – Profit Maximization with Linear Programming

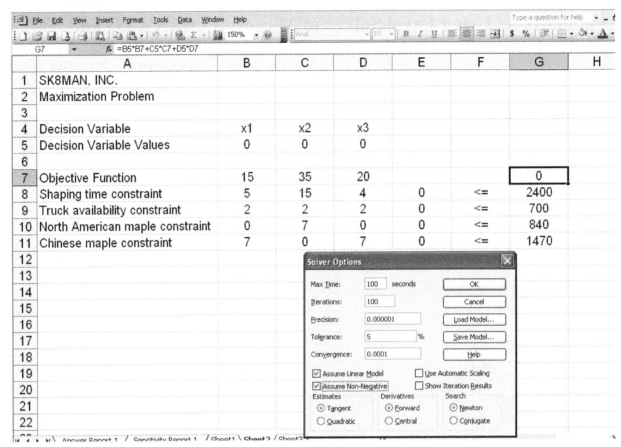

Figure 2.A.2: Solver Options window in Microsoft 2003

Finally, click "Solve" and choose the desired reports, as shown in Figure 2.A.3.

Chapter 2 — Optimize Product Mix – Profit Maximization with Linear Programming

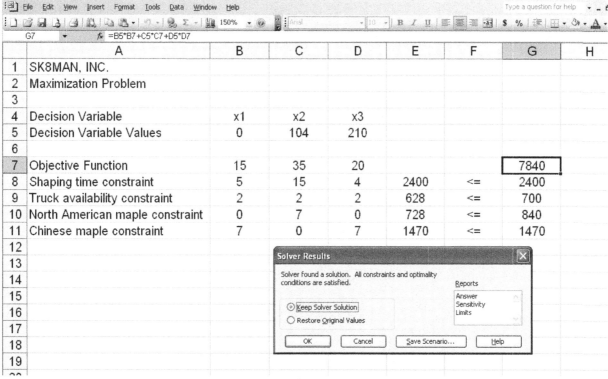

Figure 2.A.3: Solver results window

Microsoft Office 2007

To add in Solver in Microsoft Office 2007, click the Microsoft Office Button and choose "Excel Options." Click "Add-Ins." Under the "Manage" box, select "Excel Add-ins" and click "Go." Check the "Solver Add-in" box and click "OK."

Once Solver has been added, the Solver command is in the Analysis group on the Data tab, as shown in Figure 2.2.9.

All other steps for using Excel Solver in Microsoft Office 2007 are the same as for Microsoft Office 2003, given above.

CHAPTER 3:

Analyze Optimal Solutions:
Sensitivity Analysis

Section 3.0: Introduction

In addition to solving problems, operations researchers are often interested in learning how sensitive their solutions are to changes in the parameters of the problem. Consider the Computer Flips problem in the previous chapter. How sensitive is the solution to changes in the amount of profit that is made on each type of computer? What would be the effect of increasing the amount of available installation time or testing time? Questions such as these are part of what is called **sensitivity analysis**.

Section 3.1: Computer Flips, a Junior Achievement Company

Recall from Chapter 2 that Computer Flips is a Junior Achievement Company that begins producing two computer models: Simplex and Omniplex. The pertinent data from the Computer Flips problem appear in Table 3.1.1.

	Simplex	Omniplex
Profit per Computer	$200	$300
Installation Time per Computer	60 min.	120 min.

Table 3.1.1: Computer Flips information for two computer types

In addition, Computer Flips has 2,400 min of installation time available per week (five students, each working eight hours per week). They are also under two market restrictions. They estimate that they cannot sell more than 20 Simplex computers or 16 Omniplex computers per week. Gates Williams, the production manager for Computer Flips, wants to find the production rate per week for each type of computer that will maximize total profit.

3.1.1 Problem Formulation

Gates Williams writes the complete linear programming formulation for this problem. He begins with the definition of the decision variables, then he writes the objective function, and finally he lists the constraints.

Decision Variables
  Let: x_1 = the weekly production rate of Simplex computers
         x_2 = the weekly production rate of Omniplex computers

Objective Function
  Maximize: $z = 200x_1 + 300x_2$, where z = Computer Flips' weekly profit

Constraints
  Subject to:
    Installation Time (min): $60x_1 + 120x_2 \leq 2400$
    Simplex Market (#): $x_1 \leq 20$
    Omniplex Market (#): $x_2 \leq 16$
    Non-Negativity: $x_1 \geq 0$ and $x_2 \geq 0$

1. Using Excel Solver, find the optimal solution to the Computer Flips problem.
   a. How many Simplex and Omniplex computers should be produced per week?
   b. How much profit will Computer Flips earn per week?

3.1.2 Solver Answer Report

To begin to analyze this solution, Gates Williams has Solver generate the Answer and Sensitivity Reports, as shown in Figure 3.1.1. He starts by looking at the Answer Report Figure

3.1.2. He notices that the optimal solution and the value of the objective function are listed on this report.

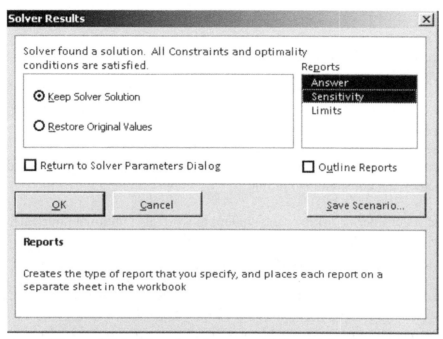

Figure 3.1.1: Choosing Answer and Sensitivity Reports

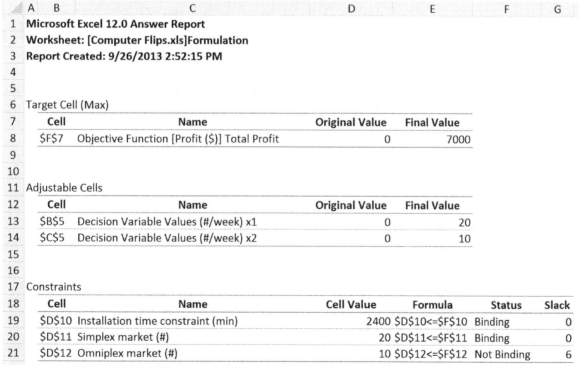

Figure 3.1.2: Computer Flips Answer Report

2. How could you determine the optimal solution and objective function value by looking *only* at the Answer Report?

Next, Gates Williams looks at the Constraints portion of the Answer Report. He notices constraints are either listed as Binding or Not Binding. He also notices that there is a column for Slack.

3. In the context of this problem, what does the information given in the Cell Value column mean?

4. Connect your response to the previous question to the amount of constraints available.
   a. Based on this information, what do you think *Binding* means?
   b. Based on this information, what do you think *Slack* means?
   c. If you were only given the slack value for a constraint, how could you determine whether that constraint is binding?

Gates Williams is considering asking each student to work an additional hour each week. This would increase the available installation time by: (5 students)(1 hour)(60 minutes) = 300. Therefore, the new total installation time available would be 2,700 minutes.

5. In your Excel worksheet, change the right hand side of the installation time constraint to 2700 minutes. Solve the problem again and open the Answer Report.
   a. What changes do you observe?
   b. Do you think it is worth it for Gates Williams to ask the students to work this extra time each week?

Gates Williams decides that he will not ask students to work this extra hour each week. Instead, he wonders if an increase in the Omniplex marketability will increase the weekly profit.

6. In your Excel worksheet, change the right hand side of the installation time constraint back to 2400 minutes. Then, increase the Omniplex marketability constraint to 17. Solve the problem again and open the Answer Report.
   a. What changes do you observe from the original problem?
   b. What happens if the Omniplex marketability constraint is 20 computers? 50 computers?
   c. Do you think Gates Williams should try to increase the marketability of the Omniplex computer? Why or why not?

3.1.3 Solver Sensitivity Report: Variable Cells

Now, Gates Williams opens the Sensitivity Report, Figure 3.1.3. He notices that the report is split into two sections: Variable Cells and Constraints. (Note: in older versions of Microsoft Office, Variable Cells are referred to as Adjustable Cells.)

In general, **Variable Cells** tell Gates Williams how the objective function coefficients may change. More specifically, Variable Cells show the increase or decrease of an objective function coefficient without changing the optimal solution. These changes only apply to one objective function coefficient at a time (all other coefficients must remain constant).

Next, the **Constraints** section tells Gates Williams how the objective function value changes. Specifically, Constraint cells give the objective function value based upon an increase or decrease of the right hand side (RHS) of a constraint. These changes only apply to one RHS constraint at a time (all other RHS constraints must remain the same).

Gates Williams feels that he only has control over how profitable each computer is. That is, he cannot change any of the constraints, but he could consider increasing the price of a computer. Therefore, he decides to only explore the Variable Cells in the Sensitivity Report.

	A	B	C	D	E	F	G	H
1	Microsoft Excel 12.0 Sensitivity Report							
2	Worksheet: [Computer Flips.xls]Formulation							
3	Report Created: 9/26/2013 2:52:15 PM							
4								
5								
6	Adjustable Cells							
7				Final	Reduced	Objective	Allowable	Allowable
8		Cell	Name	Value	Cost	Coefficient	Increase	Decrease
9		$B$5	Decision Variable Values (#/week) x1	20	0	200	1E+30	50
10		$C$5	Decision Variable Values (#/week) x2	10	0	300	100	300
11								
12	Constraints							
13				Final	Shadow	Constraint	Allowable	Allowable
14		Cell	Name	Value	Price	R.H. Side	Increase	Decrease
15		$D$10	Installation time constraint (min)	2400	2.5	2400	720	1200
16		$D$11	Simplex market (#)	20	50	20	20	12
17		$D$12	Omniplex market (#)	10	0	16	1E+30	6

Figure 3.1.3: Computer Flips Sensitivity Report

The information in the Variable Cells section of the Sensitivity Report tells how sensitive the optimal solution is to changes in the objective function coefficients of the decision variables. Solver considers changes made to one coefficient at a time. In particular, **Allowable Increase** refers to how much the objective coefficient can be increased without changing the final values. Similarly, **Allowable Decrease** tells how much the objective coefficient can be decreased without changing the final values.

For now, Gates Williams only concerns himself with the Allowable Increase column of the Variable Cells section. He considers the coefficient of x_2, which is the amount of profit generated by the sale of one Omniplex computer. Currently, that profit is $300 per computer. He sees that the Allowable Increase for the coefficient of x_2 is $100.

Gates Williams is curious about the effect of increasing the profit per Omniplex computer. He considers increasing the profit by a value below the Allowable Increase, above the Allowable Increase, and exactly at the Allowable Increase. Thus, he explores the effect of increasing the profit by $50 (below the Allowable Increase), $200 (above the Allowable Increase), and $100 (exactly the Allowable Increase). The corresponding new objective function coefficients of x_2

are $350, $500, and $400, respectively. These changes affect only the objective function in the formulation of the problem. All of the constraints remain the same.

7. Write the new objective functions for each of these three changes.

First, Gates Williams considers increasing the profit of Omniplex computers so that the profitability is now $350. However, he notices that this increase in profitability, $50, is less than the Allowable Increase.

8. In your Excel worksheet, change the objective function coefficient for Omniplex computers to $350. Solve the problem again.
    a. What changes to do you observe from the original problem?
    b. Do you think Gates Williams should try to increase the profitability of the Omniplex computer by $50? Why or why not?

Next, Gates Williams looks at the effect of increasing the profitability of Omniplex by more than the Allowable Increase. He increases the profitability by $200.

9. In your Excel worksheet, change the objective function coefficient for Omniplex computers to $500. Solve the problem again.
    a. What changes to do you observe from the original problem?
    b. Do you think Gates Williams should try to increase the profitability of the Omniplex computer by $200? Why or why not?

At this point, Gates Williams has looked at increasing the profitability of Omniplex by $50, which is less than the Allowable Increase, and by $200, which is more than the Allowable Increase. He wonders what would happen if the profitability of Omniplex increases by exactly $100, which is the Allowable Increase.

10. In your Excel worksheet, change the objective function coefficient for Omniplex computers to $400. Solve the problem again.
    a. What changes to do you observe from the original problem?
    b. Did your classmates obtain the same optimal solution as you?
    c. Do you think Gates Williams should try to increase the profitability of the Omniplex computer by $100? Why or why not?

In order to gain a better understanding of the three examples explored above, Gates Williams considers the geometry of the situation.

11. Draw a graph of the feasible region for the original problem, including the original line of constant profit ($z = 200x_1 + 300x_2$) passing through the optimal corner point (refer to Section 2.1).

12. Draw another graph of the feasible region for the original problem.
    a. Draw the line of constant profit when the profitability of the Omniplex computer has increased by $50 to $350. Which corner point maximizes profit in this situation?

b. Draw the line of constant profit when the profitability of the Omniplex computer has increased by $200 to $500. Which corner point maximizes profit in this situation?

c. Draw the line of constant profit when the profitability of the Omniplex computer has increased by $100 to $400. Which corner point maximizes profit in this situation? How is this situation different from the previous two?

13. Consider the case where the profit margin on Omniplex is increased to $400.
    a. What is the profit if 8 Simplex and 16 Omniplex computers are produced?
    b. What is the profit if 20 Simplex and 10 Omniplex computers are produced?
    c. Are there are other feasible points that produce this profit? If so, where are they? If not, why not?

14. Based on what you saw in this section, describe what you think would happen if you considered the Allowable Decrease instead.
    a. How do you think the final values would change if Gates Williams decreased the profitability of Omniplex computers by $200?
    b. How do you think the final values would change if Gates Williams decreased the profitability of Omniplex computers by $400?
    c. How do you think the final values would change if Gates Williams decreased the profitability of Omniplex computers by $300?
    d. Put these changes into Excel to see if your predictions were correct.

Finally, Gates Williams notices that the Sensitivity Report shows an Allowable Increase of 1E+30 in the coefficient of x_1 in the objective function. The number 1E+30 is Solver's way of conveying the expression $1 \cdot 10^{30}$. This very large number is the best Solver can do to indicate an infinite Allowable Increase.

To understand why the Allowable Increase is infinite, Gates Williams first needs to think about what the coefficient of x_1 in the objective function represents. It is the profit margin on Simplex computers. Solver is showing that no matter how much the profitability of Simplex computers increases, it will not change the optimal solution. In other words, increasing the profit margin on Simplex computers is not going to change the optimal number to make. This makes sense because the optimal solution shows that to maximize profits, 20 Simplex (and 10 Omniplex) computers should be made. Twenty is the most that can be made and still satisfy the market limit on weekly sales of Simplex computers. Increasing the profit margin of Simplex computers will not change the fact that no more than 20 per week can be sold, and they're already making 20 each week.

In this section, we have explored the effects on the optimal solution of increasing or decreasing the profitability of one of the computer models. In reality, the situation is much more complicated. For example, if Computer Flips increased the price of an Omniplex computer by $100-$200 to make it more profitable, doing so might affect the Omniplex market constraint. The increased price might lower the market constraint. So, in practice, a company would try to explore all of the ramifications of making changes in the important parameters of the problem.

Section 3.2: SK8MAN, Inc.

Recall from Chapter 2 that SK8MAN, Inc. manufactures skateboards. G.F. Hurley, the production manager at SK8MAN, Inc., needed to determine the production rate for each type of skateboard in order to make the most profit. Table 3.2.1 shows the relevant data from Chapter 2.

	Sporty	Fancy	Pool Runner	Amount Available
Profit per skateboard ($)	$15	$35	$20	
Shaping time required (min)	5	15	4	2,400
Truck availability (#)	2	2	2	700
North American maple veneers required (#)	0	7	0	840
Chinese maple veneers required (#)	7	0	7	1,470

Table 3.2.1: SK8MAN, Inc. data

3.2.1 Problem Formulation

In Chapter 2, the following problem formulation was developed:

Decision Variables
Let:
x_1 = the weekly production rate of Sporty boards
x_2 = the weekly production rate of Fancy boards
x_3 = the weekly production rate of Pool Runners boards

Objective Function
Maximize: $z = 15x_1 + 35x_2 + 20x_3$, where z = Computer Flips' weekly profit

Constraints
Subject to:

Shaping Time (min): $5x_1 + 15x_2 + 4x_3 \leq 2,400$

Trucks (#): $2x_1 + 2x_2 + 2x_3 \leq 700$

North American Maple (#): $7x_2 \leq 840$

Chinese Maple (#): $7x_1 + 7x_3 \leq 1,470$

Non-Negativity: $x_1, x_2, x_3 \geq 0$

3.2.2 Solver Answer Report

A spreadsheet formulation of the problem and an Answer Report showing the optimal solution appear in Figures 3.2.1 and 3.2.2.

	A	B	C	D	E	F	G
1	Chapter 3: Sensitivity Analysis						
2	3.2 SK8MAN, Inc. (3 variables)						
3	Profit Maximization Problem						
4							
5	Decision Variable	Sporty (x_1)	Fancy (x_2)	Pool-Runner (x_3)			
6	Decision Values [# to make per week]	0	104	210			
7							Total Profit
8	Objective Function [Profit ($)]	15	35	20			$7,840.00
9							
10	Constraints						
11	Shaping Time (minutes)	5	15	4	2400	≤	2400
12	Truck Availability	2	2	2	628	≤	700
13	North American Maple Veneers	0	7	0	728	≤	840
14	Chinese Maple Veneers	7	0	7	1470	≤	1470

Figure 3.2.1: Formulation for the 3-decision variable SK8MAN problem

Objective Cell (Max)

Cell	Name	Original Value	Final Value
$G$8	Objective Function [Profit ($)] Total Profit	$0.00	$7,840.00

Variable Cells

Cell	Name	Original Value	Final Value	Integer
$B$6	Decision Values [# to make per week] Sporty (x1)	0	0	Contin
$C$6	Decision Values [# to make per week] Fancy (x2)	0	104	Contin
$D$6	Decision Values [# to make per week] Pool-Runner (x3)	0	210	Contin

Constraints

Cell	Name	Cell Value	Formula	Status	Slack
$E$11	Shaping Time (minutes)	2400	$E$11<=$G$11	Binding	0
$E$12	Truck Availability	628	$E$12<=$G$12	Not Binding	72
$E$13	North American Maple Veneers	728	$E$13<=$G$13	Not Binding	112
$E$14	Chinese Maple Veneers	1470	$E$14<=$G$14	Binding	0

Figure 3.2.2: Answer Report for the 3-variable SK8MAN problem

As seen in Figure 3.2.1, the optimal solution is $x_1 = 0$, $x_2 = 104$, and $x_3 = 210$; that is, G.F. Hurley should produce no Sporty boards, 104 Fancy boards, and 210 Pool-Runner boards per week. With this product mix, SK8MAN will make a total profit of $7,840 each week. This information also appears in the Answer Report shown in Figure 3.2.2, along with some information about the constraints.

Examining the Answer Report, the first section refers to the Objective Cell (Max), and the variable name is "Objective Function [Profit ($)] Total Profit." The target cell in the spreadsheet, G8, stores the value of the objective function. Solver finds the maximum profit (because "Max" was selected during the Solver Parameters set-up) that meets all of the constraints. The maximum total profit Solver reports under the column Final Value is $7,840.00. The Original Value of $0.00 simply refers to the amount that was in the cell before Solver was run.

The second section is labeled Variable Cells and refers to the decision variables, x_1, x_2, and x_3. These are adjusted as Solver searches for the optimal solution. The Final Values of 0, 104, and 210, respectively, are the optimal solution. That is, the optimal solution is $x_1 = 0$, $x_2 = 104$, and $x_3 = 210$.

The third section of the Answer Report is labeled Constraints. The four constraints are all less than or equal to ($\leq$) constraints. The left hand side value of each constraint represents the total amount used by the production plan. These totals are stored in cells E11, E12, E13, and E14 and are reported in the column labeled "Cell Value." The right hand side values for each of the three constraints are stored in column G in cells G11, G12, G13, and G14.

G.F. Hurley notices that of the four constraints, two of them are binding and two are not binding. But, he wonders what this means.

He notices that for the two binding constraints, there is a 0 in the column labeled Slack. He looks back to the Solver solution in Figure 3.2.1 and notices that for each of the two binding constraints, the left hand side of the constraint is equal to the right hand side.

For example, the workers at SK8MAN will use $5(0) + 15(104) + 4(210) = 2{,}400$ minutes for shaping (cell E11). They have 2,400 minutes available for shaping (cell G11). In addition, they will use $7(0) + 7(210) = 1{,}470$ Chinese maple veneers (cell E14). They have 1470 Chinese maple veneers available (cell G14).

In other words, there is no slack because every bit of each of those resources is being used up by the optimal solution.

However, for the non-binding constraints in the Answer Report, the Slack values are not zero. They are listed as 72 and 112. Again returning to the Solver solution in Figure 3.1.1, G.F. Hurley notices that the left hand side of the truck availability constraint is 628, and the right hand side is 700. This time the two sides of the constraint are not equal because the optimal solution does not use up all available trucks. There is a slack of $700 - 628 = 72$ trucks. That means that SK8MAN could use 72 more trucks. They will not do that, though, because in order to use these extra trucks, they would have to make more skateboards, which is impossible due to the shaping time and the Chinese maple veneers constraints.

Similarly, G.F. Hurley notices that the left hand side of the North American maple veneers constraint is 728 and the right hand side is 840. Thus, they have an extra $840 - 728 = 112$ North American maple veneers available.

3.2.3 Solver Sensitivity Report: Variable Cells

Figure 3.2.3 contains the Variable Cells section of the Sensitivity Report SK8MAN. Use it to answer the questions that follow.

Variable Cells

Cell	Name	Final Value	Reduced Cost	Objective Coefficient	Allowable Increase	Allowable Decrease
$B$6	Decision Values [# to make per week] Sporty (x1)	0	-7.333333333	15	7.333333333	1E+30
$C$6	Decision Values [# to make per week] Fancy (x2)	104	0	35	40	35
$D$6	Decision Values [# to make per week] Pool-Runner (x3)	210	0	20	1E+30	7.333333333

Figure 3.2.3: Sensitivity Report for the 3-variable SK8MAN problem

The Sensitivity Report provides information about each of the decision variables, which it calls Variable Cells. It also provides information about each of the constraints.

Variable Cells: Allowable Increase

The information provided in the Adjustable Cells section of the Sensitivity Report tells how sensitive the optimal solution is to changes in the objective function coefficients of the decision variables. Solver considers changes made to one coefficient at a time.

G.F. Hurley notices that for x_2, there is an Allowable Increase of 40. But he wonders what this refers to. What can be increased by 40? What does such an increase "allow"? Allowable Increase refers to increasing the coefficient of the decision variable in the objective function. In this case, the coefficient of x_2 is the amount of profit generated by the sale of one Fancy skateboard. Currently, that profit is $35 per skateboard.

G.F. Hurley is curious about the effect of increasing the profit per Fancy skateboard. He first explores the effect of increasing the profit by a value below the Allowable Increase. He chooses to increase the profit of Fancy boards by $25 to $60. Doing so changes the objective function to:

$z = 15x_1 + 60x_2 + 20x_3.$

The effect of this change is shown in Figures 3.2.4a and 3.2.4b.

G.F. Hurley notices that in Figure 3.2.4, the profitability of Fancy boards has been changed to $60. However, when looking at the Answer Report for that change, he sees that the optimal solution is the same: make Sporty boards at the rate of 0 per week, make Fancy boards at the rate of 104 per week, and make Pool-Runner boards at the rate of 210 per week.

Therefore, because G.F. Hurley increased the objective function coefficient by an amount *less than* the Allowable Increase, there was no change in the optimal solution. Although the optimal solution is the same, the increase in Omniplex profitability did change the weekly profit from $7,840 to $10,440. This $2,600 increase results from the $25 increase in profit on Fancy boards, and SK8MAN makes 104 per week: ($25)(104 Fancy boards) = $2600.

Figure 3.2.4a

	A	B	C	D	E	F	G
1	Chapter 3: Sensitivity Analysis						
2	3.2 SK8MAN, Inc. (3 variables)						
3	Profit Maximization Problem						
4							
5	Decision Variable	Sporty (x_1)	Fancy (x_2)	Pool-Runner (x_3)			
6	Decision Values [# to make per week]	0	104	210			
7							Total Profit
8	Objective Function [Profit ($)]	15	60	20			$10,440.00
9							
10	Constraints						
11	Shaping Time (minutes)	5	15	4	2400	≤	2400
12	Truck Availability	2	2	2	628	≤	700
13	North American Maple Veneers	0	7	0	728	≤	840
14	Chinese Maple Veneers	7	0	7	1470	≤	1470

Figure 3.2.4a: Formulation when the profitability of Fancy boards is $60

Objective Cell (Max)

Cell	Name	Original Value	Final Value
$G$8	Objective Function [Profit ($)] Total Profit	$0.00	$10,440.00

Variable Cells

Cell	Name	Original Value	Final Value	Integer
$B$6	Decision Values [# to make per week] Sporty (x1)	0	0	Contin
$C$6	Decision Values [# to make per week] Fancy (x2)	0	104	Contin
$D$6	Decision Values [# to make per week] Pool-Runner (x3)	0	210	Contin

Constraints

Cell	Name	Cell Value	Formula	Status	Slack
$E$11	Shaping Time (minutes)	2400	$E$11<=$G$11	Binding	0
$E$12	Truck Availability	628	$E$12<=$G$12	Not Binding	72
$E$13	North American Maple Veneers	728	$E$13<=$G$13	Not Binding	112
$E$14	Chinese Maple Veneers	1470	$E$14<=$G$14	Binding	0

Figure 3.2.4b: Answer Report when the profitability of Fancy boards is $60

Next, G.F. Hurley increases the profitability of Fancy boards by an amount greater than the Allowable Increase. Figures 3.2.5a and 3.2.5b show the effect of increasing the objective function coefficient from $35 to $80 (an increase of $45), so that the objective function would now be:

$$z = 15x_1 + 80x_2 + 20x_3$$

In this case, G.F. Hurley sees that changing the profitability of Fancy boards to $80 yields a new optimal solution. This time, the optimal solution changes the production rates to 0 Sporty boards, 120 Fancy boards, and 150 Pool-Runner boards per week. Also, the amount of profit generated has now increased to $12,600 per week.

At this point, G.F. Hurley sees that increasing the profitability of Fancy boards by an amount *less than* the Allowable Increase has no impact on the optimal solution. However, increasing the profitability by an amount *greater than* the Allowable Increase changes the optimal solution. Next, he wonders what will happen if he increases the profitability by exactly the Allowable Increase.

	A	B	C	D	E	F	G
1	Chapter 3: Sensitivity Analysis						
2	3.2 SK8MAN, Inc. (3 variables)						
3	Profit Maximization Problem						
4							
5	Decision Variable	Sporty (x_1)	Fancy (x_2)	Pool-Runner (x_3)			
6	Decision Values [# to make per week]	0	120	150			
7							Total Profit
8	Objective Function [Profit ($)]	15	80	20			$12,600.00
9							
10	Constraints						
11	Shaping Time (minutes)	5	15	4	2400	≤	2400
12	Truck Availability	2	2	2	540	≤	700
13	North American Maple Veneers	0	7	0	840	≤	840
14	Chinese Maple Veneers	7	0	7	1050	≤	1470

Figure 3.2.5a: Formulation when the profitability of Fancy boards is $80

Objective Cell (Max)

Cell	Name	Original Value	Final Value
$G$8	Objective Function [Profit ($)] Total Profit	$0.00	$12,600.00

Variable Cells

Cell	Name	Original Value	Final Value	Integer
$B$6	Decision Values [# to make per week] Sporty (x1)	0	0	Contin
$C$6	Decision Values [# to make per week] Fancy (x2)	0	120	Contin
$D$6	Decision Values [# to make per week] Pool-Runner (x3)	0	150	Contin

Constraints

Cell	Name	Cell Value	Formula	Status	Slack
$E$11	Shaping Time (minutes)	2400	$E$11<=$G$11	Binding	0
$E$12	Truck Availability	540	$E$12<=$G$12	Not Binding	160
$E$13	North American Maple Veneers	840	$E$13<=$G$13	Binding	0
$E$14	Chinese Maple Veneers	1050	$E$14<=$G$14	Not Binding	420

Figure 3.2.5b: Answer Report when the profitability of Fancy boards is $80

Figures 3.2.6a and 3.2.6b show the effect of increasing the profitability of Fancy boards by exactly $40, where the objective function is:

$$z = 15x_1 + 75x_2 + 20x_3$$

	A	B	C	D	E	F	G
1	Chapter 3: Sensitivity Analysis						
2	3.2 SK8MAN, Inc. (3 variables)						
3	Profit Maximization Problem						
4							
5	Decision Variable	Sporty (x_1)	Fancy (x_2)	Pool-Runner (x_3)			
6	Decision Values [# to make per week]	0	120	150			
7							Total Profit
8	Objective Function [Profit ($)]	15	75	20			$12,000.00
9							
10	Constraints						
11	Shaping Time (minutes)	5	15	4	2400	≤	2400
12	Truck Availability	2	2	2	540	≤	700
13	North American Maple Veneers	0	7	0	840	≤	840
14	Chinese Maple Veneers	7	0	7	1050	≤	1470

Figure 3.2.6a: Formulation when the profitability of Fancy boards is $75

Objective Cell (Max)

Cell	Name	Original Value	Final Value
$G$8	Objective Function [Profit ($)] Total Profit	$0.00	$12,000.00

Variable Cells

Cell	Name	Original Value	Final Value	Integer
$B$6	Decision Values [# to make per week] Sporty (x1)	0	0	Contin
$C$6	Decision Values [# to make per week] Fancy (x2)	0	120	Contin
$D$6	Decision Values [# to make per week] Pool-Runner (x3)	0	150	Contin

Constraints

Cell	Name	Cell Value	Formula	Status	Slack
$E$11	Shaping Time (minutes)	2400	$E$11<=$G$11	Binding	0
$E$12	Truck Availability	540	$E$12<=$G$12	Not Binding	160
$E$13	North American Maple Veneers	840	$E$13<=$G$13	Binding	0
$E$14	Chinese Maple Veneers	1050	$E$14<=$G$14	Not Binding	420

Figure 3.2.6b: Answer Report when the profitability of Fancy boards is $75

G.F. Hurley notices that when the profitability of Fancy boards increases by exactly $40 (to $75), the optimal solution reported by Solver is to produce 0 Sporty boards, 120 Fancy boards, and 150 Pool-Runner boards. This solution is the same as when the profitability of Fancy boards was $80. The total profit is now $12,000:

$$15(0) + 75(120) + 20(150) = \$12{,}000.$$

However, he also notices that if the profitability of Fancy boards goes up to $75, the amount of weekly profit generated by the original optimal solution (0 Sporty boards, 104 Fancy boards, and 210 Pool-Runner boards) is also $12,000:

$$15(0) + 75(104) + 20(210) = \$12,000.$$

Both production plans lie on the same plane: $z = 15x_1 + 75x_2 + 20x_3$. In fact, any point that lies on this plane and is within the feasible region is an optimal solution. Therefore, there are infinitely many optimal solutions when the profitability of Fancy boards is $75 (i.e., when the coefficient of x_2 is increased exactly by the amount of the Allowable Increase).

Note: This idea was explored graphically in the previous section, where the objective function was a line, rather than a plane. Visualizing the SK8MAN problem graphically is much more difficult because it has three decision variables.

Finally, G.F. Hurley notices that the Sensitivity Report in Figure 3.2.3 shows an Allowable Increase of 1E+30 in the coefficient of x_3 in the objective function. The number 1E+30 is Solver's way of expressing the number 1×10^{30}. This very large number is the best Solver can do to indicate an *infinite* Allowable Increase.

To understand why the Allowable Increase is infinite, G.F. Hurley first needs to think about what the coefficient of x_3 in the objective function represents. It is the profitability of Pool-Runner boards. Solver is showing that no matter how much the profitability of Pool-Runner boards increases, it will not change the optimal solution. In other words, increasing the profitability of Pool-Runner boards is not going to change the optimal number to make.

This makes sense because the optimal solution shows that to maximize profits, 210 Pool-Runner boards (as well as 0 Sporty boards and 104 Fancy boards) should be made. Making 210 Pool-Runner boards per week consumes 1,470 Chinese maple veneers, which is exactly the number available per week. Therefore, no more than 210 Pool-Runner boards can be made per week, no matter how much their profitability increases. SK8MAN is already making all of the Pool-Runners that it possibly can!

Variable Cells: Allowable Decrease

Next, G.F. Hurley returns to the original problem and considers decreasing one of the coefficients in the objective function. Referring again to Figure 3.2.3, he notices the Sensitivity Report indicates an Allowable Decrease of approximately 7.33 for the coefficient of x_3 in the objective function.

He considers decreasing the coefficient of x_3 by 5 (a value smaller than the Allowable Decrease), 15 (a value greater than the Allowable Decrease) and 7.33 (the Allowable Increase). The corresponding new profit coefficients for each case are 15, 5, and 12.67, respectively.

Figures 3.2.7, 3.2.8, and 3.2.9 show the formulation and Answer Report for each of those decreases in the profit margin of Pool-Runner boards.

	A	B	C	D	E	F	G
1	Chapter 3: Sensitivity Analysis						
2	3.2 SK8MAN, Inc. (3 variables)						
3	Profit Maximization Problem						
4							
5	Decision Variable	Sporty (x_1)	Fancy (x_2)	Pool-Runner (x_3)			
6	Decision Values [# to make per week]	0	104	210			
7							Total Profit
8	Objective Function [Profit ($)]	15	35	15			$6,790.00
9							
10	Constraints						
11	Shaping Time (minutes)	5	15	4	2400	≤	2400
12	Truck Availability	2	2	2	628	≤	700
13	North American Maple Veneers	0	7	0	728	≤	840
14	Chinese Maple Veneers	7	0	7	1470	≤	1470

Figure 3.2.7a: Formulation when the profitability of Pool-Runner boards is $15

Objective Cell (Max)

Cell	Name	Original Value	Final Value
$G$8	Objective Function [Profit ($)] Total Profit	$0.00	$6,790.00

Variable Cells

Cell	Name	Original Value	Final Value	Integer
$B$6	Decision Values [# to make per week] Sporty (x1)	0	0	Contin
$C$6	Decision Values [# to make per week] Fancy (x2)	0	104	Contin
$D$6	Decision Values [# to make per week] Pool-Runner (x3)	0	210	Contin

Constraints

Cell	Name	Cell Value	Formula	Status	Slack
$E$11	Shaping Time (minutes)	2400	$E$11<=$G$11	Binding	0
$E$12	Truck Availability	628	$E$12<=$G$12	Not Binding	72
$E$13	North American Maple Veneers	728	$E$13<=$G$13	Not Binding	112
$E$14	Chinese Maple Veneers	1470	$E$14<=$G$14	Binding	0

Figure 3.2.7b: Answer Report when the profitability of Pool-Runner boards is $15

In Figures 3.2.7a and 3.2.7b, the profitability of Pool-Runner boards has been decreased from $20 to $15. This is a decrease of $5, which is smaller than the Allowable Decrease of $7.33. G.F. Hurley notices that the optimal solution is still to make 0 Sporty boards, 104 Fancy boards, and 210 Pool-Runner boards. But the total weekly profit has decreased by $1,050 to $6,790. This is because the profit on each of the Pool-Runner boards made has decreased by $5, and 210 · $5 = $1,050.

Chapter 3: Sensitivity Analysis
3.2 SK8MAN, Inc. (3 variables)
Profit Maximization Problem

	A	B	C	D	E	F	G
1	Chapter 3: Sensitivity Analysis						
2	3.2 SK8MAN, Inc. (3 variables)						
3	Profit Maximization Problem						
4							
5	Decision Variable	Sporty (x_1)	Fancy (x_2)	Pool-Runner (x_3)			
6	Decision Values [# to make per week]	210	90	0			
7							Total Profit
8	Objective Function [Profit ($)]	15	35	5			$6,300.00
9							
10	Constraints						
11	Shaping Time (minutes)	5	15	4	2400	≤	2400
12	Truck Availability	2	2	2	600	≤	700
13	North American Maple Veneers	0	7	0	630	≤	840
14	Chinese Maple Veneers	7	0	7	1470	≤	1470

Figure 3.2.8a: Formulation when the profitability of Pool-Runner boards is $5

Objective Cell (Max)

Cell	Name	Original Value	Final Value
$G$8	Objective Function [Profit ($)] Total Profit	$0.00	$6,300.00

Variable Cells

Cell	Name	Original Value	Final Value	Integer
$B$6	Decision Values [# to make per week] Sporty (x1)	0	210	Contin
$C$6	Decision Values [# to make per week] Fancy (x2)	0	90	Contin
$D$6	Decision Values [# to make per week] Pool-Runner (x3)	0	0	Contin

Constraints

Cell	Name	Cell Value	Formula	Status	Slack
$E$11	Shaping Time (minutes)	2400	$E$11<=$G$11	Binding	0
$E$12	Truck Availability	600	$E$12<=$G$12	Not Binding	100
$E$13	North American Maple Veneers	630	$E$13<=$G$13	Not Binding	210
$E$14	Chinese Maple Veneers	1470	$E$14<=$G$14	Binding	0

Figure 3.2.8b: Answer Report when the profitability of Pool-Runner boards is $5

In Figures 3.2.8a and 3.2.8b, the profitability of Pool-Runner boards has been decreased from $20 to $5. This is a decrease of $15, which is more than the Allowable Decrease. This time, G.F. Hurley notices that the optimal solution has changed from 0 Sporty boards, 104 Fancy boards, and 210 Pool-Runner boards to 210 Sporty boards, 90 Fancy boards, and 0 Pool-Runner boards.

This happened because the profit margin is $15 for Sporty boards, $35 for Fancy boards, and $5 for Pool-Runner boards. In this case, making 210 Sporty boards, 90 Fancy boards, and 0 Pool-Runner boards is more profitable than making 0 Sporty boards, 104 Fancy boards, and 210 Pool-Runner boards. That is, producing 0 Sporty boards, 104 Fancy boards, and 210 Pool-Runner boards generates a profit of:

15(0) + 35(104) + 5(210) = $4,690 per week.

On the other hand, producing 210 Sporty boards, 90 Fancy boards, and 0 Pool-Runner boards generates a profit of:

15(210) + 35(90) + 5(0) = $6,300 per week.

	A	B	C	D	E	F	G
1	Chapter 3: Sensitivity Analysis						
2	3.2 SK8MAN, Inc. (3 variables)						
3	Profit Maximization Problem						
4							
5	Decision Variable	Sporty (x_1)	Fancy (x_2)	Pool-Runner (x_3)			
6	Decision Values [# to make per week]	210	90	0			
7							Total Profit
8	Objective Function [Profit ($)]	15	35	12.66666667			$6,300.00
9							
10	**Constraints**						
11	Shaping Time (minutes)	5	15	4	2400	≤	2400
12	Truck Availability	2	2	2	600	≤	700
13	North American Maple Veneers	0	7	0	630	≤	840
14	Chinese Maple Veneers	7	0	7	1470	≤	1470

Figure 3.2.9a: Formulation when profitability of Pool-Runner boards is approximately $12.67

Objective Cell (Max)

Cell	Name	Original Value	Final Value
$G$8	Objective Function [Profit ($)] Total Profit	$0.00	$6,300.00

Variable Cells

Cell	Name	Original Value	Final Value	Integer
$B$6	Decision Values [# to make per week] Sporty (x1)	0	210	Contin
$C$6	Decision Values [# to make per week] Fancy (x2)	0	90	Contin
$D$6	Decision Values [# to make per week] Pool-Runner (x3)	0	0	Contin

Constraints

Cell	Name	Cell Value	Formula	Status	Slack
$E$11	Shaping Time (minutes)	2400	$E$11<=$G$11	Binding	0
$E$12	Truck Availability	600	$E$12<=$G$12	Not Binding	100
$E$13	North American Maple Veneers	630	$E$13<=$G$13	Not Binding	210
$E$14	Chinese Maple Veneers	1470	$E$14<=$G$14	Binding	0

Figure 3.2.9b: Answer Report when profitability of Pool-Runner boards is approximately $12.67

In Figures 3.2.9a and 3.2.9b, the profitability of Pool-Runner boards has been decreased from $20 to $12.67. This is a decrease of $7.33, which is the Allowable Decrease. Examining the Answer Report for this case, G.F. Hurley sees that Solver reports 210 Sporty boards, 90 Fancy boards, and 0 Pool-Runner boards as the optimal solution. This yields a total profit of $6,300:

15(210) + 35(90) + 12.67(0) = $6,300 per week.

However, the original optimal solution of 0 Sporty boards, 104 Fancy boards, and 210 Pool-Runner boards also yields a total profit of $6,300:

15(0) + 35(104) + 12.67(210) = $6,300 per week.

This is another case where the plane representing the objective function coincides with one of the boundaries of the feasible region. Once again, there are infinitely many possible optimal solutions along that boundary.

G.F. Hurley turns his attention to the Allowable Decrease in the objective coefficient of x_2. He notices that Solver reports a value of $35. Since the profitability of Fancy boards is currently $35, an Allowable Decrease of $35 means that no matter how small the profit margin is, as long as Fancy boards generate a positive profit margin, the optimal solution will not change.

Variable Cells: Reduced Cost

Next, G.F. Hurley notices that there is a Reduced Cost of approximately -7.33 listed for Sporty boards (x_1). To see what this means, G.F. Hurley forces the production of one Sporty board by adding the constraint $x_1 = 1$. Figure 3.2.10 shows the new formulation after Solver has found the optimal solution.

	A	B	C	D	E	F	G
1	Chapter 3: Sensitivity Analysis						
2	3.2 SK8MAN, Inc. (3 variables)						
3	Profit Maximization Problem						
4							
5	Decision Variable	Sporty (x_1)	Fancy (x_2)	Pool-Runner (x_3)			
6	Decision Values [# to make per week]	1	103.933333	209			
7							Total Profit
8	Objective Function [Profit ($)]	15	35	20			$7,832.67
9							
10	Constraints						
11	Shaping Time (minutes)	5	15	4	2400	≤	2400
12	Truck Availability	2	2	2	627.867	≤	700
13	North American Maple Veneers	0	7	0	727.533	≤	840
14	Chinese Maple Veneers	7	0	7	1470	≤	1470
15	Force One Sporty Board	1	0	0	1	=	1

Figure 3.2.10: Forcing the production of one Sporty board (x_1)

The new optimal solution is 1 Sporty board, approximately 103.933 Fancy boards, and 209 Pool-Runner boards. (There is nothing wrong with having a non-integer solution since the decision variables are production rates per week, not number of skateboards sold.) This production mix generates a total profit of $7,832.67 per week. This is a decrease of $7840 – $7832.67 = $7.33 per week. Therefore, **Reduced Cost** refers to the change in the Final Value of the objective function that is caused by increasing a decision variable by one unit.

1. SK8MAN has a regular customer who wants to special order 10 Sporty boards. If SK8MAN manufactures those boards, how will that affect profit for that week?

Alternatively, one could think of Reduced Cost as the amount by which the objective function coefficient would have to increase to before it would profitable to make that item. For example, G.F. Hurley notices that the Allowable Increase for Sporty boards is approximately 7.33 and the Reduced Cost for Sporty boards is approximately -7.33. This is not a coincidence! If G.F. Hurley increases the profitability of Sporty boards by more than $7.33, then Sporty boards would be profitable to produce.

To experiment with this idea, G.F. Hurley changes the profitability of Sporty boards to $23. The result is shown in Figure 3.2.11. In this case, it is now profitable to produce Sporty boards.

	A	B	C	D	E	F	G
1	Chapter 3: Sensitivity Analysis						
2	3.2 SK8MAN, Inc. (3 variables)						
3	Profit Maximization Problem						
4							
5	Decision Variable	Sporty (x_1)	Fancy (x_2)	Pool-Runner (x_3)			
6	Decision Values [# to make per week]	210	90	0			
7							Total Profit
8	Objective Function [Profit ($)]	23	35	20			$7,980.00
9							
10	Constraints						
11	Shaping Time (minutes)	5	15	4	2400	≤	2400
12	Truck Availability	2	2	2	600	≤	700
13	North American Maple Veneers	0	7	0	630	≤	840
14	Chinese Maple Veneers	7	0	7	1470	≤	1470
15							

Figure 3.2.11: Formulation when the profitability of Sporty boards is $23

In a maximization problem, the Reduced Cost value will always be less than or equal to zero. If the Reduced Cost is less than zero, then it is not profitable to make the product. G.F. Hurley considers what it means to have a Reduced Cost value of zero.

The Sensitivity Report in Figure 3.2.3 shows that the Reduced Cost values for the other two decision variables, x_2 and x_3, are both zero. That is because both of those decision variables are already part of the optimal solution. If a decision variable is not part of the optimal solution, its final value is zero. Its Reduced Cost measures the amount the final value of the objective function would be reduced if the value of the decision variable were increased by just one unit.

Table 3.2.2 summarizes the impact of changes in the objective function coefficients. The table differentiates between changes within the allowable range and changes outside the allowable range. It also distinguishes between decision variables that are non-zero and those that are zero. If a decision variable is zero, changing the objective function coefficient within the allowable range has no impact on anything including the value of the objective function. Conversely, it is important to note that when the change exceeds the range, the decision variable becomes an non-zero element of the optimal solution.

	Changes to objective function coefficients			
Impact on…	**Decision Variables with non-zero values**		**Decision variables with zero values**	
	Within range of allowable changes	**Outside range** of allowable changes	**Within range** of allowable changes	**Outside range** of allowable changes
Decision Variables	Values do not change	Values change	Values do not change	Values change and decision variable becomes non-zero
Objective Function	Objective function increases or decreases with the change of the coefficient	Increases or decreases with the change of the coefficient and the changes to the decision variables	No change in objective function	Increases or decreases with the change of the coefficient and the changes to the decision variables

Table 3.2.2: Summary of the impact of changes in the objective function coefficients

3.2.4 Solver Sensitivity Report: Constraints

Shadow Price

G.F. Hurley then moves to the Constraints section of the Sensitivity Report (see Figure 3.2.12), where the most useful piece of information is the Shadow Price. The **Shadow Price** tells the effect on the value of the objective function of increasing the resource that is constraining the solution by 1 unit. In other words, the Shadow Price refers to the amount by which the objective function value changes given a 1-unit increase or decrease in one right hand side of a constraint.

Variable Cells

Cell	Name	Final Value	Reduced Cost	Objective Coefficient	Allowable Increase	Allowable Decrease
$B$6	Decision Values [# to make per week] Sporty (x1)	0	-7.333333333	15	7.333333333	1E+30
$C$6	Decision Values [# to make per week] Fancy (x2)	104	0	35	40	35
$D$6	Decision Values [# to make per week] Pool-Runner (x3)	210	0	20	1E+30	7.333333333

Constraints

Cell	Name	Final Value	Shadow Price	Constraint R.H. Side	Allowable Increase	Allowable Decrease
$E$11	Shaping Time (minutes)	2400	2.333333333	2400	240	1560
$E$12	Truck Availability	628	0	700	1E+30	72
$E$13	North American Maple Veneers	728	0	840	1E+30	112
$E$14	Chinese Maple Veneers	1470	1.523809524	1470	343.6363636	420

Figure 3.2.12: Sensitivity Report for the three-variable SK8MAN problem

For example, the shaping time constraint shows a Shadow Price of approximately 2.33. That means if the amount of shaping time were increased by 1 unit, the value of the objective function would increase by 2.33 units.

In the context of the SK8MAN problem, the units of shaping time are minutes, and the units of the objective function are dollars. Suppose the workers agree to work a total of 100 minutes longer each week. Doing so would increase the available shaping time by 100 minutes.

According to the Sensitivity Report, that should increase the value of the objective function for the optimal solution by approximately 100 · $2.3333 = $233.33.

Figures 3.2.13a and 3.2.13b show that the objective has increased to $8,073.33. The new production plan is 0 Sporty boards, approximately 110.67 Fancy boards, and 210 Pool-Runner boards.

	A	B	C	D	E	F	G
1	Chapter 3: Sensitivity Analysis						
2	3.2 SK8MAN, Inc. (3 variables)						
3	Profit Maximization Problem						
4							
5	Decision Variable	Sporty (x_1)	Fancy (x_2)	Pool-Runner (x_3)			
6	Decision Values [# to make per week]	0	110.666667	210			
7							Total Profit
8	Objective Function [Profit ($)]	15	35	20			$8,073.33
9							
10	Constraints						
11	Shaping Time (minutes)	5	15	4	2500	≤	2500
12	Truck Availability	2	2	2	641.333	≤	700
13	North American Maple Veneers	0	7	0	774.667	≤	840
14	Chinese Maple Veneers	7	0	7	1470	≤	1470

Figure 3.2.13a: Formulation when available shaping time is increased by 100 minutes

Objective Cell (Max)

Cell	Name	Original Value	Final Value
$G$8	Objective Function [Profit ($)] Total Profit	$0.00	$8,073.33

Variable Cells

Cell	Name	Original Value	Final Value	Integer
$B$6	Decision Values [# to make per week] Sporty (x1)	0	0	Contin
$C$6	Decision Values [# to make per week] Fancy (x2)	0	110.6666667	Contin
$D$6	Decision Values [# to make per week] Pool-Runner (x3)	0	210	Contin

Constraints

Cell	Name	Cell Value	Formula	Status	Slack
$E$11	Shaping Time (minutes)	2500	$E$11<=$G$11	Binding	0
$E$12	Truck Availability	641.3333333	$E$12<=$G$12	Not Binding	58.66666667
$E$13	North American Maple Veneers	774.6666667	$E$13<=$G$13	Not Binding	65.33333333
$E$14	Chinese Maple Veneers	1470	$E$14<=$G$14	Binding	0

Figure 3.2.13b: Answer Report when available shaping time is increased by 100 minutes

On the other hand, suppose that the available shaping time is reduced by 150 minutes. That reduction should then reduce the final value of the objective function by approximately 150 · $2.333 = $350. Figures 3.2.14a and 3.2.14b demonstrate that the objective function has decreased to $7,490. The new production plan is 0 Sporty boards, 94 Fancy boards, and 210 Pool-Runner boards.

	A	B	C	D	E	F	G
1	Chapter 3: Sensitivity Analysis						
2	3.2 SK8MAN, Inc. (3 variables)						
3	Profit Maximization Problem						
4							
5	Decision Variable	Sporty (x_1)	Fancy (x_2)	Pool-Runner (x_3)			
6	Decision Values [# to make per week]	0	94	210			
7							Total Profit
8	Objective Function [Profit ($)]	15	35	20			$7,490.00
9							
10	Constraints						
11	Shaping Time (minutes)	5	15	4	2250	≤	2250
12	Truck Availability	2	2	2	608	≤	700
13	North American Maple Veneers	0	7	0	658	≤	840
14	Chinese Maple Veneers	7	0	7	1470	≤	1470

Figure 3.2.14a: Formulation when available shaping time is reduced by 150 minutes

Objective Cell (Max)

Cell	Name	Original Value	Final Value
$G$8	Objective Function [Profit ($)] Total Profit	$0.00	$7,490.00

Variable Cells

Cell	Name	Original Value	Final Value	Integer
$B$6	Decision Values [# to make per week] Sporty (x1)	0	0	Contin
$C$6	Decision Values [# to make per week] Fancy (x2)	0	94	Contin
$D$6	Decision Values [# to make per week] Pool-Runner (x3)	0	210	Contin

Constraints

Cell	Name	Cell Value	Formula	Status	Slack
$E$11	Shaping Time (minutes)	2250	$E$11<=$G$11	Binding	0
$E$12	Truck Availability	608	$E$12<=$G$12	Not Binding	92
$E$13	North American Maple Veneers	658	$E$13<=$G$13	Not Binding	182
$E$14	Chinese Maple Veneers	1470	$E$14<=$G$14	Binding	0

Figure 3.2.14b: Answer Report when available shaping time is reduced by 150 minutes

Thus, the Shadow Price shows how much the value of the objective function will increase or decrease for each unit of increase or decrease in the availability of one of the constraining resources.

Constraints: Allowable Increase and Allowable Decrease
Returning once again to the Sensitivity Report in Figure 3.2.12, G.F. Hurley puts his attention towards the columns for an **Allowable Increase** and **Allowable Decrease** for each of the constraints. These refer to increases or decreases in the right hand side of a constraint (i.e., increasing or decreasing the availability of one of the constraining resources, such as shaping time). If an increase or decrease falls within the range determined by the Allowable Increase and Allowable Decrease, then the Shadow Price will remain the same.

For example, from the Sensitivity Report, the Allowable Increase in shaping time is 240 minutes, and the Allowable Decrease is 1,560 minutes. So, if a change in the availability of shaping time falls in the range between an increase of 240 minutes and a decrease of 1,560 minutes, the Shadow Price will stay constant at approximately $2.33.

G.F. Hurley wonders what happens if a change in the available shaping time falls outside this range. He supposes that the shaping time increases by 241 minutes. Since 241 is greater than the Allowable Increase, there should be an effect on the Shadow Price. Figures 3.2.15a and 3.2.15b show the spreadsheet formulation and sensitivity report, respectively, for this change.

	A	B	C	D	E	F	G
1	Chapter 3: Sensitivity Analysis						
2	3.2 SK8MAN, Inc. (3 variables)						
3	Profit Maximization Problem						
4							
5	Decision Variable	Sporty (x_1)	Fancy (x_2)	Pool-Runner (x_3)			
6	Decision Values [# to make per week]	0	120	210			
7							Total Profit
8	Objective Function [Profit ($)]	15	35	20			$8,400.00
9							
10	Constraints						
11	Shaping Time (minutes)	5	15	4	2640	≤	2641
12	Truck Availability	2	2	2	660	≤	700
13	North American Maple Veneers	0	7	0	840	≤	840
14	Chinese Maple Veneers	7	0	7	1470	≤	1470

Figure 3.2.15a: Formulation when the shaping time constraint increases by 241 minutes

Variable Cells

Cell	Name	Final Value	Reduced Cost	Objective Coefficient	Allowable Increase	Allowable Decrease
$B$6	Decision Values [# to make per week] Sporty (x1)	0	-5	15	5	1E+30
$C$6	Decision Values [# to make per week] Fancy (x2)	120	0	35	1E+30	35
$D$6	Decision Values [# to make per week] Pool-Runner (x3)	210	0	20	1E+30	5

Constraints

Cell	Name	Final Value	Shadow Price	Constraint R.H. Side	Allowable Increase	Allowable Decrease
$E$11	Shaping Time (minutes)	2640	0	2641	1E+30	1
$E$12	Truck Availability	660	0	700	1E+30	40
$E$13	North American Maple Veneers	840	5	840	0.466666667	840
$E$14	Chinese Maple Veneers	1470	2.857142857	1470	1.75	1470

Figure 3.2.15b: Sensitivity Report when the shaping time constraint increases by 241 minutes

G.F. Hurley notices that the Shadow Price has changed to zero. A Shadow Price of zero means that there is no value in increasing the availability of installation time any further. Therefore, increasing the availability of a resource beyond the Allowable Increase *decreases* the Shadow Price.

He also notices that increasing the available shaping time changes the optimal solution to 0 Sporty boards, 120 Fancy boards, and 210 Pool-Runner boards per week. Furthermore, the Allowable Increase for shaping time has changed to infinity (1E+30). This means there is no value in increasing the available shaping time any more.

Next, G.F. Hurley investigates what happens if the available shaping time decreases below the Allowable Decrease of 1,560. Suppose he decreases it by 1,561 minutes. The available shaping time becomes 839 minutes. Figures 3.2.16a and 3.2.16b show the spreadsheet formulation and sensitivity report for this change.

	A	B	C	D	E	F	G
1	Chapter 3: Sensitivity Analysis						
2	3.2 SK8MAN, Inc. (3 variables)						
3	Profit Maximization Problem						
4							
5	Decision Variable	Sporty (x_1)	Fancy (x_2)	Pool-Runner (x_3)			
6	Decision Values [# to make per week]	0	0	209.75			
7							Total Profit
8	Objective Function [Profit ($)]	15	35	20			$4,195.00
9							
10	Constraints						
11	Shaping Time (minutes)	5	15	4	839	≤	839
12	Truck Availability	2	2	2	419.5	≤	700
13	North American Maple Veneers	0	7	0	0	≤	840
14	Chinese Maple Veneers	7	0	7	1468.25	≤	1470

Figure 3.2.16a: Formulation when the shaping time constraint decreases by 1,561 minutes

Variable Cells

Cell	Name	Final Value	Reduced Cost	Objective Coefficient	Allowable Increase	Allowable Decrease
$B$6	Decision Values [# to make per week] Sporty (x1)	0	-10	15	10	1E+30
$C$6	Decision Values [# to make per week] Fancy (x2)	0	-40	35	40	1E+30
$D$6	Decision Values [# to make per week] Pool-Runner (x3)	209.75	0	20	1E+30	8

Constraints

Cell	Name	Final Value	Shadow Price	Constraint R.H. Side	Allowable Increase	Allowable Decrease
$E$11	Shaping Time (minutes)	839	5	839	1	839
$E$12	Truck Availability	419.5	0	700	1E+30	280.5
$E$13	North American Maple Veneers	0	0	840	1E+30	840
$E$14	Chinese Maple Veneers	1468.25	0	1470	1E+30	1.75

Figure 3.2.16b: Sensitivity Report when the shaping time constraint decreases by 1,561 minutes

G.F. Hurley notices that this change increases the value of the Shadow Price to $5. That is, decreasing the availability of a resource beyond the Allowable Decrease *increases* the Shadow Price. This makes economic sense, because decreasing the availability of a resource, as he did, increases the value per unit of that resource.

Table 3.2.4 summarizes the impact of changes in the right hand side values of the constraints. The table differentiates between changes within the allowable range and changes outside the allowable range. It also distinguishes between binding and non-binding constraints. Unlike objective coefficient changes, RHS changes of binding constraints always change the values of the decision variables. This is true even if the change is within the allowable range. The only difference between within the range and outside the range is whether or not the shadow price changes. This is a complex concept and beyond the scope of this text. If a constraint is non-binding, its shadow price is zero. Thus, any changes within the range have no impact on the decision variables or the objective function. However, when the change exceeds the allowable range for the rhs value, the constraint becomes binding and the shadow price becomes non-zero.

Impact on...	Changes to right hand side values of constraints			
	Binding Constraints		**Non-binding constraints**	
	Within range of allowable changes	**Outside range** of allowable changes	**Within range** of allowable changes	**Outside range** of allowable changes
Decision variables	Values change	Values change	Values do not change	Values change
Objective function	Objective function change predicted by multiplying shadow price by change in RHS	Objective function change cannot be predicted	No changes	Objective function change cannot be predicted
Shadow price	Stays the same	Changes – Marginal value of resource changes	No changes. Stays at zero.	Shadow price increases from 0 as constraint becomes binding

Table 3.2.4: Summary of the impact of changes in the right hand side values of constraints

3.2.5 Adding a Fourth Product—Is it profitable?

The managers at SK8MAN, Inc. are now considering adding a fourth line of skateboards to their portfolio of products. The EasyRider skateboard will be made from seven North American maple veneers, require 12 minutes of shaping time, and, of course, require 2 trucks. The managers believe that each EasyRider skateboard manufactured will earn $25 profit. They are excited by the prospect of adding a new product to their line, but the key question is whether it will be profitable to do so. Figures 3.2.17 and 3.2.18 display the problem formulation with a fourth decision variable and the optimal solution, as well as the Sensitivity Report.

Chapter 3 — Analyze Optimal Solutions—Sensitivity Analysis

	A	B	C	D	E	F	G	H
1	Chapter 3: Sensitivity Analysis							
2	3.2 SK8MAN, Inc. (4 variables)							
3	Profit Maximization Problem							
4								
5	Decision Variable	Sporty (x_1)	Fancy (x_2)	Pool-Runner (x_3)	EasyRider (x_4)			
6	Decision Values [# to make per week]	0	104	210	0			
7								Total Profit
8	Objective Function [Profit ($)]	15	35	20	25			$7,840.00
9								
10	Constraints							
11	Shaping Time (minutes)	5	15	4	12	2400	≤	2400
12	Truck Availability	2	2	2	2	628	≤	700
13	North American Maple Veneers	0	7	0	7	728	≤	840
14	Chinese Maple Veneers	7	0	7	0	1470	≤	1470

Figure 3.2.17: Formulation for the 4-decision variable SK8MAN problem

Variable Cells

Cell	Name	Final Value	Reduced Cost	Objective Coefficient	Allowable Increase	Allowable Decrease
$B$6	Decision Values [# to make per week] Sporty (x1)	0	-7.333333333	15	7.333333333	1E+30
$C$6	Decision Values [# to make per week] Fancy (x2)	104	0	35	40	3.75
$D$6	Decision Values [# to make per week] Pool-Runner (x3)	210	0	20	1E+30	7.333333333
$E$6	Decision Values [# to make per week] EasyRider (x4)	0	-3	25	3	1E+30

Constraints

Cell	Name	Final Value	Shadow Price	Constraint R.H. Side	Allowable Increase	Allowable Decrease
$F$11	Shaping Time (minutes)	2400	2.333333333	2400	240	1560
$F$12	Truck Availability	628	0	700	1E+30	72
$F$13	North American Maple Veneers	728	0	840	1E+30	112
$F$14	Chinese Maple Veneers	1470	1.523809524	1470	343.6363636	420

Figure 3.2.18: Sensitivity Report for the 4-variable SK8MAN problem

Notice that the optimal solution has not changed, despite the addition of a new product. That means it is not profitable to make the new product. SK8MAN, Inc. will earn more profit by continuing to make only Fancy and Pool-Runner skateboards. Now the question is what, if anything, can be done so that making the new EasyRider boards would be part of SK8MAN's optimal production plan.

To answer that question, G.F. Hurley turns his attention to the Sensitivity Report. Considering the information on the EasyRider board (product x_4), he sees that the Allowable Increase in the objective coefficient is three. That means that the profitability of EasyRider boards would have to increase by at least $3 (to $28) per board before they would become part of the optimal solution.

The shadow prices on the constraints help explain why the profit margin would need to be at least $28 for each EasyRider skateboard. Each EasyRider board requires seven North American maple veneers and two trucks. The related resource constraints have zero shadow prices because not all of these resources are currently being used. However, each EasyRider requires 12 minutes of installation. Each minute has a shadow price of $2.333. If G.F. Hurley multiplies 12 by

$2.333, he obtains $28. Thus the resources needed to produce an EasyRider board are valued at $28 with the current optimal production plan.

Now, suppose the marketing division at SK8MAN, Inc. has just signed a contract with Allie Loop, the top female skateboarder in the world. She will endorse the new EasyRider board. Taking into consideration the cost of Allie Loop's endorsement contact, the marketing division estimates that the retail price of an EasyRider can be increased by $5. This would then increase the profitability of EasyRider to $30 per board. Since the increase in profitability is larger than the Allowable Increase, this should be enough to make it profitable to produce EasyRider boards.

Figures 3.2.19a and 3.2.19b shows the problem formulation and the Sensitivity Report after increasing the profitability of the EasyRider board (x_4) to $30 per board.

	A	B	C	D	E	F	G	H
1	Chapter 3: Sensitivity Analysis							
2	3.2 SK8MAN, Inc. (4 variables)							
3	Profit Maximization Problem							
4								
5	Decision Variable	Sporty (x_1)	Fancy (x_2)	Pool-Runner (x_3)	EasyRider (x_4)			
6	Decision Values [# to make per week]	0	40	210	80			
7								Total Profit
8	Objective Function [Profit ($)]	15	35	20	30			$8,000.00
9								
10	Constraints							
11	Shaping Time (minutes)	5	15	4	12	2400	≤	2400
12	Truck Availability	2	2	2	2	660	≤	700
13	North American Maple Veneers	0	7	0	7	840	≤	840
14	Chinese Maple Veneers	7	0	7	0	1470	≤	1470

Figure 3.2.19a: Formulation when the profitability of EasyRider boards is $30

Variable Cells

Cell	Name	Final Value	Reduced Cost	Objective Coefficient	Allowable Increase	Allowable Decrease
$B$6	Decision Values [# to make per week] Sporty (x1)	0	-6.666666667	15	6.666666667	1E+30
$C$6	Decision Values [# to make per week] Fancy (x2)	40	0	35	2.5	5
$D$6	Decision Values [# to make per week] Pool-Runner (x3)	210	0	20	1E+30	6.666666667
$E$6	Decision Values [# to make per week] EasyRider (x4)	80	0	30	5	2

Constraints

Cell	Name	Final Value	Shadow Price	Constraint R.H. Side	Allowable Increase	Allowable Decrease
$F$11	Shaping Time (minutes)	2400	1.666666667	2400	240	120
$F$12	Truck Availability	660	0	700	1E+30	40
$F$13	North American Maple Veneers	840	1.428571429	840	70	112
$F$14	Chinese Maple Veneers	1470	1.904761905	1470	140	420

Figure 3.2.19b: Sensitivity Report when the profitability of EasyRider boards is $30

When the profitability of the EasyRider board is increased to $30 per board, it becomes profitable to produce 80 of them per week. G.F. Hurley compares this optimal production plan with the optimal production plan before SK8MAN got the Allie Loop endorsement (see Figure 3.2.17). He notices that 210 Pool-Runner boards (x_3) will still be produced, but only 40 Fancy boards will be produced. So, in order to produce 80 EasyRider boards, 64 *fewer* Fancy boards would have to be made.

G.F. Hurley wonders why this is more profitable to produce 64 fewer Fancy boards while producing 80 more EasyRider boards. He considers the profit margins on each of the boards. Making 64 fewer Fancy boards would decrease the total profit by (64)($35) = $2,240. At the same time, making 80 EasyRider boards that were not being made before would increase the total profit by (80)($30) = $2,400. Thus, the total profit is being increased by $2,400 − $2,240 = $160 per week.

Section 3.3: The Pallas Sport Shoe Company

Recall from Chapter 2 that the Pallas Sport Shoe Company manufactures six different lines of sport shoes: High Rise, Max-Riser, Stuff It, Zoom, Sprint, and Rocket. Table 3.3.1 displays the amount of daily profit generated by each pair of shoes for each of these six products. It also lists the amount of time each line of shoes requires for the six steps of production. The last line of the table shows the total amount of time per day available for each of the six production steps. Sue Painter, the production manager of the company would like to determine the daily production rates for each line of shoes that will maximize profit.

	High Rise	Max-Riser	Stuff It	Zoom	Sprint	Rocket	Total Time Available (minutes per day)
Profit ($)	18	23	22	20	18	19	
Stamping (min)	1.25	2	1.5	1.75	1	1.25	420
Upper Finishing (min)	3.5	3.75	5	3	4	4.25	1,260
Insole Stitching (min)	2	3.25	2.75	2.25	3	2.5	840
Molding (min)	5.5	6	7	6.5	8	5	2,100
Sole-to-Upper Joining (min)	7.5	7.25	6	7	6.75	6.5	2,100
Inspecting (min)	2	3	2	3	2	3	840

Table 3.3.1: Profit and production detail per pair for six lines of sport shoes

3.3.1 Problem Formulation

The formulation of the problem is given below.

Decision Variables
Let:
x_1 = the daily production rate of High Rise
x_2 = the daily production rate of Max-Riser
x_3 = the daily production rate of Stuff It
x_4 = the daily production rate of Zoom
x_5 = the daily production rate of Sprint
x_6 = the daily production rate of Rocket

Objective Function
Maximize: $z = 18x_1 + 23x_2 + 22x_3 + 20x_4 + 18x_5 + 19x_6$,
where z = the amount of profit Pallas Sport Shoe Company earns per day.

Constraints
Subject to:

Stamping (min):	$1.25x_1$	$+ 2x_2$	$+ 1.5x_3$	$+ 1.75x_4$	$+ 1x_5$	$+ 1.25x_6$	≤ 420
Upper Finishing (min):	$3.5x_1$	$+ 3.75x_2$	$+ 5x_3$	$+ 3x_4$	$+ 4x_5$	$+ 4.25x_6$	$\leq 1,260$
Insole Stitching (min):	$2x_1$	$+ 3.25x_2$	$+ 2.75x_3$	$+ 2.25x_4$	$+ 3x_5$	$+ 2.5x_6$	≤ 840
Molding (min):	$5.5x_1$	$+ 6x_2$	$+ 7x_3$	$+ 6.5x_4$	$+ 8x_5$	$+ 5x_6$	$\leq 2,100$
Sole-to-Upper Joining (min):	$7.5x_1$	$+ 7.25x_2$	$+ 6x_3$	$+ 7x_4$	$+ 6.75x_5$	$+ 6.5x_6$	$\leq 2,100$
Inspecting (min):	$2x_1$	$+ 3x_2$	$+ 2x_3$	$+ 3x_4$	$+ 2x_5$	$+ 3x_6$	≤ 840
Non-Negativity (#):	x_1	, x_2	, x_3	, x_4	, x_5	, x_6	≥ 0

This formulation as it appears in a spreadsheet is presented in Figure 3.3.1. Solver has been run, and the optimal solution also appears in the spreadsheet.

	A	B	C	D	E	F	G	H	I	J
1	Chapter 3: Sensitivity Analysis									
2	3.3 Pallas Sport Shoe Company									
3	Profit Maximization									
4										
5	Decision Variable	High Rise (x_1)	Max-Riser (x_2)	Stuff It (x_3)	Zoom (x_4)	Sprint (x_5)	Rocket (x_6)			
6	Decision Values [= to make per day]	0	4.28294434	45.12172	72.3275	104.902	89.82118			
7										Total Profit
8	Objective Function [Profit ($)]	18	23	22	20	18	19			$6,132.57
9										
10	Constraints							Used		Available
11	Stamping (minutes)	1.25	2	1.5	1.75	1	1.25	420	$\leq$	420
12	Upper Finishing (minutes)	3.5	3.75	5	3	4	4.25	1260	$\leq$	1260
13	Insole Stitching (minutes)	2	3.25	2.75	2.25	3	2.5	840	$\leq$	840
14	Molding (minutes)	5.5	6	7	6.5	8	5	2100	$\leq$	2100
15	Sole-to-Upper Joining (minutes)	7.5	7.25	6	7	6.75	6.5	2100	$\leq$	2100
16	Inspecting (minutes)	2	3	2	3	2	3	799.34219	$\leq$	840

Figure 3.3.1: Pallas Sport Shoes Spreadsheet Formulation and Optimal Solution

3.3.2 Interpreting the Solution

1. Without referring to an Answer or Sensitivity Report, which of the constraints in the spreadsheet in Figure 3.3.1 are binding and which are non-binding? How do you know?

2. Similarly, which one of the constraints will show a Shadow Price of zero in the Sensitivity Report, and why does that make sense?

Sue Painter has seen the Answer and Sensitivity Reports. She wonders, "How do I go about implementing this optimal solution?" In order to answer this question, the production manager must understand what the optimal solution means.

3. The optimal solution given in the spreadsheet from Figure 3.3.1 lists $x_1 = 0$. What does that mean? What does it mean that $x_2 \approx 4.2829$?

Recalling that the decision variables in the problem were defined as daily production rates, $x_2 \approx$ 4.2829 means that on most days, 4 Max-Riser shoes will be produced. Then, approximately every fourth day, 5 Max-Riser shoes will be produced. This production plan would yield 4.25 Max-Riser shoes every four days.

Similarly, a daily production rate for $x_3 \approx 45.1217$ means that on most days 45 will be produced, but on about every eighth day, 46 will be produced. This production plan would yield 45.125 Stuff It shoes every eight days.

4. How might the production rate of 72.3275 for product x_4 be implemented?

So, in order to implement the optimal production plan, the production manager will have to allocate production resources in such a way that the optimal production rates are achieved.

Figure 3.3.2 shows the Sensitivity Report for the optimal solution to the Pallas Sport Shoe problem. The production manager notices that it reports an Allowable Increase of about $0.0507 in the coefficient of x_1 in the objective function.

Variable Cells

Cell	Name	Final Value	Reduced Cost	Objective Coefficient	Allowable Increase	Allowable Decrease
$B$6	Decision Values [# to make per day] High Rise (x1)	0	-0.05070018	18	0.05070018	1E+30
$C$6	Decision Values [# to make per day] Max-Riser (x2)	4.282944345	0	23	0.083039285	0.239247312
$D$6	Decision Values [# to make per day] Stuff It (x3)	45.12172352	0	22	0.100842737	1.120253165
$E$6	Decision Values [# to make per day] Zoom (x4)	72.32746858	0	20	0.25648415	0.055429065
$F$6	Decision Values [# to make per day] Sprint (x5)	104.9019749	0	18	6.260393168	0.529761905
$G$6	Decision Values [# to make per day] Rocket (x6)	89.82118492	0	19	0.572583906	0.040500229

Constraints

Cell	Name	Final Value	Shadow Price	Constraint R.H. Side	Allowable Increase	Allowable Decrease
$H$11	Stamping (minutes) Used	420	5.580179533	420	113.5815474	8.859180036
$H$12	Upper Finishing (minutes) Used	1260	1.793895871	1260	15.4507772	111.8007117
$H$13	Insole Stitching (minutes) Used	840	0.255655296	840	72.56195965	4.008064516
$H$14	Molding (minutes) Used	2100	0.203375224	2100	29.90972919	134.3521595
$H$15	Sole-to-Upper Joining (minutes) Used	2100	0.422262118	2100	20.14864865	179.7068966
$H$16	Inspecting (minutes) Used	799.3421903	0	840	1E+30	40.65780969

Figure 3.3.2: Sensitivity Report for the Pallas Sport Shoe problem

5. Suppose Pallas Shoes was able to increase the profit margin on x_1 to $18.05. Would this change affect the optimal solution? Why or why not?

6. Suppose Pallas Shoes was able to increase the profitability of x_1 to $18.10. What would be the effect on the optimal solution of this increase?

3.3.3 Using the Sensitivity Report to Make Decisions

Pallas Shoes is considering adding an hour of overtime to one of the workers. Sue Painter must decide to which of the production tasks the overtime should go.

7. Using the Sensitivity Report in Figure 3.3.2 to guide the decision, to which of the six production tasks should the extra time be added? Why?

Suppose that the union contract mandates that any overtime work be paid at double the normal rate of $28 per hour.

8. Would it be profitable to add to one hour of overtime? If so, how much larger than the cost of the overtime would the increase in profits be? If not, at what hourly pay rate would it be profitable?

Finally, the managers at Pallas Sport Shoes are considering adding another line of shoes. The data for the new Pro-Go model is shown in Table 3.3.2.

	Pro-Go
Profit	$20
Stamping	1.5
Upper Finishing	3.9
Insole Stitching	2.6
Molding	6.3
Sole-to-Upper Joining	6.8
Inspecting	2.5

Table 3.3.2: Profit and production detail per pair of Pro-Go sport shoes

At present, there are no plans to increase the total amount of time available for each of the six steps of production. Figure 3.3.3 contains the new spreadsheet and optimal solution with the information for Pro-Go as decision variable x_7. Figure 3.3.4 shows the Sensitivity Report.

	A	B	C	D	E	F	G	H	I	J	K
1	Chapter 3: Sensitivity Analysis										
2	3.3 Pallas Sport Show Company										
3	Profit Maximization										
4											
5	Decision Variable	High Rise (x_1)	Max-Riser (x_2)	Stuff It (x_3)	Zoom (x_4)	Sprint (x_5)	Rocket (x_6)	Pro-Go (x_7)			
6	Decision Values [= to make per day]	0	4.2829443	45.1217	72.3275	104.902	89.8212	0			
7											Total Profit
8	Objective Function [Profit ($)]	18	23	22	20	18	19	20			$6,132.57
9											
10	Constraints								Used		Available
11	Stamping (minutes)	1.25	2	1.5	1.75	1	1.25	1.5	420	≤	420
12	Upper Finishing (minutes)	3.5	3.75	5	3	4	4.25	3.9	1260	≤	1260
13	Insole Stitching (minutes)	2	3.25	2.75	2.25	3	2.5	2.6	840	≤	840
14	Molding (minutes)	5.5	6	7	6.5	8	5	6.3	2100	≤	2100
15	Sole-to-Upper Joining (minutes)	7.5	7.25	6	7	6.75	6.5	6.8	2100	≤	2100
16	Inspecting (minutes)	2	3	2	3	2	3	2.5	799.34219	≤	840

Figure 3.3.3: Formulation with seven decision variables

Variable Cells

Cell	Name	Final Value	Reduced Cost	Objective Coefficient	Allowable Increase	Allowable Decrease
$B$6	Decision Values [# to make per day] High Rise (x1)	0	-0.05070018	18	0.05070018	1E+30
$C$6	Decision Values [# to make per day] Max-Riser (x2)	4.282944345	0	23	0.083039285	0.239247312
$D$6	Decision Values [# to make per day] Stuff It (x3)	45.12172352	0	22	0.100842737	1.120253165
$E$6	Decision Values [# to make per day] Zoom (x4)	72.32746858	0	20	0.25648415	0.055429065
$F$6	Decision Values [# to make per day] Sprint (x5)	104.9019749	0	18	6.260393168	0.529761905
$G$6	Decision Values [# to make per day] Rocket (x6)	89.82118492	0	19	0.572583906	0.040500229
$H$6	Decision Values [# to make per day] Pro-Go (x7)	0	-0.183813285	20	0.183813285	1E+30

Constraints

Cell	Name	Final Value	Shadow Price	Constraint R.H. Side	Allowable Increase	Allowable Decrease
$I$11	Stamping (minutes) Used	420	5.580179533	420	113.5815474	8.859180036
$I$12	Upper Finishing (minutes) Used	1260	1.793895871	1260	15.4507772	111.8007117
$I$13	Insole Stitching (minutes) Used	840	0.255655296	840	72.56195965	4.008064516
$I$14	Molding (minutes) Used	2100	0.203375224	2100	29.90972919	134.3521595
$I$15	Sole-to-Upper Joining (minutes) Used	2100	0.422262118	2100	20.14864865	179.7068966
$I$16	Inspecting (minutes) Used	799.3421903	0	840	1E+30	40.65780969

Figure 3.3.4: Sensitivity Report with seven decision variables

9. Why was it not profitable to produce the new product?

10. How much would its profit margin have to increase to make it profitable enough to produce?

Chapter 3 (Sensitivity Analysis) Homework Questions

1. Recall in Section 2.1.6. The Computer Flips Junior Achievement Company produces four models. Four JA students do the installation work and each of them works 10 hours per week. Another student does the testing work and he also works for 10 hours per week. Market research indicates that the combined sales of Simplex and Multiplex cannot exceed 20 computers per week, and the combined sales of Omniplex and Megaplex cannot exceed 16 computers per week. The table below contains the relevant data.

	Simplex	Omniplex	Multiplex	Megaplex
Profit	200	300	250	400
Installation Time (minutes)	60	120	90	150
Testing Time (minutes)	20	24	24	30

The sensitivity report for the Computer Flips problem is given below. The total profit is $7,000.
Microsoft Excel 12.0 Sensitivity Report
Worksheet: [Computer_Flips.xls]Sheet1

Adjustable Cells

Cell	Name	Final Value	Reduced Cost	Objective Coefficient	Allowable Increase	Allowable Decrease
$B$5	x1 Simplex	15	0	200	66.67	26.67
$C$5	x2 Omniplex	0	-20	300	20.00	1E+30
$D$5	x3 Multiplex	0	-20	250	20.00	1E+30
$E$5	x4 Megaplex	10	0	400	100	25.00

Constraints

Cell	Name	Final Value	Shadow Price	Constraint R.H. Side	Allowable Increase	Allowable Decrease
$F$11	Installation Time	2400	1.67	2400	360	200
$F$12	Testing Time	600	5	600	40	120
$F$13	Market Restriction 1	15	0	20	1E+30	5
$F$14	Market Restriction 2	10	0	16	1E+30	6

a. Computer Flips made its reputation in the 1990s with the launch of Simplex and Omniplex brands. It wants to keep these two classic computers in its product line. If the profit on each Omniplex is increased from $300 to $325, would that be a large enough increase to add Omniplex to the optimal product mix?

b. The five students are not paid per hour. They simply split the weekly profits equally. However, the student in charge of doing testing has expressed am immediate need for extra cash. He requested an opportunity to work a half hour more this week. He agrees to accept an hourly wage of $12/hour but not share in the extra profit. Should the team approve his request?

c. Because of the poor market conditions, the combined demand for Omniplex and Megaplex has decreased from 16 to 8 computers (Market Restriction 2). Computer Flips has decided it cannot sell more than eight of these types of computers this week. How would this affect the optimal solution? Explain.

2. Refer back to Section 2.3.3 SK8MAN Inc. This example includes two decision variables and three constraints as formulated below with a corresponding graphical representation as presented earlier in Figure 2.3.7.

<u>Objective Function</u>
  Maximize: $z = 15x_1 + 35x_2$
<u>Constraints</u>
  Shaping Time: $5x_1 + 15x_2 \leq 2400$
  Trucks: $2x_1 + 2x_2 \leq 700$
  North American Maple: $7x_2 \leq 840$
  Non-Negativity: $x_1 \geq 0$ and $x_2 \geq 0$

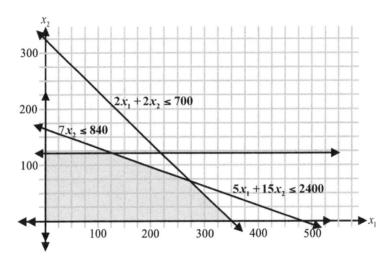

Figure 2.3.7: The SK8man feasible region after adding the North American maple constraint

Answer Report

Cell	Name	Original Value	Final Value
$E$3	Decision variables	0	6550

Cell	Name	Original Value	Final Value
$C$3	Decision variables Sporty	0	285
$D$3	Decision variables Fancy	0	65

Cell	Name	Cell Value	Formula	Status	Slack
$E$6	Shaping Time	2400	$E$6<=$G$6	Binding	0
$E$7	Trucks	700	$E$7<=$G$7	Binding	0
$E$8	North American Maple	455	$E$8<=$G$8	Not Binding	385

Sensitivity Report

Adjustable Cells

Cell	Name	Final Value	Reduced Cost	Objective Coefficient	Allowable Increase	Allowable Decrease
$C$3	Decision variables Sporty	285	0	15	20	3.33333
$D$3	Decision variables Fancy	65	0	35	10	20

Constraints

Cell	Name	Final Value	Shadow Price	Constraint R.H. Side	Allowable Increase	Allowable Decrease
$E$6	Shaping Time	2400	2	2400	550	650
$E$7	Trucks	700	2.5	700	260	220
$E$8	North American Maple	455	0	840	1E+30	385

a. There is a rumor that a competitor will soon start selling a skateboard directly comparable to Fancy. This will cause the profit margin to decline by $6 to $29. If this rumor comes true, how would that impact the optimal solution and total profit?

b. There is a rumor that a competitor will soon start selling a skateboard directly comparable to Sporty. This will cause the profit margin to decline by $3.50 to $11.50. If this rumor comes true, how would that impact the optimal solution and total profit?

c. SK8man is considering enhancements to their Sporty product that will increase the profit margin by $4 to $19. How would that impact the optimal solution and total profit?

d. Explain graphically what happens to the objective function when the profit margin on Fancy increases by $10 to a $45. What information in the above reports suggests this would happen?

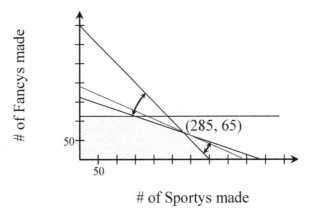

e. Describe a corresponding change in profit for the Sporty product that would have the same impact.

f. In general, maintenance on the shaping equipment is done before work starts each day. Occasionally workers must turn off the equipment and realign critical components. This takes 30 minutes. What is the impact on the total profit of this loss of production time? How would this change the optimal production plan?

g. Management is considering extending the use of shaping equipment by an hour. How would this affect total profit?

h. The local supplier of trucks is willing to provide 100 extra trucks per shipment. How would this impact the optimal production plan?

i. Assume now that these extra trucks come at a premium price that adds $3 to the cost. Should management be willing to purchase these extra trucks? What if the premium price were only $1.50 what would be the decision? What information in the sensitivity report helps resolve these questions?

j. The supplier of North American Maple is prepared to offer an additional 100 veneers at a discount price. Should management purchase these extra veneers? Explain.

3. Recall from Chapter 2 homework that Anderson Cell Phone Company produces smart phones and standard phones. There are 10 workers available who are each available to work 7 hours per day on assembly. The table below gives the assembly times (in minutes) required to assemble each type of phone and their associated profits. 2000 LCD screens are available per day.

	Smart Phone	Standard Phone
Profit	$40	$30
Assembly Time	2.5	1.5

Below are the Answer and the Sensitivity Reports for the problem.

Target Cell (Max)

Cell	Name	Original Value	Final Value
$D$2	OBJ. FNC.	$0	$72,000

Adjustable Cells

Cell	Name	Original Value	Final Value
$B$3	D.V. Smart	0	1200
$C$3	D.V. Standard	0	800

Constraints

Cell	Name	Cell Value	Formula	Status	Slack
$D$5	Assembly Time	4200	$D$5<=$F$5	Binding	0
$D$6	Screen	2000	$D$6<=$F$6	Binding	0

Adjustable Cells

Cell	Name	Final Value	Reduced Cost	Objective Coefficient	Allowable Increase	Allowable Decrease
$B$3	D.V. Smart	1200	0	40	10	10
$C$3	D.V. Standard	800	0	30	10	6

Constraints

Cell	Name	Final Value	Shadow Price	Constraint R.H. Side	Allowable Increase	Allowable Decrease
$D$5	Assembly Time	4200	10	4200	800	1200
$D$6	Screen	2000	15	2000	800	320

a. What are the binding constraints?

b. What would be the daily profit if one assembly worker did not come to work due to illness and there were only 9 workers available?

c. Check your answer to part (c) by reusing Solver to determine the optimal solution with 9 workers.

4. Because of intense market competition, the selling price for standard phones has decreased by $5 with a similar decrease in the profit for a standard phone.

   a. Graph the feasible region and show the line of constant profit passing through the optimal corner point. Does the optimal solution change? What is the objective function value?

   b. What if the price decrease is $6? $7? What are the new values of the objective function?

   c. What information in the sensitivity report helps explain why a $5 and $7 decrease do not similarly impact the optimal production plan?

5. Below is the sensitivity report for the GA Sports Soccer Ball homework problem in Chapter 2. The thermal molding machine is down for 20 minutes.

   GA Soccer Ball Sensitivity Report
   Adjustable Cells

Cell	Name	Final Value	Reduced Cost	Objective Coefficient	Allowable Increase	Allowable Decrease
$B$3	D.V. Professional	15	0	15	5	15
$C$3	D.V. Practice	30	0	10	1E+30	2.5

   Constraints

Cell	Name	Final Value	Shadow Price	Constraint R.H. Side	Allowable Increase	Allowable Decrease
$D$5	Cutting Time	420	1.071	420	14	210
$D$6	Hand-stitching	15	0	16	1E+30	1
$D$7	Thermal Molding	360	0.208	360	360	24

   a. The thermal molding machine needs emergency repairs and could not be used for 20 minutes. How much did it cost to the company in lost profit?

   b. The company is considering hiring more stitchers. Is this a good idea?

   c. The net profit per ball already includes the salary of workers during the regular day. However, the cutter informed management that he was available to work one hour overtime with overtime pay that is an extra $8 per hour more than regular pay. Should management increase the cutter's hours? Justify your answer.

d. GA is concerned that increasing competition may force them to cut their profit margins on practice balls by $2 per ball. Would this cause them to change their production plan? How much would their profits decrease?

6. Consider the four product version of the Family Cow dairy in chapter 2. The owner of the dairy can make and sell cream, butter, plain yogurt and cheese. The prices for these products are $7/quart, $12/lb, $3.5/lb, and $5.5/lb, respectively.

   a. Create the Answer and Sensitivity Reports for the problem.

   b. The farmer's nephew is considering helping out after school. He is available two hours a day but wants to be paid $10 per hour for his time? Would it be worth hiring him? If so what would be the net impact on the daily profit?

   c. The farmer was surprised that the optimal plan did not include cream. He wonders if a $0.25 increase in the sale price for cream would change his production plan. Would the plan be affected by this price increase?

   d. The farmer's best customer insists she needs two quarts of cream tomorrow. What will be the impact on his total profit? Rerun Solver to determine his new optimal plan for that day.

   e. There is a holiday coming up. The local store believes it can sell 10 pounds of butter per day. What will be the new objective value?

7. In Chapter 2 homework, Katia considered investing her money in a bond fund and a domestic stock fund.

Sensitivity Analysis
Adjustable Cells

Cell	Name	Final Value	Reduced Cost	Objective Coefficient	Allowable Increase	Allowable Decrease
$B$3	D.V. bond	133333.33	0	0.08	0.07	0.0425
$C$3	D.V. stock	66666.67	0	0.15	0.17	0.07

Constraints

Cell	Name	Final Value	Shadow Price	Constraint R.H. Side	Allowable Increase	Allowable Decrease
$D$5	Max Stock Investment	66666.67	0	75000	1E+30	8333.333
$D$6	Max Risk	20000	0.467	20000	1250	10000
$D$7	Max Investment	200000	0.057	200000	200000	25000

   a. What are the binding and non-binding constraints for her decision?

b. Why is the shadow price for the maximum stock investment constraint 0?

c. Which would have a greater impact on total revenue, adding $1,000 to the total investment or increasing the allowable risk by $1,000?

d. Increase the total investment by $1,000 and describe how the optimal investment strategy changed.

8. In chapter 2 Katia also considered investing in an international stock fund. She is concerned that the forecasted return is overly optimistic.

Adjustable Cells

Cell	Name	Final Value	Reduced Cost	Objective Coefficient	Allowable Increase	Allowable Decrease
$B$3	D.V. bond	150000	0	0.08	0.14	0.036
$C$3	D.V. domestic stock	0	-0.035	0.15	0.035	1E+30
$D$3	D.V. international stoc	50000	0	0.22	0.18	0.046666667

Constraints

Cell	Name	Final Value	Shadow Price	Constraint R.H. Side	Allowable Increase	Allowable Decrease
$E$5	Max Stock Investment	$50,000	$0.00	75000	1E+30	25000
$E$6	Max Risk	$20,000	$0.70	20000	5000	10000
$E$7	Max Investment	$200,000	$0.05	200000	200000	100000

a. Would the optimal investment strategy change if the return on investment for the international fund were only 20%.

b. What change to the projected annual return for the domestic stock will result in it being included in the optimal investment plan?

c. Katia is considering setting aside $5,000 to pay off some bills. How much revenue would she lose as a result?

d. How much increased revenue would she gain if she were willing to accept a $1,000 increase in risk? Rerun Solver to determine the new optimal solution and show how the new optimal investment plan confirms your answer.

9. Remember John Farmer from Chapter 2's set of homework problems? He would like to help his father find the optimal crop mix to plant in the 640-acre farm under cultivation. Mr. Farmer had budgeted $60,000 to cover all expenses. Use the information given in the table to answer the sensitivity questions.

	Corn	Soybean	Wheat
Price/bushel	$2.90	$7.85	$3.75
Yield/acre	155.8 bushels	41.4 bushels	60 bushels
Seed cost/acre	$45.50	$34	$24
Fertilizer cost/acre	$88.10	$37.80	$69.25
Fuel cost/acre	$24.40	$10	$40
Worker cost/acre	$13	$9.50	$9

The sensitivity report for the crop mix problem is given below:

Adjustable Cells

Cell	Name	Final Value	Reduced Cost	Objective Coefficient	Allowable Increase	Allowable Decrease
$B$17	Corn	19.67	0	280.82	156.87	47.13
$C$17	Soybean	620.33	0	233.69	47.13	83.76
$D$17	Wheat	0	-181.07	82.75	181.07	1E+30

Constraints

Cell	Name	Final Value	Shadow Price	Constraint R.H. Side	Allowable Increase	Allowable Decrease
$F$19	Land	640	179.70	640	17.17	289.12
$F$20	Budget	60000	0.59	60000	49440	1568

a. A farmers' association announced it will grant low interest credit to local farmers. The interest rate is 7%. Mr. Farmer is considering increasing his budget by borrowing $10,000. How much will the net income increase with the new budget?

b. In order to encourage corn production of corn, the USDA will provide a subsidy that will increase the corn price per bushel to $3.50. Does it affect Mr. Farmer's decision?

c. Mr. Farmer is considering diversifying his crop by planting 10 acres of wheat. How would this impact his total profit?

d. Mr. Farmer's daughter wants her father to delay planting wheat until the price increases to the point that wheat is part of the optimal solution. At what price should Mr. Farmer consider producing wheat?

e. The owner of the neighboring farm suggested leasing his small farm to Mr. Farmer. He is asking for $1000 for this 10 acre farm. Is this a good offer for Mr. Farmer?

10. In chapter 2 homework, Elegant Fragrances, Ltd developed an optimal solution for its 4 product lineup.

    a. The company is concerned about frequent changes in the market for their perfumes. These changes can cause prices to drop by 10% or even 20%. Would a decline of 10% in any perfume price cause the optimal solution to change? What about a 20% decline?

Sensitivity Report

Adjustable Cells

Cell	Name	Final Value	Reduced Cost	Objective Coefficient	Allowable Increase	Allowable Decrease
$E$3	Decision X1	14285.71	0.00	324.25	1E+30	59.47
$F$3	Decision X2	5595.24	0.00	191.52	103.12	41.19
$G$3	Decision X3	0.00	-76.46	436.3	76.46	1E+30
$H$3	Decision X4	9880.95	0.00	210.46	57.67	70.20

Constraints

Cell	Name	Final Value	Shadow Price	Constraint R.H. Side	Allowable Increase	Allowable Decrease
$J$6	Mango Pulp 4 Perfumes	1400.0	2148.4	1400	51.43	338.78
$J$7	Tea Leaves 4 Perfumes	1600.0	1201.4	1600	257.14	191.84
$J$8	Juniper Berry 4 Perfumes	1000.0	2853.2	1000	470.00	90.00
$J$9	White Rose 4 Perfumes	1435.7	0.0	1500	1E+30	64.29

    b. The company is considering buying more raw materials. Which raw materials does it need to buy? Which of these will have the greatest impact per pound? How can a shadow price per pound of raw material be more than $1,000 when the profit margins are always less than $500?

    c. The optimal solution included only three of the four products. The optimal does not include Evergreen. However, management wants to offer a full range of products. It is thinking of requiring a minimum production of 1,000 lbs. of each perfume. They know this will reduce profits but wonder if the reduction in profit will be more than 2%. Explain why the impact of this requirement is going to be small relative to the total profit which is more than $8.45 million. Rerun the model with this requirement and describe the impact on production

Chapter 3 Summary

What have we learned?

The process of **sensitivity analysis** is used to explore the robustness of the optimal solution to a linear programming problem. Oftentimes there is uncertainty or variability in the parameters of a problem. Sensitivity analysis allows the decision maker to understand how sensitive the optimal solution is to changes in these parameters. Knowing the implications of changes to parameters also allows for managers to make better decisions about how operations can be improved.

1. Solve the problem and generate reports.
    - Set up spreadsheet formulation of problem.
    - Set up Solver Parameters and Options.
    - Solve and generate Answer and Sensitivity Reports.

2. Interpret the Answer Report.
    - Know what information is contained in the Answer Report.
    - Know how the same information is represented in the solved spreadsheet.

3. Interpret the Sensitivity Report.
    - Know what information is contained in the Sensitivity Report.
    - Investigate effects of changing objective function coefficients.
    - Investigate effects of changing constraint right hand sides.
    - Understand what changes affect optimal solution.
    - Understand what changes affect the optimal value of the objective function.

Terms

Answer Report A report generated by Solver after it has found the optimal solution. This report summarizes the optimal solution, the optimal value of the objective function, and the status of the constraints.

Binding When a constraint models the consumption of a resource, it is binding if the solution to the problem uses all of the resource that is available. This is the case when the left hand side and the right hand side of a constraint are equal.

Parameters The data that define a problem. These include objective function coefficients, coefficients within constraints, and constraint right hand side values.

Sensitivity Analysis The process of exploring how changing the parameters of a linear programming problem affects the optimal solution and optimal value of the objective function.

Sensitivity Report A report generated by Solver after it has found the optimal solution. This report provides information to predict how the optimal solution or the optimal value of the objective function will change in response to varying specific parameters of the problem.

	A	B	C	D	E	F	G
1	2.2 SK8MAN, Inc.						
2	LP - Max						
3		Sporty	Fancy	Pool-Runner			
4	Decision Variables	x1	x2	x3			
5	Decision Variable Values (#/week)	0	104	210			
6					Total Profit		
7	Objective Function [Profit ($/week)]	15	35	20	7840		
8							
9	Constraints						
10	Shaping time (min/week)	5	15	4	2400	≤	2400
11	Truck availability (#/week)	2	2	2	628	≤	700
12	North American maple veneers (#/week)	0	7	0	728	≤	840
13	Chinese maple veneers (#/week)	7	0	7	1470	≤	1470

Spreadsheet formulation showing optimal solution

	A	B	C	D	E	F	G
6		Target Cell (Max)		**E**	**C**		
7		**D** Cell	Name	Original Value	Final Value		
8		$E$7	Objective Function [Profit ($/week)] Total Profit	0	7840		
9							
10							
11		Adjustable Cells		**E**	**C**		
12		Cell	Name	Original Value	Final Value		
13		$B$5	Decision Variable Values (#/week) x1	0	0		
14		$C$5	Decision Variable Values (#/week) x2	0	104		
15		$D$5	Decision Variable Values (#/week) x3	0	210		
16							
17		**B**					
18		Constraints		**A**		**G**	**F**
19		Cell	Name	Cell Value	Formula	Status	Slack
20		$E$10	Shaping time (min/week) Total Profit	2400	$E$10<=$G$10	Binding	0
21		$E$11	Truck availability (#/week) Total Profit	628	$E$11<=$G$11	Not Binding	72
22		$E$12	North American maple veneers (#/week) Total Profit	728	$E$12<=$G$12	Not Binding	112
23		$E$13	Chinese maple veneers (#/week) Total Profit	1470	$E$13<=$G$13	Binding	0

Answer report

A: **Cell Value** — Values in this column show the total for the left hand side of each constraint based on the optimal solution.

B: **Constraints** — The section of the Answer Report dealing with the system of constraints.

C: **Final Value** — The value that appeared in the indicated cell after the "Solve" button was pushed.

D: **Objective/Target Cell** — The section of the Answer Report dealing with the objective function.

E: **Original Value** — The value that appeared in the indicated cell before the "Solve" button was pushed.

F: **Slack** — The difference between the right hand side and the left hand side values of a constraint.

G: **Status** — Shows whether a constraint is binding or not binding.

H: **Variable/Adjustable Cells** — The section of the Answer Report dealing with the decision variable values.

	A	B	C	D	E	F	G	H
6	Adjustable Cells				**H**	**G**	**D**	**B**
7	**J**			Final	Reduced	Objective	Allowable	Allowable
8		Cell	Name	Value	Cost	Coefficient	Increase	Decrease
9		$B$5	Decision Variable Values (#/week) x1	0	-7.33333333	15	7.33333333	1E+30
10		$C$5	Decision Variable Values (#/week) x2	104	0	35	40	35
11		$D$5	Decision Variable Values (#/week) x3	210	0	20	1E+30	7.33333333
12	**F**							
13	Constraints				**I**	**E**	**C**	**A**
14				Final	Shadow	Constraint	Allowable	Allowable
15		Cell	Name	Value	Price	R.H. Side	Increase	Decrease
16		$E$10	Shaping time (min/week) Total Profit	2400	2.333333333	2400	240	1560
17		$E$11	Truck availability (#/week) Total Profit	628	0	700	1E+30	72
18		$E$12	North American maple veneers (#/week) Total Profit	728	0	840	1E+30	112
19		$E$13	Chinese maple veneers (#/week) Total Profit	1470	1.523809524	1470	343.636364	420

Sensitivity report

A:	**Allowable Decrease (Constraints)**	The amount the constraint right hand side can decrease without affecting the shadow price.
B:	**Allowable Decrease (Variable/Adjustable)**	The amount the objective function coefficient can decrease without affecting the optimal solution.
C:	**Allowable Increase (Constraints)**	The amount the constraint right hand side can increase without affecting the shadow price.
D:	**Allowable Increase (Variable/Adjustable)**	The amount the objective function coefficient can increase without affecting the optimal solution.
E:	**Constraint R.H. Side**	When a constraint models the consumption of a resource, the right hand side value shows the amount of that resource that is available.
F:	**Constraints**	The section of the Sensitivity Report dealing with the system of constraints.
G:	**Objective Coefficient**	Lists the objective function coefficients for each decision variable.
H:	**Reduced Cost**	The amount the optimal value of the objective function will change in order to increase a decision variable value from zero to one.
I:	**Shadow Price**	The amount of change in the optimal value of the objective function if the right hand side of the constraint were increased by one unit.
J:	**Variable/Adjustable Cells**	The section of the Sensitivity Report dealing with the decision variable values.

Chapter 3 (Sensitivity Analysis) Objectives

You should be able to:

- Enter the problem formulation into Excel

- Set up Solver Parameters and Options

- Interpret the optimal solution in the context of the problem

- Analyze the Answer Report

- Analyze the Sensitivity Report

- Know the type of information contained in the sections of the Answer and Sensitivity Reports

- Differentiate between Allowable Increase/Decrease in adjustable cells and in constraints

- Explain meaning of Shadow Price being positive, negative, or zero

- Understand meaning of Reduced Cost

Chapter 3 Study Guide

1. What information have we used from the Answer Reports?

2. Write a definition for each in your own words.
   a. Binding constraint

   b. Non-binding constraint

3. What is the "Final Value" on an Answer Report?

4. What is slack? What does the cell value tell us?

5. What information have we used from the Sensitivity Reports?

6. For the decision variables, what do the allowable increase and decrease tell us about the variables?

7. What does the reduced cost tell us?

8. Complete each statement describing what happens to the objective function:
   a. If the shadow price for a constraint is 0, then

   b. If the shadow price for a constraint is 100, then

9. How does "Constraint R. H. Side" relate to the word problem?

10. What information is listed in the final value column?

11. Which report would you use to find the final value of the objective function?

12. In 5 or more COMPLETE, GRAMMATICALLY CORRECT sentences compare and contrast the answer report and sensitivity report. Tell how they can be used to analyze problems. **Use at least one example** of how we have used them with the problems done in class.

CHAPTER 4:

Minimize Calories or Cost with Linear Programming

Section 4.0: Introduction

The last two chapters focused on maximizing profit. First, Chapter 2 explained how to explore maximization linear programming problems by hand and then with Excel. Then, Chapter 3 was dedicated to interpreting and analyzing the solutions and constraints of these maximization problems. In both of these chapters, the goal was to find the largest value of the objective function, given a set of constraints.

In this chapter, minimization linear programming problems are introduced. Excel Solver is again used, but this time, the intent is to obtain the smallest value of the objective function for a given set of constraints. The chapter begins by exploring the problem of finding a food program for Malawian children. The food program needs to meet daily nutritional requirements while minimizing calories.

Next, a group of planners is trying to reduce the amount of water pollution coming into two watersheds in Wisconsin. They need to keep in mind a number of constraints as they attempt to minimize the cost of the project. Excel Solver is used to solve this problem.

Finally, the chapter ends with a linear programming problem involving a gasoline distributor. The distributor needs to determine the optimal gasoline blend while minimizing the cost of this blend. Again, Excel Solver is used to solve this problem.

In each of these problems, the goal is to minimize the value of the objective function, representing either calories or cost. The methods used to formulate and solve these problems are very similar to the methods used to solve maximization problems. The largest difference is simply the way one thinks about the set-up of the problem. There will also be a greater mix of constraint types. In the previous chapters all of the constraints were of the form, "less than or equal to" and involved a resource or market demand limit. In this chapter many of the constraints will be of the form, "greater than or equal to."

Section 4.1: Nutrition in Malawi

Malawi is a landlocked country in southern Africa (see Figures 4.1.1 and 4.1.2). Its population of over 13,000,000 lives in an area about the size of the state of Pennsylvania. Malawi's economy is largely based on agriculture. Much of its population is impoverished. As a result, the diets of Malawian children are frequently deficient in essential nutrients.

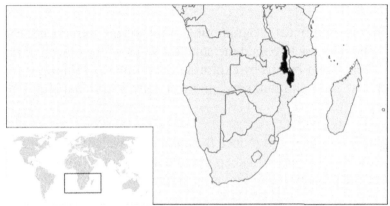

Figure 4.1.1: Map of southern Africa with Malawi in black

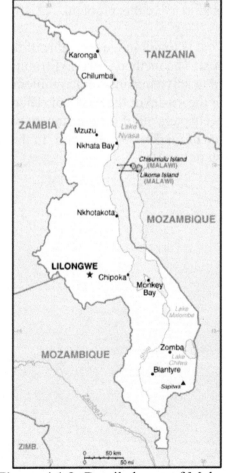

Figure 4.1.2: Detailed map of Malawi

Malawi is an impoverished nation, so the financial aspect of any food program is a vital concern. For that reason, Dr. Corr, an administrator at the World Health Organization, needs to determine an optimal food program for this country. In particular, he needs to minimize the total number of calories while meeting the minimum requirements for key nutrients, using the highest nutrient concentrated food combinations. This will be the most efficient way to meet the children's minimum nutritional requirements. A common problem in poor regions is that their diet is dominated by low cost high caloric foods with little other nutritional value. All of the foods under consideration are readily available in Malawi at low cost. To simplify the example, we chose a subset of foods available in Malawi. (For a more detailed example see the 2002 article by Darmon, Ferguson, and Briend entitled "Linear and nonlinear programming to optimize the nutrient density of a population's diet: An example based on diets of preschool children in rural Malawi" in *The American Journal of Clinical Nutrition*, volume 75, number 2, pages 245-253.)

The key nutrients Dr. Corr takes into account and the minimum daily requirements recommended by the World Health Organization are listed in Table 4.1.1. The foods available to the Malawians are shown in Figure 4.1.3. Notice that the units of measurement are not all the same in Table 4.1.1. For example, protein is measured in grams. Calcium and iron are measured in milligrams. Vitamins B_9 and B_{12} are measured in micrograms. Nutritional facts per gram for these foods appear in Table 4.1.2. These nutrients are all scaled based on the same units of measurement in Table 4.1.1. For example, let's look at a gram of maize flour. Each gram of maize flour contains 0.08120 grams of protein. It also contains 0.0612 milligrams of calcium and 0.03450 milligrams of iron. Each gram of maize flour contains 0.2450 micrograms of vitamin B_9 and so forth. The caloric content per gram of each of the food sources is given in Table 4.1.3.

Nutrient	Minimum daily requirement
Protein	20 grams (g)
Calcium (Ca)	400 milligrams (mg)
Iron (Fe)	7 mg
Folate (Vitamin B_9)	50 micrograms (μg)
Cyanocobalamin (Vitamin B_{12})	0.5 μg
Ascorbic acid (Vitamin C)	20 mg
Thiamine (Vitamin B_1)	0.7 mg
Riboflavin (Vitamin B_2)	1.1 mg
Niacin (Vitamin B_3)	12.1 mg
Retinol (Vitamin A)	400 μg

Table 4.1.1: Nutrients and minimum daily requirements

Figure 4.1.3: Malawian foods

Food	Protein (g)	Ca (mg)	Fe (mg)	Vit. B_9 (µg)	Vit. B_{12} (µg)	Vit. C (mg)	Vit. B_1 (mg)	Vit. B_2 (mg)	Vit. B_3 (mg)	Vit. A (µg)
Maize flour	0.08120	0.0612	0.03450	0.2450	0	0	0.00385	0.00201	0.03630	0.112
Tangerines	0.00805	0.3640	0.00156	0.1560	0	0.268	0.00058	0.00036	0.00377	0.338
Pigeon peas	0.06760	0.4290	0.10000	1.1000	0	0	0.00148	0.00571	0.00781	0
Matemba	0.20100	0.1000	0.00556	0.2440	0.0158	0	0.00041	0.00063	0.03900	0
Potatoes	0.01960	0.0507	0.00350	0.0922	0	0.128	0.00105	0.00021	0.01390	0
Chinese cabbage	0.01500	1.0500	0.00800	0.6630	0	0.450	0.00040	0.00070	0.00500	2.230

Table 4.1.2: Nutritional content per gram of foods

Food	Energy content (cal/g)
Maize flour	3.620
Tangerines	0.532
Pigeon peas	1.190
Matemba	0.956
Potatoes	0.931
Chinese cabbage	0.131

Table 4.1.3: Energy content of foods

Dr. Corr uses linear programming to design a diet that meets all nutritional requirements while keeping the intake of calories at a minimum. In this case, each nutrient under consideration acts as a constraint. The total number of calories is the objective function. Note that the objective is to *minimize* the number of calories. Therefore, the method used to solve this problem must be different from the previous chapters, where the objective was maximization.

1. How do you think this minimization problem differs from the maximization problems in the previous two chapters?

Minimization Linear Programming Problems

Minimization linear programming problems are solved very similarly to maximization problems. However, instead of the optimal solution being the *largest* value of the objective function, it is now the *smallest*.

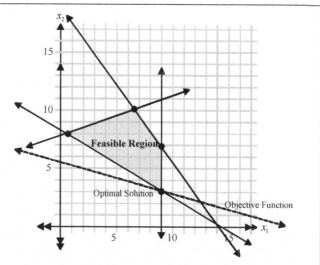

Figure 4.1.4: Example of a feasible region and optimal solution in a minimization problem

4.1.1 Linear Programming Formulation

The first step in the formulation of a linear programming problem is to define the decision variables in the problem. The decision variables are then used to define the objective function. In the Malawian diet problem, Dr. Corr seeks to minimize the daily intake of calories. Therefore, the objective function must represent the total calories in the diet per day. These calories come from the available foods. Thus, each decision variable represents the amount of each type of food in the diet each day.

2. How many decision variables should be defined?

3. How should the decision variables be defined?

4. Use the decision variables and the values in Table 4.1.3 to write the objective function, where z represents the number of calories consumed per day.

The last step in the formulation is writing the system of constraints for the problem. The diet must meet or exceed the minimum daily allowances for each of the key nutrients listed in Table 4.1.1. Therefore, Dr. Corr multiplies the number of grams of the food by its corresponding nutritional content to obtain the amount of that nutrient consumed. For example, if a person eats x_1 grams of maize flour, he/she consumes $0.0612 * x_1$ mg of calcium (see Table 4.1.2). Dr. Corr multiplies these two numbers together to show the number of milligrams of calcium the maize flour serving provides:

(x_1 grams of maize flour)(0.0612 mg of calcium per gram) = $0.0612x_1$ mg of calcium

Dr. Corr needs to determine the total number of milligrams of calcium in the food intake for an entire day. To do so, he must repeat the above process for each decision variable and add the results together. This yields the following expression:

$$0.0612x_1 + 0.3640x_2 + 0.4290x_3 + 0.1000x_4 + 0.0507x_5 + 1.0500x_6$$

This represents the total number of milligrams of calcium in the food intake for an entire day. Then, Dr. Corr completes this constraint by making the above expression greater than or equal to 400:

$$0.0612x_1 + 0.3640x_2 + 0.4290x_3 + 0.1000x_4 + 0.0507x_5 + 1.0500x_6 \geq 400$$

In this equation both the left hand side of the equation and right hand side are in milligrams.

5. Why is this constraint greater than or equal to 400?

6. Continue this process to find a constraint inequality for each of the nutrients in Table 4.1.1.

Next, Dr. Corr builds the notion of a balanced diet into the model. To do this, he takes into account what nutritionists recommend as the minimum and maximum number of calories for a typical Malawian child aged 6-9 years. This recommendation is also broken down into a range of calories for each of the various food groups. Table 4.1.4 contains these recommended minimums and maximums.

Food Group	Food	Minimum (cal/day)	Maximum (cal/day)
Cereals	Maize flour	900	1,100
Fruits	Tangerines	15	45
Legumes	Pigeon peas	45	150
Fish, meat, eggs	Matemba	30	90
Roots	Potatoes	60	240
Vegetables	Chinese cabbage	15	45

Table 4.1.4: Minimum and maximum calories per day by food group for children in Malawi

For example, the table shows that the recommended number of calories per day from cereals is between 900 and 1,100 calories. In the Malawian children's diets, the food Dr. Corr chooses to represent this group is maize flour, because it is the most available grain in Malawi. The number of grams of maize flour consumed per day is represented by x_1. From Table 4.1.3, its energy content is 3.620 calories per gram. Thus, the total number of calories coming from cereals would be $3.620x_1$. Now, it is recommended that the total be greater than or equal to 900 and less than or equal to 1,100 per day. This is really two constraints:

$$3.620x_1 \geq 900 \text{ and } 3.620x_1 \leq 1,100$$

In other words, the total number of calories from cereals must be *between* 900 and 1,100.

7. Continue this process to find a constraint inequality for each of the remaining food groups.

8. Using your responses to question 3 through question 7, write the complete problem formulation.

9. Based on this problem formulation, write a general prediction for the results of this problem. For example, are there any foods that must be consumed? Are there any foods that may not need to be consumed? Explain your answer.

In the Homework Exercises for this chapter, you will revisit this problem. You will be asked to enter your problem formulation into a spreadsheet solver to find the optimal solution and then to interpret this solution using answer and sensitivity reports.

Section 4.2: Minimizing Cost to Reduce Phosphorus in Watersheds

Water pollution comes from many sources. Water that runs off construction sites following rainstorms—known as *construction runoff*—contributes to water pollution. As water runs off construction sites, it picks up harmful sediment. The sediment might contain lead or mercury, nutrients like nitrogen and phosphorus, as well as oil, grease, and pesticides.

In urban and suburban areas, rain or snowfall that does not evaporate or soak into the ground is called *urban storm runoff*. Urban storm runoff also carries nutrients, sediment, and chemicals as it flows eventually into our waterways.

Fertilizers, pesticides, manure, and tilled soil are beneficial to crops. However, they can become harmful to our water as rains and irrigation wash them away. This is referred to as water pollution from *agricultural sources*.

Finally, industries, such as factories, release water back into the environment that has not been completely relieved of its nutrients and sediments. Much of this water runoff from *industrial sources* contains the chemical phosphorus, which potentially can harm the environment.

Phosphorus is one of the key nutrients necessary for the growth of plants and animals. However, in large amounts, it leads to excessive plant growth and decay. Phosphorus also favors certain weedy species over others. Too much phosphorus is likely to cause severe reductions in water quality. This is known as *excessive phosphorus loading*. Each form of water pollution discussed here (construction runoff, urban storm runoff, agricultural sources, and industrial sources) contributes to the amount of phosphorus in the water.

A challenge for many environmental groups is how to reduce the amount of phosphorus in the water at a minimal cost. Linear programming can be used to determine how to minimize the cost while achieving a specified phosphorus level for the best environmental quality.

In northeastern Wisconsin, there is a watershed system that is an area of concern because of the amount of phosphorus loading. A *watershed system* is the various land areas which drain into a certain lake or river. The communities in northeastern Wisconsin are planning to reduce the pollution in their watersheds. In this region, water drains into either Lake Winnebago or Green Bay (see Figure 4.2.1). In the rest of the state, water drains into the Mississippi River.

Figure 4.2.1: Map of Wisconsin showing Lake Winnebago and Green Bay

Nadia Manning is leading a group of planners who are focusing on pollution from the four sources discussed above. These are construction site runoff, urban storm runoff, agricultural sources, and industrial sources including municipal treatment plants. Nadia wants to reduce the amount of phosphorus in each watershed. Her goal is to reduce the amount of phosphorus by exactly 40,000 kilograms in Lake Winnebago (approximately 44 tons) and exactly 85,000 kilograms of phosphorus in Green Bay (approximately 93.5 tons), as seen in Table 4.2.1.

	Watershed	Target Reduction Amount of Phosphorus
1	Lake Winnebago	40,000 kg
2	Green Bay	85,000 kg

Table 4.2.1: Amount of phosphorus by watershed

The cost of the phosphorus reduction varies for each source. These costs are found in Table 4.2.2.

	Source	Cost of Phosphorus Reduction per kg
1	Construction runoff	$770
2	Urban storm runoff	$2,025
3	Agricultural sources	$26
4	Industrial sources	$75

Table 4.2.2: Phosphorus reduction costs by source

Table 4.2.1 shows the target reduction goal for each of the two watersheds. Nadia wants to exactly meet these reduction goals. She understands that exceeding the target will always increase the cost. Since the objective of this problem is to minimize the cost, there is no reason to have any more phosphorus than the specified amounts.

Table 4.2.2 shows the cost by source of reducing phosphorus in the water. It is important to notice those quantities are fixed and, thus, not variable. The solution to the water pollution problem requires reducing various sources of pollution in each of the watersheds to meet the reduction goals. Nadia and her team of planners are deciding how to reach those goals while keeping costs to a minimum.

4.2.1 Linear Programming Formulation

Nadia develops the complete linear programming formulation. First, she needs to define the decision variables. One way would be to let g_1, g_2, g_3, and g_4 represent the amount of phosphorus reduction in the Green Bay watershed from each of the four sources. Similarly, w_1, w_2, w_3, and w_4 could represent the amount of phosphorus reduction in Lake Winnebago from each source. Nadia notices she needs a different letter for the variables relating to each watershed.

Nadia and her team decide they need a system that uses just one letter for both watersheds. Therefore, she employs **double-subscripted variables**. As the name suggests, these variables have two subscripts. In this case, the first subscript refers to the watershed (Lake Winnebago or Green Bay). The second subscript refers to the source (construction runoff, urban storm runoff, agricultural sources, or industrial sources).

To help visualize this, Nadia creates a **matrix** (Table 4.2.3). In the matrix, the first subscript refers to the row in which the variable is written. The second subscript refers to its column. In general, $x_{i,j}$ represents the element in row i and column j of the matrix.

In the water pollution example, one decision variable represents the amount of reduction in phosphorus going into the Lake Winnebago watershed from agricultural sources. A single double-subscripted variable ($x_{1,3}$) may be used for this decision variable. The first subscript indicates the first watershed, Lake Winnebago, and the second subscript indicates the third pollution source, agriculture. Table 4.2.3 contains eight decision variables arranged in rows by the two watersheds and in columns by the four sources of pollution.

(kg of phosphorus)		Source (*j*)			
		Construction runoff	Urban storm runoff	Agricultural sources	Industrial sources
Watershed (*i*)	Lake Winnebago	$x_{1,1}$	$x_{1,2}$	$x_{1,3}$	$x_{1,4}$
	Green Bay	$x_{2,1}$	$x_{2,2}$	$x_{2,3}$	$x_{2,4}$

Table 4.2.3: Definition of decision variables

Now that Nadia has defined the decision variables, she writes the objective function. Since the goal is to minimize cost, she uses the values in Table 4.2.2 to develop the following objective function.

Minimize:
$$z = \$770(x_{1,1} + x_{2,1}) + \$2025(x_{1,2} + x_{2,2}) + \$26(x_{1,3} + x_{2,3}) + \$75(x_{1,4} + x_{2,4})$$
$$= \$770x_{1,1} + \$770x_{2,1} + \$2025x_{1,2} + \$2025x_{2,2} + \$26x_{1,3} + \$26x_{2,3} + \$75x_{1,4} + \$75x_{2,4}$$

Notice that each cost appears twice in the objective function. That makes sense, because we are assuming that it costs the same amount per kilogram to remove the pollution from either of the two watersheds. There is, however, a difference in cost to reduce each type of pollution. These cost differences can be an order of magnitude. For example, the cost to remove one kilogram of construction runoff is more than ten times as expensive as the cost to reduce a kilogram of industrial pollution. The reason for this is that the construction runoff arrives from widely dispersed areas while industrial pollution is more concentrated.

1. What is the magnitude of difference in cost for urban runoff and agricultural sources? Why do you think it might be less costly to control agricultural pollution as compared to urban pollution?

There are also some constraints Nadia needs to consider. First, the amount of phosphorus must be reduced by the values in Table 4.2.1. Thus, Nadia develops the following constraints:

Lake Winnebago target reduction: $x_{1,1} + x_{1,2} + x_{1,3} + x_{1,4} = 40,000$
Green Bay target reduction: $x_{2,1} + x_{2,2} + x_{2,3} + x_{2,4} = 85,000$

In addition to meeting the target reductions, Nadia and her team want to ensure that the pollution reductions over all the sources are evenly distributed. Nadia notices that pollution from agricultural sources is the least expensive to reduce. Without setting stipulations, agricultural polluters would be overburdened with phosphorus reductions. On the other hand, urban storm runoff would likely not be reduced at all, because it is by far the most costly. The team of planners agrees to requirements for reductions per source as shown in Tables 4.2.4 and 4.2.5.

Source	Minimum percent reduction of these sources into each watershed
Construction runoff	30%
Urban storm runoff	30%

Table 4.2.4: Minimum proportions of reductions per source

Source	Maximum percent reduction of these sources into each watershed
Agricultural sources	15%
Industrial sources	15%

Table 4.2.5: Maximum proportions of reductions per source

Based on this information, Nadia and her team develop the following constraints.

Construction runoff for Lake Winnebago:	$x_{1,1} \geq 12{,}000 = (0.3)(40{,}000)$ kg
Construction runoff for Green Bay:	$x_{2,1} \geq 25{,}500 = (0.3)(85{,}000)$ kg
Urban storm runoff for Lake Winnebago:	$x_{1,2} \geq 12{,}000 = (0.3)(40{,}000)$ kg
Urban storm runoff for Green Bay:	$x_{2,2} \geq 25{,}500 = (0.3)(85{,}000)$ kg
Agricultural sources for Lake Winnebago:	$x_{1,3} \leq 6{,}000 = (0.15)(40{,}000)$ kg
Agricultural sources for Green Bay:	$x_{2,3} \leq 12{,}750 = (0.15)(85{,}000)$ kg
Industrial sources for Lake Winnebago:	$x_{1,4} \leq 6{,}000 = (0.15)(40{,}000)$ kg
Industrial sources for Green Bay:	$x_{2,4} \leq 12{,}750 = (0.15)(85{,}000)$ kg

2. Why do the first four constraints use "$\geq$"?

3. Why do the last four constraints use "$\leq$"?

4. Why is the right-hand side of the first constraint 12,000 kg?

5. Why is the right-hand side of the last constraint 12,750 kg?

6. Without looking at any table or chart explain how you could tell that the constraint containing the decision variable $x_{2,3}$ refers to reducing pollution from agricultural sources in Green Bay?

7. Which constraint sets a limit on pollution reduction from urban storm runoff in Lake Winnebago?

Now, Nadia and her team have the complete linear programming formulation:

Decision Variables

Let: $x_{1,1}$ = phosphorus reduction from construction runoff in Lake Winnebago (in kg)
$x_{1,2}$ = phosphorus reduction from urban storm runoff in Lake Winnebago (in kg)
$x_{1,3}$ = phosphorus reduction from agricultural sources in Lake Winnebago (in kg)
$x_{1,4}$ = phosphorus reduction from industrial sources in Lake Winnebago (in kg)
$x_{2,1}$ = phosphorus reduction from construction runoff in Green Bay (in kg)
$x_{2,2}$ = phosphorus reduction from urban storm runoff in Green Bay (in kg)
$x_{2,3}$ = phosphorus reduction from agricultural sources in Green Bay (in kg)
$x_{2,4}$ = phosphorus reduction from industrial sources in Green Bay (in kg)

Objective Function
Minimize: $z = \$770x_{1,1} + \$770x_{2,1} + \$2025x_{1,2} + \$2025x_{2,2} + \$26x_{1,3} + \$26x_{2,3} + \$75x_{1,4} + \$75x_{2,4}$
where z = the total cost of reducing the amount of phosphorus in the watersheds.

Constraints
Subject to:

Lake Winnebago target reduction:	$x_{1,1} + x_{1,2} + x_{1,3} + x_{1,4} = 40{,}000$
Construction runoff for Lake Winnebago:	$x_{1,1} \geq 12{,}000$
Urban storm runoff for Lake Winnebago:	$x_{1,2} \geq 12{,}000$
Agricultural sources for Lake Winnebago:	$x_{1,3} \leq 6{,}000$
Industrial sources for Lake Winnebago:	$x_{1,4} \leq 6{,}000$
Green Bay target reduction:	$x_{2,1} + x_{2,2} + x_{2,3} + x_{2,4} = 85{,}000$
Construction runoff for Green Bay:	$x_{2,1} \geq 25{,}500$
Urban storm runoff for Green Bay:	$x_{2,2} \geq 25{,}500$
Agricultural sources for Green Bay:	$x_{2,3} \leq 12{,}750$
Industrial sources for Green Bay:	$x_{2,4} \leq 12{,}750$
Non-Negativity:	$x_{1,1} \geq 0, x_{1,2} \geq 0, x_{1,3} \geq 0, x_{1,4} \geq 0,$
	$x_{2,1} \geq 0, x_{2,2} \geq 0, , x_{2,3} \geq 0,$ and $x_{2,4} \geq 0$

4.2.2 Using the Excel Solver

To solve this problem, Nadia and her team rely on Excel Solver. The spreadsheet is shown in Figure 4.2.2.

	A	B	C	D	E	F	G	H	I	J	K	L	M
1	LP Minimization		Lake Winnebago (LW)				Green Bay (GB)						
2	Wisconsin Watershed Problem		x1,1 Contruction	x1,2 Urban Storm	x1,3 Agricultural	x1,4 Industrial	x2,1 Contruction	x2,2 Urban Storm	x2,3 Agricultural	x2,4 Industrial			
3		Decision Variable Values [kg of phosphorous]									Total Cost		
4		Objective Function [Cost ($/kg)]	$770	$2,025	$26	$75	$770	$2,025	$26	$75	$0		
5													
6		Contraints									Reduction		
7	Lake Winnebago	Target Reduction for LW (kg)	1	1	1	1					0	=	40,000
8		Total Reduction for Construction Runoff_LW	1								0	≥	12,000
9		Total Reduction for Urban Storm Runoff_LW		1							0	≥	12,000
10		Total Reduction for Agricultural Sources_LW			1						0	≤	6,000
11		Total Reduction for Industrial Sources_LW				1					0	≤	6,000
12	Green Bay	Target Reduction for GB (kg)					1	1	1	1	0	=	85,000
13		Total Reduction for Construction Runoff_GB					1				0	≥	25,500
14		Total Reduction for Urban Storm Runoff_GB						1			0	≥	25,500
15		Total Reduction for Agricultural Sources_GB							1		0	≤	12,750
16		Total Reduction for Industrial Sources_GB								1	0	≤	12,750

Figure 4.2.2: Complete linear programming formulation in Excel

Nadia sets up this linear programming formulation in Excel with the eight in row 3. She added labels in row 1 that grouped the decision variables into two sets of four decision variables as in Figure 4.2.2. She also added labels in column A to group the constraints.

Solving a minimization problem in Excel is very similar to solving a maximization problem. Nadia simply needs to tell Solver to minimize the objective function. She follows the same procedures developed in earlier problems to set up the parameters of the model. The critical difference for this problem is that it is a minimization problem. Nadia uses Excel Solver (See Figure 4.2.3) to come up with the optimal solution in Figure 4.2.4.

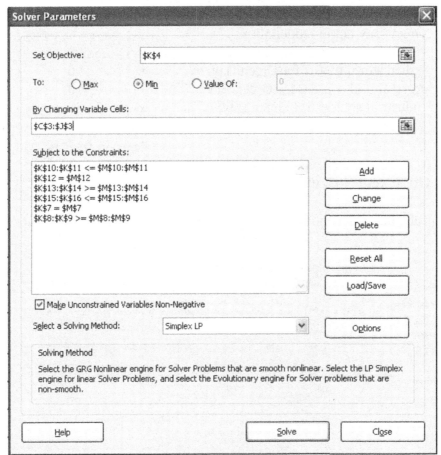

Figure 4.2.3: Solver parameters for Watershed Pollution Reduction - Cost Minimization

	A	B	C	D	E	F	G	H	I	J	K	L	M
1		LP Minimization	\multicolumn{4}{c}{Lake Winnebago (LW)}								
2		Wisconsin Watershed Problem	x1,1 Contruction	x1,2 Urban Storm	x1,3 Agricultural	x1,4 Industrial	x2,1 Contruction	x2,2 Urban Storm	x2,3 Agricultural	x2,4 Industrial			
3		Decision Variable Values [kg of phosphorous]	16,000	12,000	6,000	6,000	34,000	25,500	12,750	12,750	Total Cost		
4		Objective Function [Cost ($/kg)]	$770	$2,025	$26	$75	$770	$2,025	$26	$75	$116,331,250		
5													
6		Contraints									Reduction		
7	Lake Winnebago	Target Reduction for LW (kg)	1	1	1	1					40,000	=	40,000
8		Total Reduction for Construction Runoff_LW	1								16,000	≥	12,000
9		Total Reduction for Urban Storm Runoff_LW		1							12,000	≥	12,000
10		Total Reduction for Agricultural Sources_LW			1						6,000	≤	6,000
11		Total Reduction for Industrial Sources_LW				1					6,000	≤	6,000
12	Green Bay	Target Reduction for GB (kg)					1	1	1	1	85,000	=	85,000
13		Total Reduction for Construction Runoff_GB					1				34,000	≥	25,500
14		Total Reduction for Urban Storm Runoff_GB						1			25,500	≥	25,500
15		Total Reduction for Agricultural Sources_GB							1		12,750	≤	12,750
16		Total Reduction for Industrial Sources_GB								1	12,750	≤	12,750

Figure 4.2.4: Spreadsheet with optimal solution

The optimal solution shown in Figure 4.2.4 indicates the kilograms of phosphorus reduction from each source in each watershed. These values meet all of the constraints for the reduction of phosphorus. The solution achieves the goal of meeting these constraints for reduction while keeping cost to a minimum.

8. What is the optimal solution?

9. What is the value of the objective function for that optimal solution?

10. Which of the constraints are binding?

4.2.3 Interpreting Results

Nadia and her team examine the Answer Report shown in Figure 4.2.5 and the Sensitivity Report shown in Figure 4.2.7.

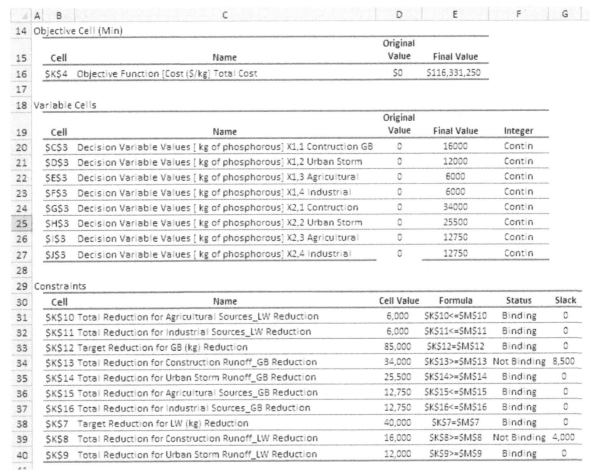

Figure 4.2.5: Answer Report for optimal solution

	A	B	C	D	E	F	G	H
6		Variable Cells						
7				Final	Reduced	Objective	Allowable	Allowable
8		Cell	Name	Value	Cost	Coefficient	Increase	Decrease
9		$C$3	Decision Variable Values [kg of phosphorous] X1,1 Contruction GB	16000	0	770	1255	695
10		$D$3	Decision Variable Values [kg of phosphorous] X1,2 Urban Storm	12000	0	2025	1E+30	1255
11		$E$3	Decision Variable Values [kg of phosphorous] X1,3 Agricultural	6000	0	26	744	1E+30
12		$F$3	Decision Variable Values [kg of phosphorous] X1,4 Industrial	6000	0	75	695	1E+30
13		$G$3	Decision Variable Values [kg of phosphorous] X2,1 Contruction	34000	0	770	1255	695
14		$H$3	Decision Variable Values [kg of phosphorous] X2,2 Urban Storm	25500	0	2025	1E+30	1255
15		$I$3	Decision Variable Values [kg of phosphorous] X2,3 Agricultural	12750	0	26	744	1E+30
16		$J$3	Decision Variable Values [kg of phosphorous] X2,4 Industrial	12750	0	75	695	1E+30
17								
18		Constraints						
19				Final	Shadow	Constraint	Allowable	Allowable
20		Cell	Name	Value	Price	R.H. Side	Increase	Decrease
21		$K$10	Total Reduction for Agricultural Sources_LW Reduction	6000	-744	6000	4000	6000
22		$K$11	Total Reduction for Industrial Sources_LW Reduction	6000	-695	6000	4000	6000
23		$K$12	Target Reduction for GB (kg) Reduction	85000	770	85000	1E+30	8500
24		$K$13	Total Reduction for Construction Runoff_GB Reduction	34000	0	25500	8500	1E+30
25		$K$14	Total Reduction for Urban Storm Runoff_GB Reduction	25500	1255	25500	8500	25500
26		$K$15	Total Reduction for Agricultural Sources_GB Reduction	12750	-744	12750	8500	12750
27		$K$16	Total Reduction for Industrial Sources_GB Reduction	12750	-695	12750	8500	12750
28		$K$7	Target Reduction for LW (kg) Reduction	40000	770	40000	1E+30	4000
29		$K$8	Total Reduction for Construction Runoff_LW Reduction	16000	0	12000	4000	1E+30
30		$K$9	Total Reduction for Urban Storm Runoff_LW Reduction	12000	1255	12000	4000	12000

Figure 4.2.6: Sensitivity Report for optimal solution

11. From the Sensitivity Report, which of the constraints have a shadow price?

12. Explain how the two previous answers are related.

Recall that *shadow price* refers to the amount by which the objective function value changes given a unit increase or decrease in one right-hand side (RHS) value of a constraint.

13. Interpret the meaning of the shadow price for the "Target Reduction for GB" constraint.

14. Interpret the meaning of the shadow price for the "Total Reduction for Agricultural Souces_GB" constraint. Why is it negative?

15. How does the interpretation of the shadow price in a minimization problem differ from the the interpretation of the shadow price in a maximization problem?

While the answer and sensitivity reports give Nadia and her team a lot of important information, they do not give them the entire picture. They must use their knowledge of the problem. It is important to remember the units for the various constraints and decision variables. For instance, the decision variables are in kilograms of phosphorus.

16. Which of the sources provides the highest amount of phosphorus reduction? Why does this make sense?

17. Which of the sources provides the least amount of phosphorus reduction? Why does this make sense?

Nadia's team notices that the Sensitivity Report for the Watersheds Problem lists an Allowable Decrease in the objective coefficients of four of the decision variables as 1E+30. In other words, these coefficients may be decreased as much as they want without affecting the optimal solution.

They also notice that these four decision variables (that have an Allowable Decrease of 1E+30) are the decision variables for agricultural and industrial sources of pollution. These two sources are the least costly of the four sources from which to reduce pollution. As a result, Nadia's team sees why the four constraints related to agricultural and industrial sources are binding. They want to reduce as much pollution as possible from the least expensive sources in order to minimize the total cost.

Then they notice that the two constraints for urban storm runoff are also binding, but reducing urban storm runoff is the *most* expensive source to reduce. Nadia calls the team's attention to the direction of the constraint inequalities. For urban storm runoff, both constraint inequalities are "greater than or equal to." One way to think of "greater than or equal to" is "at least." Since the goal is to keep costs at a minimum, the most expensive source of pollution should be reduced as little as required. For example, reducing pollution from urban storm runoff by "at least 25,500 kg" means reduce it by exactly 25,500 kg. To reduce it more while staying within the total target reduction would be more costly. Now they understand why those two constraints are also binding.

Returning to the Sensitivity Report, the team notices that Solver reports a Shadow Price of 1255 for both of the urban storm runoff constraints. That means that if the right-hand side of either of those constraints increases by 1 kg, the total cost of the project would increase by $1,255. They understand why the cost would increase, but they do not see where the 1255 came from. To see why that makes sense, Nadia changes the spreadsheet formulation, as shown in Figure 4.2.8. The spreadsheet shows an increase to 12,001 kg in the right-hand side of the Lake Winnebago urban storm runoff constraint, as well as the decision variable values for the resulting optimal solution.

	A	B	C	D	E	F	G	H
1	Chapter 4: LP Minimization							
2	4.2 WI Watershed							
3	Reduction Cost Minimization							
4								
5	Decision Variable Values [= of kg of Phosphorus]	Construction Runoff	Urban Storm Runoff	Agricultural Sources	Industrial Sources			
6								
7	Lake Winnebago (LW)	15999	12001	6000	6000			
8	Green Bay (GB)	34000	25500	12750	12750			
9								Total Cost
10	Objective Function [Cost ($/kg)]	770	2025	26	75			$116,332,505
11								
12	Constraints							
13	Target Reduction for LW (kg)	1	1	1	1	40000	=	40,000
14	Total Reduction for Construction Runoff_LW	1				15999	≥	12,000
15	Total Reduction for Urban Storm Runoff_LW		1			12001	≥	12,001
16	Total Reduction for Agricultural Sources_LW			1		6000	≤	6,000
17	Total Reduction for Industrial Sources_LW				1	6000	≤	6,000
18	Target Reduction for GB (kg)	1	1	1	1	85000	=	85,000
19	Total Reduction for Construction Runoff_GB	1				34000	≥	25,500
20	Total Reduction for Urban Storm Runoff_GB		1			25500	≥	25,500
21	Total Reduction for Agricultural Sources_GB			1		12750	≤	12,750
22	Total Reduction for Industrial Sources_GB				1	12750	≤	12,750

Figure 4.2.8: Increasing the amount of reduction in Lake Winnebago urban storm runoff by 1 kg

By increasing the amount of reduction in urban storm runoff in the Lake Winnebago watershed by 1 kg, the final value of only two of the decision variables changes. Urban storm runoff in Lake Winnebago increases from 12,000 kg to 12,001 kg, and construction runoff in Lake Winnebago decreases from 16,000 kg to 15,999 kg.

The increase in urban storm runoff in Lake Winnebago by 1 kg causes the total cost of the project to increase by $2,025 (see Table 4.2.2). In addition, reducing the construction runoff in Lake Winnebago by 1 kg causes the total cost of the project to decrease by $770. Thus, Nadia saw that the net effect of the change she made is $2,025 – $770 = $1,255, which is exactly the amount of increase in the total cost of the project.

Next, Nadia and her team notice that four of the Shadow Prices in the Sensitivity Report are negative numbers (e.g., the Shadow price for the constraint for agricultural sources in Lake Winnebago is -$744). This means that increasing the right-hand side of the constraint will decrease the total cost of the project. The negative shadow prices are all linked to less than or equal to constraints. These constraints limit the use of lower cost reduction strategies. Increasing the right hand side of any of these constraints expands the size of the feasible region. As a result, it is possible to improve the optimal solution. In a cost minimization problem improvements result in a reduction in total cost. This reduction appears as a negative shadow price. For example, Figure 4.2.9 shows the spreadsheet Nadia has after increasing the right-hand side of the Lake Winnebago agricultural sources constraint by 1 kg to 6,001. (Note: the right-hand side of the constraint for urban storm runoff for Lake Winnebago has been changed back to 12,000 kg.)

	A	B	C	D	E	F	G	H
1	Chapter 4: LP Minimization							
2	4.2 WI Watershed							
3	Reduction Cost Minimization							
4								
5	Decision Variable Values [# of kg of Phosphorus]	Construction Runoff	Urban Storm Runoff	Agricultural Sources	Industrial Sources			
6								
7	Lake Winnebago (LW)	15999	12000	6001	6000			
8	Green Bay (GB)	34000	25500	12750	12750			
9								Total Cost
10	Objective Function [Cost ($/kg)]	770	2025	26	75			$116,330,506
11								
12	Constraints							
13	Target Reduction for LW (kg)	1	1	1	1	40000	=	40,000
14	Total Reduction for Construction Runoff_LW	1				15999	≥	12,000
15	Total Reduction for Urban Storm Runoff_LW		1			12000	≥	12,000
16	Total Reduction for Agricultural Sources_LW			1		6001	≤	6,001
17	Total Reduction for Industrial Sources_LW				1	6000	≤	6,000
18	Target Reduction for GB (kg)	1	1	1	1	85000	=	85,000
19	Total Reduction for Construction Runoff_GB	1				34000	≥	25,500
20	Total Reduction for Urban Storm Runoff_GB		1			25500	≥	25,500
21	Total Reduction for Agricultural Sources_GB			1		12750	≤	12,750
22	Total Reduction for Industrial Sources_GB				1	12750	≤	12,750

Figure 4.2.9: Increasing the amount of reduction from Lake Winnebago agricultural sources by 1 kg

The team notices that this change in the constraint causes two changes in the final values of the decision variables. The amount of pollution reduction from agricultural sources in the Lake Winnebago watershed increases by 1 kg, from 6,000 to 6,001. At the same time, the amount of pollution reduction from construction runoff in the Lake Winnebago watershed decreases by 1 kg, from 16,000 to 15,999. The net effect of these two changes is $26 – $770 = -$744, which matches the reported Shadow Price.

Finally, after a lot of hard work, Nadia and her team of planners know how much reduction in phosphorus should be coming from each source and how much it will cost to do the entire reduction process. Based on this information, they are now ready to move forward on this project.

Section 4.3: Disk Gasoline Distributors, Inc.

Disk Gasoline Distributors, Inc. obtains gasoline wholesale from three refineries. It then blends this gasoline and introduces additives. These additives are designed to improve vehicle performance. Disk delivers the finished product to various gasoline retailers.

During the first quarter of the year, the management at Disk wants to produce a blend of gasoline meeting a particular set of specifications. The product will be delivered to retailers in the Southeast. Table 4.3.1 contains those specifications. To meet the company's goals for profitability, it must produce 500,000 gallons per week of the blend.

Octane rating:	Greater than or equal to 87
Vapor pressure:	Less than 7.2 pounds per square inch (psi)
Sulfur content:	Less than 75 parts per million (ppm)
Olefins (a family of toxic pollutants):	Less than 10% by volume (%v)

Table 4.3.1: Gasoline blend specifications at Disk Gasoline Distributors

The octane rating is a performance measure of a gasoline. The higher the octane rating, the better the gasoline performs. Some vehicles require gasoline with an octane rating higher than 89. Vapor pressure is a measure of the extent to which a gasoline is subject to evaporation. Sulfur content and olefin content determine how cleanly a gasoline blend burns in a vehicle. A lower content of either produces cleaner gasoline.

Disk buys gasoline in 100-gallon units called hectogallons (hgal) directly from three refineries in the southeast:
- Vicksburg, MS,
- Norco, LA, and
- Mobile, AL.

The cost per hgal from the refineries is $274.90 from Vicksburg, $265.90 from Norco, and $249.90 from Mobile. These costs include delivery. The characteristics of the gasoline that Disk can obtain from these three refineries are contained in Table 4.3.2.

	Refinery		
Characteristic	**Vicksburg**	**Norco**	**Mobile**
Octane rating	89	88	85
Vapor pressure	7.23	7.09	7.32
Sulfur content	72	86	58
Olefins	7.52	8.97	13.38

Table 4.3.2: Characteristics of gasoline from three refineries

The managers at Disk would like to minimize the cost of the gasoline used in this blend. However, they are further constrained by the capacities of the three refineries. They can obtain no more than 210,000 gallons per week from Vicksburg, no more than 190,000 gallons per week from Norco, and no more than 200,000 gallons per week from Mobile. Edward Thompson, the production manager at Disk, wants to know how much gasoline to purchase from each refinery in order to keep the cost of the blend at a minimum.

4.3.1 Linear Programming Formulation

Edward Thompson begins by defining the decision variables. He must decide how much gasoline to obtain from each of the three refineries per week. So he lets:
x_1 = gasoline purchased from Vicksburg (hgal per week),
x_2 = gasoline purchased from Norco (hgal per week), and
x_3 = gasoline purchased from Mobile (hgal per week).

Now he must define a function that represents the objective of the problem in terms of the decision variables. Because the goal is to minimize the cost of the gasoline in the blend, Edward Thompson lets:
$z = 274.9x_1 + 265.9x_2 + 249.9x_3$.

Edward Thompson wants to minimize z, subject to all of the constraints. First, there is a production constraint. The company wants to produce 500,000 gallons of the blend per week. However, Edward Thompson notices that the units of purchase are hundred-gallons. There are 5,000 hgal in 500,000 gallons. Therefore, he writes the following constraint:
$x_1 + x_2 + x_3 = 5{,}000$.

Next, he must account for the capacities at each of the refineries:
$x_1 \leq 2{,}100$,
$x_2 \leq 1{,}900$, and
$x_3 \leq 2{,}000$.

Notice that these constraints have also been expressed in hgal.

Next, Edward Thompson must include all of the constraints that derive from the specifications of the blend.

He knows the octane rating of the blend must be greater than or equal to 87. He also knows the octane ratings of the gasoline coming from all three refineries. However, he cannot just average those numbers. The amount of gasoline from each of the refineries might not be the same. To account for this possibility, he must use a **weighted average** to calculate the octane rating of the blend.

To compute this weighted average, Edward Thompson first multiplies the octane rating of the gasoline from each of the refineries by the amount purchased from that refinery. Then, he adds the three products and divides the sum by the total amount of the blend:

$$\frac{89x_1 + 88x_2 + 85x_3}{x_1 + x_2 + x_3} = \text{the octane rating of the blend.}$$

Now the octane rating of the blend must be greater than or equal to 87, so he writes:
$$\frac{89x_1 + 88x_2 + 85x_3}{x_1 + x_2 + x_3} \geq 87$$

Then, Edward Thompson manipulates the original inequality to remove the rational expression on the left-hand side. Doing so replaces the fraction with a linear expression:

$89x_1 + 88x_2 + 85x_3 \geq 87(x_1 + x_2 + x_3)$
$89x_1 + 88x_2 + 85x_3 \geq 87x_1 + 87x_2 + 87x_3$
$2x_1 + x_2 - 2x_3 \geq 0$

Similarly, the constraints on vapor pressure, sulfur content, and olefin content can be found using this weighted average approach.

Vapor pressure: $\dfrac{7.23x_1 + 7.09x_2 + 7.32x_3}{x_1 + x_2 + x_3} \leq 7.2$

$7.23x_1 + 7.09x_2 + 7.32x_3 \leq 7.2(x_1 + x_2 + x_3)$
$7.23x_1 + 7.09x_2 + 7.32x_3 \leq 7.2x_1 + 7.2x_2 + 7.2x_3$
$0.03x_1 - 0.11x_2 + 0.12x_3 \leq 0$

Sulfur content: $\dfrac{72x_1 + 86x_2 + 58x_3}{x_1 + x_2 + x_3} \leq 75$

$72x_1 + 86x_2 + 58x_3 \leq 75(x_1 + x_2 + x_3)$
$72x_1 + 86x_2 + 58x_3 \leq 75x_1 + 75x_2 + 75x_3$
$-3x_1 + 11x_2 - 17x_3 \leq 0$

Olefin content: $\dfrac{7.52x_1 + 8.97x_2 + 13.38x_3}{x_1 + x_2 + x_3} \leq 10$

$7.52x_1 + 8.97x_2 + 13.38x_3 \leq 10(x_1 + x_2 + x_3)$
$7.52x_1 + 8.97x_2 + 13.38x_3 \leq 10x_1 + 10x_2 + 10x_3$
$-2.48x_1 - 1.03x_2 + 3.38x_3 \leq 0$

Finally, Edward Thompson includes the non-negativity constraints: $x_1 \geq 0$, $x_2 \geq 0$, and $x_3 \geq 0$.

Complete Linear Programming Formulation
Now, Edward Thompson writes the complete formulation.

<u>Decision Variables</u>
Let: x_1 = hgal of gasoline purchased from Vicksburg per week,
x_2 = hgal of gasoline purchased from Norco per week, and
x_3 = hgal of gasoline purchased from Mobile per week.

Objective Function
Minimize: $z = 274.9x_1 + 265.9x_2 + 249.9x_3$,
where z = the cost of the gasoline blend

Constraints
Subject to:
Production (hgal): $x_1 + x_2 + x_3 = 5,000$
Vicksburg Refinery Capacity (hgal): $x_1 \leq 2,100$
Norco Refinery Capacity (hgal): $x_2 \leq 1,900$
Mobile Refinery Capacity (hgal): $x_3 \leq 2,000$.
Octane Rating: $2x_1 + x_2 - 2x_3 \geq 0$
Vapor Pressure: $0.03x_1 - 0.11x_2 + 0.12x_3 \leq 0$
Sulfur Content: $-3x_1 + 11x_2 - 17x_3 \leq 0$
Olefin Content: $-2.48x_1 - 1.03x_2 + 3.38x_3 \leq 0$
Non-Negativity: $x_1 \geq 0, x_2 \geq 0$, and $x_3 \geq 0$

4.3.2 Excel Solver

Figure 4.3.1 contains a spreadsheet formulation for the Disk Gasoline Distributors problem. Figures 4.3.2 and 4.3.3 contain the Answer Report and the Sensitivity Report, respectively, which were generated by Solver.

	Section 4.3	Disk Gasoline Blending				
	Decision Variables	Hundred Gal. Vicksburg	Hundred Gal. Norco	Hundred Gal. Mobile		
		X1	X2	X3		
	Decision Values	1811.11	1900.00	1288.89		
					Total Cost	
	Costs/100 gal.	274.9	265.9	249.9	$1,325,178	
	Constraints				LHS	RHS
	Production	1	1	1	5000	=	5000
Capacity Constraints	Vicksburg Capacity	1			1811.11	<=	2100
	Norco Capacity		1		1900	<=	1900
	Mobile Capacity			1	1288.89	<=	2000
Content and Performance Constraints	Octane rating (≥87)	2	1	-2	2944.44	>=	0
	Vapor Pressure (≤7.2)	0.03	-0.11	0.12	0	<=	0
	Sulfur Content (≤75)	-3	11	-17	-6444.44	<=	0
	Olefin Content (≤10)	-2.48	-1.03	3.38	-2092.11	<=	0

Figure 4.3.1: Spreadsheet formulation of the Disk Gasoline Distributors problem

Objective Cell (Min)

Cell	Name	Original Value	Final Value
$G$8	Objective Function [Cost (100-gal)] Total Cost	$0.00	$1,325,177.78

Variable Cells

Cell	Name	Original Value	Final Value	Integer
$B$6	Decision Value [# of 100-gal per week] Vicksburg (x1)	0	1811.111111	Contin
$C$6	Decision Value [# of 100-gal per week] Norco (x2)	0	1900	Contin
$D$6	Decision Value [# of 100-gal per week] Mobile (x3)	0	1288.888889	Contin

Constraints

Cell	Name	Cell Value	Formula	Status	Slack
$E$11	Production (100-gal)	5000	$E$11=$G$11	Binding	0
$E$12	Vicksburg Capacity (100-gal)	1811.111111	$E$12<=$G$12	Not Binding	288.8888889
$E$13	Norco Capacity (100-gal)	1900	$E$13<=$G$13	Binding	0
$E$14	Mobile Capacity (100-gal)	1288.888889	$E$14<=$G$14	Not Binding	711.1111111
$E$15	Octane Rating	2944.444444	$E$15>=$G$15	Not Binding	2944.444444
$E$16	Vapor Pressure (psi)	0	$E$16<=$G$16	Binding	0
$E$17	Sulfur Content (ppm)	-6444.444444	$E$17<=$G$17	Not Binding	6444.444444
$E$18	Olefin (%Vol)	-2092.111111	$E$18<=$G$18	Not Binding	2092.111111

Figure 4.3.2: Solver Answer Report for the Disk Gasoline Distributors problem

Variable Cells

Cell	Name	Final Value	Reduced Cost	Objective Coefficient	Allowable Increase	Allowable Decrease
$B$6	Decision Value [# of 100-gal per week] Vicksburg (x1)	1811.111111	0	274.9	1E+30	18.73913043
$C$6	Decision Value [# of 100-gal per week] Norco (x2)	1900	0	265.9	47.88888889	1E+30
$D$6	Decision Value [# of 100-gal per week] Mobile (x3)	1288.888889	0	249.9	25	1E+30

Constraints

Cell	Name	Final Value	Shadow Price	Constraint R.H. Side	Allowable Increase	Allowable Decrease
$E$11	Production (100-gal)	5000	283.2333333	5000	216.6666667	471.9047619
$E$12	Vicksburg Capacity (100-gal)	1811.111111	0	2100	1E+30	288.8888889
$E$13	Norco Capacity (100-gal)	1900	-47.88888889	1900	198.0124093	113.0434783
$E$14	Mobile Capacity (100-gal)	1288.888889	0	2000	1E+30	711.1111111
$E$15	Octane Rating	2944.444444	0	0	2944.444444	1E+30
$E$16	Vapor Pressure (psi)	0	-277.7777778	0	32.13139932	26
$E$17	Sulfur Content (ppm)	-6444.444444	0	0	1E+30	6444.444444
$E$18	Olefin (%Vol)	-2092.111111	0	0	1E+30	2092.111111

Figure 4.3.3: Solver Sensitivity Report for the Disk Gasoline Distributors problem

4.3.3 Interpreting Results

Edward Thompson now knows the optimal gasoline blend and how much it will cost. However, he still needs to spend some time interpreting these results.

Carefully examine the Answer and Sensitivity Reports in Figures 4.3.2 and 4.3.3. Then answer each of the following questions about the solution to this problem.
1. What is the optimal solution, and what is the cost of producing that blend?

2. Calculate the cost per hundred gallons of the optimal blend. How does the cost per hundred gallons of the blend compare to the cost per hundred gallons of each of the components?

3. Why does the Sensitivity Report list the Reduced Cost for each of the decision variables as 0?

Notice that the Allowable Increase in the objective coefficient (the cost per hundred gallons) for x_3 (the number of hgal of gasoline purchased from the Mobile refinery per week) is 25. Suppose that this objective coefficient increases by 24.9 to 274.8.

4. What do you think would happen to the optimal solution?

5. What do you think would happen to the optimal solution if the price per hundred-gallon at the Mobile refinery increased to $275.90?

6. Similarly, assume the objective coefficient of x_2 decreased by 18 to 247.9. What do you think would happen to the optimal solution?

The capacity of each of the three refineries forms a constraint in the problem formulation. According to the Answer Report, the constraint is binding for the Norco refinery, but not for the Vicksburg and Mobile refineries.

7. In the context of the problem, what does it mean that the Mobile refinery constraint is nonbinding?

8. The slack value for the Mobile refinery is given as 711.1111.
   a. What does that slack value tell you about the optimal solution?

   b. Based on other information in the problem, why do you think this has happened?

9. Sometimes Solver reports an allowable increase or decrease of 1E+30. How should that number be interpreted?

10. Why does an allowable increase in the right-hand side of the Mobile capacity constraint of 1E+30 make sense?

Next, the shadow price for the production constraint is given as 283.2333333. To see the effect of this shadow price, Edward Thompson increases the production constraint by 100 gallons. Figure 4.3.4 shows the new optimal solution when the production is 5,001 hectogallons (i.e., 500,100 gallons) instead of 5,000 hectogallons.

Section 4.3	Disk Gasoline Blending			
Decision Variables	Hundred Gal. Vicksburg	Hundred Gal. Norco	Hundred Gal. Mobile	
	X1	X2	X3	
Decision Values	1812.44	1900.00	1288.56	
				Total Cost
Costs/100 gal.	274.9	265.9	249.9	$1,325,461

	Constraints				LHS		RHS
	Production	1	1	1	5001	=	5001
Capacity Constraints	Vicksburg Capacity	1			1812.44	<=	2100
	Norco Capacity		1		1900	<=	1900
	Mobile Capacity			1	1288.56	<=	2000
Content and Performance Constraints	Octane rating (>87)	2	1	-2	2947.78	>=	0
	Vapor Pressure (<7.2)	0.03	-0.11	0.12	0	<=	0
	Sulfur Content (<75)	-3	11	-17	-6442.78	<=	0
	Olefin Content (<10)	-2.48	-1.03	3.38	-2096.54	<=	0

Figure 4.3.4: Spreadsheet formulation with one more unit of production

11. What do you observe about the total cost now?

12. What other changes do you observe?

Finally, the Shadow Prices for the capacity of each refinery are reported as zero for Vicksburg, -47.89 for Norco, and zero for Mobile.

13. Why are the Vicksburg and Mobile Shadow Prices zero?

14. How do you interpret the negative Shadow Price for Norco?

15. Why should Disk try to obtain more gasoline per week from Norco to use in the blend?

Based on these results and his interpretation of the results, Edward Thompson feels confident in moving forward with this project. He knows the optimal amounts of gasoline to obtain from each refinery, but he is also aware of the changes that could occur in the constraints and the objective function that could result in different amounts of gasoline and/or a different total cost.

Chapter 4 (LP Minimization) Homework Questions

1. Use Solver to optimize the diet problem in Malawi that you formulated in section 4.1.

    a. What is the optimal solution? Which foods are used and how much?

    b. How many calories per day are there in the optimal diet?

    c. Which nutrient constraints are binding or non-binding? If a nutrient constraint is binding/non-binding, what does this mean?

    d. What does a slack amount for a nutrient mean?

    e. Which food group calorie constraints are binding or non-binding?

    Consider the ranges on the objective function coefficients.

    f. For which foods would a small percent change in the coefficient produce a different optimal solution?

    g. What might cause the coefficient to increase or decrease in the objective function coefficient?

    h. Identify a food you eat that could have two different coefficients.

2. Use Solver to optimize the water pollution problem you formulated in section 4.2.

    a. How much does the optimal cost change if they decide to increase the target reduction of Lake Winnebago by 500 kg? How about the same increase for Green Bay?

3. Each week, the DeeLite Milk Company gets milk from three dairies and then blends the milk to get the desired amount of butterfat for the company's premier product. The average price of milk and milk fat percentage varies from year to year. DeeLite is currently planning to produce 1500 cwt per week with at least 3.7% butterfat. Dairy A can supply at most 600 cwt of milk averaging 3.9% butterfat and costing $17.0 per cwt. Dairy B can supply milk averaging 3.5% butterfat costing $13.0 per cwt. There is no practical limit on the amount they can provide DeeLite. Dairy C can supply at most 700 cwt of milk averaging 3.75% butterfat costing $15.45 per cwt.

    a. How much milk should DeeLite use from each supplier to produce 1500 cwt of milk with at least 3.7% butterfat?

    b. If DeeLite were to produce 1501 cwt what would be the total cost? How much would each additional cwt cost? Explain where you can find this value in the sensitivity analysis report.

c. DeeLite is considering increasing the butterfat content slightly to 3.71%. How much would this increase the cost? How does it change the optimal solution? Explain why it is not easy to determine this incremental cost directly from the sensitivity analysis report.

4. The Ferris Mining Company owns three iron mines in Michigan, Minnesota, and Wisconsin. Each produces three grades of iron ore - high, medium, and low. Due to a rush order, the company has to quickly supply at least 250 tons of high grade, 150 tons of medium grade, and 300 tons of low grade. Each mine produces a certain amount of each grade of iron ore each day. Each mine operates 2 shifts per day for a period of 16 hours. This mix is a function of the type of ore in each mine. The Michigan mine produces 40, 20, and 30 tons respectively of high, medium, and low grade per day. The Minnesota mine produces 25, 30 and 50 tons, respectively of high, medium and low grade per day. The Wisconsin mine produces 20, 20 and 70 tons, respectively of high, medium and low grade per day. It costs the company $15,000 per day to operate the Michigan mine, $16,000 per day to operate the Minnesota mine, and $20,000 per day for the Wisconsin mine. The company wants to determine the number of days it needs to operate each mine to complete the rush order at the lowest cost.

   a. What are the decision variables?

   b. Write the objective function.

   c. The cost of operation is given as a daily cost. What assumption about cost of operation must be made in order to use linear programming to model this decision context?

   d. Formulate the constraints.

   e. Use Excel Solver and find the optimum solution.

   f. Which constraint is non-binding and what does this mean in this decision context? How should management deal with this non-binding constraint?

   g. There is the possibility that the rush order would be increased. Use sensitivity analysis to determine the added cost of providing 10 more tons of high grade iron ore? How does this compare to the cost of providing 10 more tons of low grade ore?

   h. The local union in Wisconsin is concerned that their mine is not winning a share of these rush orders. They are considering a temporary reduction in pay that would reduce the daily cost to operate the mine. How much would the daily cost have to drop before the Wisconsin mine would win a share of the order? If the reduction exceeded this value by $10 per day, how would it change the optimal operating solution and total cost? Which mine would suffer the most by Wisconsin's actions? Surprisingly, another mine benefits from the reduced cost in Wisconsin. Explain why this happens.

5. Sue's uncle Bob grows strawberries on his 500-acre farm. Sue is an agricultural engineer and she tested the soil on his farm and found the level of phosphorus (P) to be 35 parts per million (ppm) and of potassium (K) 80 ppm. The two tables below show the recommended usage of fertilizer for growing strawberries in his climate based on the existing soil concentration.

Property within Phosphorus concentration range (ppm)	Recommended additional phosphorus (lbs/acre)
0-15	46-55
16-45	28-46
over 45	0-28

Table 1: Phosphorus fertilization rates for strawberries

Property within Potassium concentration range (ppm)	Recommended additional potassium (lbs/acre)
0-75	60-72
76-175	48-60
over 175	0-48

Table 2: Potassium fertilization rates for strawberries

Sue also suggests applying between 30 and 40 pounds of nitrogen (N) per acre and between 15 and 20 pounds of sulfur (S) per acre. Uncle Bob has six commercially available fertilizers from which to choose. An analysis of the contents and costs of these fertilizers is given in Table 3.

Product	Nutrient (%/lb)				Cost ($/lb)
	P	K	N	S	
Ammonium sulfate (AS)	0	0	21	24	0.1275
Potassium sulfate (PS)	0	52	0	12	0.1045
Urea (U)	0	0	46	0	0.2025
Urea ammonium phosphate (AP)	17	0	34	0	0.1530
Potash chloride (PC)	0	60	0	0	0.1375
Ammonium phosphate sulfate (APS)	20	0	16	14	0.1615

Table 3: Analysis of fertilizer products

Uncle Bob needs to choose a combination of the above fertilizers that will meet the engineer's recommendations while minimizing the cost per acre.

   a. Formulate the problem to help Uncle Bob make this decision. What is the equation for the total cost of fertilizer for his farm?

   b. What is his optimal purchase plan and how much will it cost for his 500 acre farm?

   c. He recently noticed that a local farm supply company was offering sales on Ammonium sulfate and Urea. Should Bob consider changing his plans for fertilizing his farm?

d. Bob asked his niece to review the upper bound constraints on the various chemicals. Which upper bound constraints should she consider increasing? How much money would be saved by increasing the upper bound by 3 lbs?

e. Sue explored the possibility of reducing the lower bound on one of the chemical constraints. Which lower bound constraints should she consider decreasing? How much money would be saved by decreasing the lower bound by 3 lbs?

6. A risk management team of the Futures Investment Company is considering investment opportunities. The company's budget is $200 million. It has identified the opportunities presented in the Table 4.

	Energy	Agric.	Manuf.	Pharma.	Transp.
Maximum Investment ($ millions)	45	15	60	90	50
Average percentage profit	20%	11%	8%	15%	10%
Value at Risk (% of investment)	15%	8%	5%	10%	7%

Table 4: Investment Opportunities

The maximum investment amount is the maximum they can invest in a specific project. For example, they cannot invest more than $45 million in energy projects. Value at risk (VaR) is the maximum percent loss they might incur in the worst case scenario. For example, value at risk for the energy project in the worst case is a loss of 15% of the initial investment. Assume that partial investment is possible. For example if they invest $4.5 million in energy, they have $675,000 VaR and $900,000 expected profit.

a. Their investment goal is to make at least $25 million in expected profit. The team wants to minimize total value at risk for their investment while meeting this goal. Formulate the decision problem

b. Determine the optimal investment strategy. What is the total Value at Risk with the optimal solution?

c. Generate a sensitivity report in Solver. If Futures Investment were to add $1 million to the budget, what is the impact on the Value at Risk? Just use the sensitivity report to answer this question?

d. Explain how investing more money can reduce the overall value at risk. If necessary rerun Solver to understand what happens.

e. Management is considering increasing the minimum expected profit by one million dollars. How much does total VaR change?

f. Management is considering raising the upper limit on one area of investment. Which maximum investment should be increased? What is the impact on VaR for each one million dollar increase in this maximum investment?

g. Which decision variable has a non-zero reduced cost? Why? Future Investment is considering a more diversified investment plan. They want to invest at least $5 million in each industry. What will be the impact of VaR? Determine this using sensitivity analysis and confirm your answer by rerunning Solver.

7. Ben Efficient has a major end-of-term set of homework assignments in mathematics. The problems are labeled A through G. Each part has a maximum "value" (in points earned) and a "time" (time in hours to complete).

	A	B	C	D	E	F	G
Point value	7	9	5	12	14	6	11
Time (hrs)	3	4	2	6	8	2	5

Table 5: Homework points and time

Assume Ben will earn partial credit that is proportional to the amount of work done (e.g., one hour spent on problem C earns 2.5 points).

a. Ben will be satisfied to earn 40 points on this assignment. This would give him a B+ grade for this homework. How do you determine the coefficients of each decision variable in this constraint?

b. Ben wants to minimize his work time and earn at least 40 points (a B grade). Define the decision variables and constraints. Why are there constraints of the type $A \leq 3$.

c. You would not spend more time on a problem than the amount required to earn all of the points. There is no value in spending more time. There are no bonuses.

d. Determine the optimal work plan and time spent.

e. Which problems have the highest rate of point return per hour of work? How is this reflected in the optimal solution? Which has the lowest? How is this reflected in the optimal solution?

f. Describe a procedure Ben could have used to determine the optimal work schedule without using linear programming.

g. How much time would it take Ben to earn 1 more point? How would this change his investment of time?

h. Problems D and E have the highest potential points. Ben has decided that his study plan must include earning at least half credit for problems D and E. What should be his new study plan?

i. Explain why this problem might in reality violate the proportionality assumption that is part of linearity?

8. All steel manufactured by Malco must meet the following standard requirements:

Item	Range Minimum	Maximum
carbon	6.4%	7%
nickel	1.8%	2.4%
tensile strength	90,000	100,000

Table 6: Steel requirements

These standards are enforced by a federal agency and they should be satisfied. Malco manufactures steel by combining three alloys. The cost and properties of each alloy are given in following Table 7. Assume that the tensile strength of a mixture of the two alloys can be determined by finding the weighted average of the tensile strengths of the alloys that are mixed together. For example, a one-ton mixture that is 20% alloy 1, 45% alloy 2, and 35% alloy 3 has a tensile strength of 0.2(84,000) + 0.45(100,000)+.35(90000) = 93,300 .[1]

	Alloy 1	Alloy 2	Alloy 3
Cost per ton	$190	$230	200$
Percent silicon	4%	5%	6%
Percent nickel	2%	3%	1%
Percent carbon	6%	8%	6.5%
Tensile strength	84,000	100,000	90,000

Table 7: Alloy Composition

Malco wants to manufacture steel that meets all of the requirements at the lowest cost possible.

- a. Identify the decision variables, objective function, and constraints. Clearly define the unit of measurement for decision variables. The unit of measurement can be fraction (or percent) or imagine it as a fraction of a ton.

- b. Use linear programming to determine how to minimize the cost of producing a ton of steel.

- c. What are the allowable increase and decrease on the cost coefficients? Explain the logic behind the ranges.

- d. If the standard agency increases the minimum of tensile strength by 1,000 pounds per square inch, how much does the optimal cost of the mix increase? How about the same increase for maximum of tensile strength?

9. Hardwood Corporation makes expensive bookcases. Each week, Hardwood Corp. needs 11,000 board feet of finished lumber to produce its furniture. The company may obtain lumber in two ways. First, it may purchase raw lumber from outside suppliers and then dry it

---

[1] Adapted from Winston book

in its kiln. Second, it may chop down trees on its own land and cut them into logs. Hardwood Corp. processes their logs into lumber at its small sawmill. It then dries the lumber in its own kiln. Hardwood incurs a cost of $0.50 per board foot to dry any type of raw lumber to convert it into finished lumber.

Hardwood can purchase grade A or grade B raw lumber to be dried in its kiln. Grade A raw lumber costs $1 a board foot. When dried, one board foot of Grade A raw lumber yields 0.9 board feet of finished lumber of the quality that can be used to make bookcases. Grade B lumber costs $0.82 a board foot and when dried one board foot of Grade B raw lumber yields 0.75 board feet of useful lumber. Each week, Hardwood can buy up to 4,000 board feet of grade A lumber and up to 7,000 board feet of grade B lumber.

It costs the company $60 to chop down a tree, create logs, and haul them to a sawmill. After the logs are cut, a tree yields 200 board feet of lumber on average. When a board foot of their lumber is dried, it produces 0.8 board feet of finished lumber. It costs $0.45 a board foot to process logs through the sawmill. Each week, the sawmill is staffed to process up to 5,000 board feet of lumber for bookcases.

On average, the kiln is available 80 hours per week for drying lumber for use in bookcases. The rate of drying one <u>hundred</u> board feet of different types of lumber is as follows: 40 minutes per <u>hundred</u> board feet of grade A lumber, 30 minutes per <u>hundred</u> board feet of grade B lumber, and 35 minutes per <u>hundred</u> board feet of lumber derived from their own logs. (Note: Lumber is dried in huge batches in large kilns. This drying can take as long as days. When you divide the total capacity of the kiln by the number of hours of operation, the rate of drying is converted into the number of minutes per hundred board feet. However, no piece of lumber stays just a few minutes in a kiln.)

    a. Identify the decision variables and use the same unit of measurement.

    b. What is the cost associated with each decision variable? Write the objective function.

    c. Formulate the lumber demand constraint. What are the coefficients associated with the decision variables?

    d. Formulate the kiln capacity constraint. What is the unit of measurement?

    e. Formulate all of the other constraints.

    f. Solve the problem using a spreadsheet solver. How many trees would be cut down each week?

    g. Hardwood Corporation is purchasing the maximum available supply of Grade A. This is impacting its cost. How much money could they save if they were able to purchase 100 more board feet of Grade A lumber? They are also considering paying a little extra to obtain more Grade B lumber. What is the maximum premium they can pay and still save money?

h. The company is considering expanding its bookcase production by 1%. This would increase its need for processed lumber by 1%. What is the cost to Hardwood of increasing its demand by 110 board feet? What would be the impact of a 3% increase in production? Which constraints would be binding?

i. Hardwood Corporation is considering adding hours to the operation of the sawmill. It will cost them $37 for each extra hour of operation. Should they increase the hours of operation?

10. Water supply[2] and the discharge of waste water in the river basins has become a major concern of many states in recent years. The Cape Fear River Basin Water Supply Plan in North Carolina evaluated the water system supply and demand needs for municipal water systems that use water from the Haw River and the Deep River. The plan included 94 water systems in 19 counties across the state and looked at water demands through 2050. Although the numbers in this problem are fictitious, this type of analysis is being done throughout the United States as we consider how to best use one of our natural resources, water. Consider a regional water supplier developing a plan for the next year. The cost of supplying water from Haw and Deep Rivers to the counties of Rockingham, Guilford, Standolph and Alamance is given in the table below. The rivers have a supply limit, and the counties have demand for water that must be satisfied. The table below includes daily demand and supply. Daily demand and supply is measured in mgd (millions of gallons of water used daily). The cost is dollars per mgd delivered.

		Destination				
		Rockingham	Guilford	Standolph	Alamance	Supply (mgd)
Source	Haw River	$3,200	$7,100	$6,700	$4,300	7.5
	Deep River	$2,600	$4,100	$3,700	$2,100	6.1
	Demand (mgd)	3.6	3.1	2.4	4.2	

Table 8: Water supply and demand

a. Define decision variables and use double subscripts.

b. Formulate the problem in order to decide on the amount of water supplied from each river to each county. Write the objective function and all constraints.

c. What is the optimal allocation and how much does it cost?

---

[2] The scenario is based on the North Carolina Cape Fear River Basin Water Supply Plan (http://www.ncwater.org/Reports_and_Publications/Jordan_Lake_Cape_Fear_River_Basin/CFR BWSPdraft2.pdf)
http://www.newsobserver.com/opinion/story/961916.html article on water debates in Raleigh

d. There is an existing contract that requires Haw River to deliver 1 mgd water to Guilford County. If this contract were to continue into the future what is the extra cost per mgd incurred by the region each day?

e. Standolph County is a growing community and the demand may increase by 0.2 mgd. How much will this increase the overall cost of supplying water? If the demand were to increase by 0.4 mgd Solver cannot find a solution? Explain why this happens.

f. There has been a recent drought in the area surrounding the Deep River. There is a concern that the available water will decrease by 0.2 mgd. What will be the impact of this decrease on total cost and the allocation plan?

11. Kathy has an adult cat, named Purrl. It weighs nine lbs. Kathy can choose from seven different foods to create Purrl's daily diet. She wants to keep her cat healthy but control cost. She gathered information about the daily nutritional requirements of an adult cat. Most of the nutritional requirements are defined in grams or mg. However, two are defined as the percent of the diet. She also has the nutritional content of those seven foods, their costs, and calories (See Tables 9, 10, and 11.) Her cat requires 250 cal per day. The food group includes both dry foods and canned foods, which contain some level of moisture. Cats don't require drinking water while eating canned food since the canned food is 78-82% water. But, they do require drinking approximately two milliliters of water for every gram of dry food they eat. Kathy should keep this in mind.

Food	Type	Energy	Cost	Weight
Salmon & Brown Rice (SBR)	Dry	1.8 cal/g	$20.00/bag	3.2 kg - bag
Adult Cat Formula (ACF)	Dry	1.6 cal/g	$23.68/bag	8.2 kg - bag
Duck À La Veggie (DLV)	Dry	1.9 cal/g	$33.52/bag	6.8 kg - bag
Complete Formula (CF)	Dry	1.5 cal/g	$24.96/bag	8.2 kg - bag
Chicken & Turkey Puree (CTP)	Canned	1.3 cal/g	$1.69/can	156 g - can
Beef & Egg Skillet (BES)	Canned	1.1 cal/g	$0.65/can	85 g - can
Turkey & Salmon (T&S)	Canned	1.2 cal/g	$1.21/can	156 g - can

Table 9: Cost and Calorie Information for Available Dry and Canned Foods

Nutrients	Recommended Allowances
Crude Protein	At least 12.5 g
Crude Fat	At least 5.5 g
Crude Fiber	At least 3%
Taurine	At least 0.1%
Vitamin A	At least 0.063 mg
Vitamin E	At least 2.5 mg
Magnesium	At least 25 mg
Calcium	Greater than 0.18 g
Zinc	At least 4.6 mg
Iodine	At least 0.088 mg
Iron	At least 5 mg

Table 10: Recommended Nutrition Levels

Nutrients	SBR	ACF	DLV	CF	CTP	BES	T&S
Crude Protein (grams)	0.400	0.300	0.340	0.320	0.090	0.100	0.100
Crude Fat (grams)	0.160	0.200	0.200	0.140	0.070	0.055	0.060
Crude Fiber (%)	5.0	6.0	5.9	6.5	1.9	1.5	1.8
Taurine (%)	0.13	0.14	0.12	0.11	0.07	0.06	0.09
Vitamin A (mg)	0.003	-	-	-	0.013	0.005	0.007
Vitamin E (mg)	-	0.01	-	0.05	0.025	0.012	0.010
Magnesium (mg)	-	1	1	-	.3	.15	.2
Calcium (grams)	0.013	-	-	-	0.0037	0.0061	0.0026
Zinc (mg)	-	0.100	-	0.150	0.0567	0.0270	0.0375
Iodine (mg)	0.0003	0.0002	0.0003	-	0.0009	0.0006	0.0007
Iron (mg)	0.012	0.018	0.01	-	0.052	0.063	0.047

Table 11: Nutritional Content of Available Cat Foods Per Gram

In the above tables, there are some blank spaces. Assume that those values are zero. Help Kathy satisfy the daily requirements of her cat at the lowest cost. A daily diet should be measured in grams. There are seven decision variables:

x_1:	Grams of dry Salmon & Brown Rice
x_2:	Grams of dry Adult Cat Formula
x_3:	Grams of dry Duck À La Veggie
x_4:	Grams of dry Complete Formula
x_5:	Grams of canned Chicken & Turkey Puree
x_6:	Grams of canned Beef & Egg Skillet
x_7:	Grams of canned Turkey & Salmon

Kathy wants to minimize the cost of her cat's diet. The cost coefficients in Table 9 are shown as the costs of a large bag or can of each cat food. However, the decisions variables are in grams and cost must be as well. Kathy determined the cost per gram of dry Salmon & Brown Rice as follows:

3.2 kg-bag of x_1 is $20. Then cost of a kg of is $20 / 3.2 = $6.25 per kg.
The cost per gram is $.00625.
If we use only three significant digits, the cost is $0.006 per gram.

- a. Determine the cost per gram for every other food to three significant digits and write the objective function.

- b. Formulate the constraints in which the unit of measurement is grams or mg. Treat the calorie constraint as an exact requirement.

- c. Formulate the constraints that are defined in percents. Rewrite the equation as a linear equation with a right hand side value that is numeric.

- d. What is the least cost diet? For the canned food, what fraction of a can is used per day?

- e. Which product included in the optimal solution makes up the smallest portion of the diet? Why do you think it is included in the optimal diet?

- f. Which nutritional constraints are binding? Experts are considering increasing the Iodine minimum constraint from 0.088 to 0.090. How would that impact daily cost?

- g. Purrl's diet contains 250 calories. What would be the impact on cost of a 10 calorie reduction?

Kathy is concerned that her cat will become bored with the same diet each day. She is considering some variations to the daily diet. She wonders if these changes will be costly.

- h. Write a constraint a constraint that allows Kathy to exclude a specific product from the diet. If she were to exclude the canned product found in part d, how would the optimal solution change? How much more would it cost?

- i. If she were to exclude just the primary dry food found in part d, how would the optimal solution change? How much more would it cost?

Chapter 4 Summary

What have we learned?

As we learned in chapter 2, Linear Programming is used to make decisions that will lead to the optimal solution for a situation. However, this does not always mean making something as big as possible. Often, we want to minimize something undesirable such as cost, risk, or pollution.

The process to follow is similar to the one we used in the previous chapter:

Formulate the linear model representing the situation
- Identify the decision variables
- Write the objective function
- Define the constraints including maximum, minimum, and combination of variable constraints.

Use a spreadsheet program such as Excel
- Specify the decision variables, objective function, and constraints
- Use Solver to find the optimal solution to the problem – change the default selection for "Equal To:" from Max to Min

Analyze the results
- What is the optimal plan?
- What is the minimum value of the objective function?
- What values of the decision values will result in this optimal solution?
- What effect do changes to the situation have on the optimal solutions? Changes that make the situation worse will increase the final value rather than decrease it as in the previous chapter.

Terms

Matrix A rectangular arrangement of values and/or variables in a table

Double-Subscripted Variable A variable with two subscripts to represent the location in a matrix, where the first subscript refers to the row in the matrix and the second subscript refers to the column in the matrix. Often the rows and columns can be matched with key features of a problem, such as in the water pollution problem where the rows were the watersheds and the columns were the pollution sources.

Weighted Average An average in which each quantity to be in the average is assigned a weight; the weighted average is found by first adding together the products of each variable and its weight value and then by dividing this sum by the total of the weights

Chapter 4 (LP Minimization) Objectives

You should be able to:

- Use Solver to optimize a situation in which the goal is to minimize a value

- Use a matrix and double-subscripts to represent the combination of two factors into one variable.

- Manipulate an inequality to remove variables from the denominator of a fraction and move all the decision variables to the left hand side.

Chapter 4 Study Guide

1. What is the objective function? Why are we trying to minimize the objective function in this chapter?

2. Why are some constraints shown with <= and others with >=? Give examples.

3. What is a matrix? Give an example.

4. What is a double-subscripted variable? Give an example.

5. What is a weighted average? Give an example.

6. What is shadow price?

7. What does it mean if a shadow price is negative?

CHAPTER 5:

Optimize Effectiveness or Cost with Integer Programming

Section 5.0: Introduction

In the previous chapters on linear programming, the decision variables were things that can be measured continuously such as production rates, grams of a food source, or 100-gallons of gasoline. To say that a variable can be measured continuously simply means that it can take on the value of any real number. However, in some contexts, the decision variable could be restricted. For example, a manager might need to know how many of each type of worker to hire.

1. Explain why a decision variable that represents how many workers to hire cannot be measured continuously.

In this chapter, we will discuss how to solve mathematical programming problems in which the decision variables must be restricted to integer values. We will also discuss why such a restriction makes a difference in how the problem is solved and how its solution is analyzed.

Chapter 5 — Optimize Effectiveness or Cost with Integer Programming

Section 5.1: Political Advertising

Cynthia Brown is running for the office of governor in Michigan. She is considering purchasing ads in several newspapers costing a total of $80,000 per day or TV ads costing $60,000 per day. She decides she does not want to order more than one ad per day on any medium (i.e., no more than seven newspaper ads per week and no more than seven TV ads per week). She wonders how many of each type of ad she could purchase using $600,000.

1. Suppose Cynthia Brown lets x_1 represent the number of newspaper ads purchased and x_2 represent the number of TV ads purchased. Explain why the inequality $80x_1 + 60x_2 \leq 600$ models this situation.

2. Why is 80 used instead of 80,000, 60 instead of 60,000, and 600 instead of 600,000?

First, Cynthia Brown wants to know how many of each type of ad she can purchase for *exactly* $600,000. Thus, she uses algebra to find a solution to the equation $80x_1 + 60x_2 = 600$. She starts by choosing any number of newspaper ads and solving for the number of TV ads. For example, if she chooses 5 newspaper ads (i.e., she lets $x_1 = 5$), then she has $80(5) + 60x_2 = 600$. Then she solves for x_2:

$$80(5) + 60x_2 = 600$$
$$400 + 60x_2 = 600$$
$$60x_2 = 200$$
$$x_2 = \frac{200}{60} = 3\frac{1}{3}$$

3. Why is this solution not feasible?

4. Does this one infeasible solution mean that there are no solutions at all?

5. Find a feasible solution to this equation.

The key to understanding this problem is identifying the need for the solutions to be integers. The candidate must purchase a whole number of ads; she cannot purchase half an ad. Although newspaper ads may be purchased in fractions of a page, not all fractions are possible. For example, a half-page ad could be purchased, but not a 0.341-page ad. Furthermore, a half-page ad does not cost half as much as a full-page ad. Thus, the decision variable is not continuous. If Cynthia Brown wanted to include the possibility of half page and quarter page ads, she would need to introduce another integer decision variable for each sized ad. She chose not to increase the problem size by adding these decision variables.

Recall that the coordinates of a point on a line represent a solution to the linear equation. From the example above, the point (5, 3⅓) is on the graph of $80x_1 + 60x_2 = 600$. That point, though, does not represent a feasible decision. In an integer programming problem, the only feasible

Chapter 5   Optimize Effectiveness or Cost with Integer Programming

solutions are ones in which all decision variable values are integers. Points with integer coefficients are called **lattice points**.

Some equations have integer solutions and some do not. If an equation has no integer solutions, then its graph will not pass through any lattice points.

6. If a constraint in an integer programming problem has no lattice points on its boundary, what are the implications for the corner point principle?

Consider again the equation $80x_1 + 60x_2 = 600$. If Cynthia Brown decides to purchase 3 newspaper ads and she lets $x_1 = 3$, then

$$80(3) + 60x_2 = 600$$
$$240 + 60x_2 = 600$$
$$60x_2 = 360$$
$$x_2 = 6$$

So the equation has the solution $x_1 = 3$ and $x_2 = 6$. This means the graph of the equation contains the point (3, 6). It also shows that purchasing three newspaper ads and six TV ads is a feasible solution. Now, suppose the candidate decides that she can afford to spend $650,000 each week on advertising.

7. How does that change the equation?

8. Do you think there is an integer solution to this equation? Why or why not?

Cynthia Brown wonders under what circumstances an equation will have integer solutions. The question she ponders is, "Is it possible to determine why $80x_1 + 60x_2 = 600$ has integer solutions while $80x_1 + 60x_2 = 650$ does not?"

Since the left hand sides of the two equations are identical, the key lies in the value on the right hand side and its relationship to the left hand side. Cynthia notices that the coefficients on the left hand side can be simplified by pulling out a common factor. She can then divide both sides of the equation by the greatest common factor of the coefficients.

9. What is the greatest common factor of the coefficients of x_1 and x_2?

10. Divide both sides of the equation $80x_1 + 60x_2 = 600$ by the greatest common factor found in question 9. Then divide both sides of the equation $80x_1 + 60x_2 = 650$ by the greatest common factor found in question 9.

11. What do you notice about the right hand sides (i.e., the constant term) of the two equations in question 10?

12. Remember that we are limiting the possible values of the decision variables to integers. The coefficients of both simplified equations from question 9 are also integers. With

Lead Authors: Thomas Edwards and Kenneth Chelst

these two facts in mind, what can you conclude about whether there are feasible solutions to the two equations?

The test to determine if there are not any integer solutions to an equation is to check if the greatest common factor of the coefficients is also a factor of the constant term. If the greatest common factor of the coefficients does not also divide the constant term, then the equation has no integer solutions.

(Note: The inverse of the previous statement is: If an equation has integer solutions, then the greatest common factor of the coefficients also divides the constant term. Equations where this is true are called *linear Diophantine equations*. The relationship between the greatest common divisor and integer solutions includes the assumption that the variables can be negative. If we restrict the variables to the positive domain as in our contexts, then the fact that the greatest common divisor divides the constant term does not guarantee the equation has an integer solution.)

13. Our constraints are typically inequalities, not equations. If an equation like $80x_1 + 60x_2 = 650$ does not have any integer solutions; does that mean that the corresponding inequality $80x_1 + 60x_2 \leq 650$ also has no integer solutions? Why or why not?

14. If the graph of the budget constraint in an integer programming problem does not pass through any lattice points, is it possible to spend all the available money? Explain.

At this point, Cynthia Brown has simplified the original equation, but she has not yet found an actual solution. Once she finds a solution, she can ask many more questions: Will there be more than one integer solution? What is the largest number of ads that could be purchased? What is the least number of ads that could be purchased? Is it possible to purchase the same number of TV ads as newspaper ads?

Restricting the decision variables to take on only integer values transforms the linear programming (LP) problem into an **integer programming** (IP) problem. An integer programming problem is more difficult to solve, because as was just seen, a linear equation does not need to have any integer solutions. Geometrically, that means there might not be any points on the graph of a boundary line that has integer coordinates. If such a line happens to be part of the boundary of an integer programming formulation, that might lead to an optimal solution that does not lie at a corner point, nor even on the boundary of the feasible region, but in the interior of the feasible region.

How we think of the feasible region also needs to change. The points that are considered for a feasible solution to an integer programming problem must have integer coordinates. Recall that points with integer coordinates are called lattice points. Lattice points can also be thought of as the points of intersection of the grid lines of a coordinate plane. Figure 5.1.1 shows the lattice points in the interior of the feasible region for this integer programming problem. This makes the search for the optimal solution much more complicated, especially as problems grow in their number of decision variables and constraints.

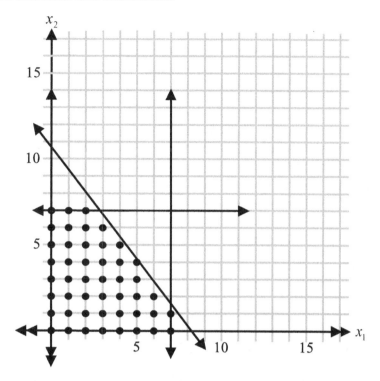

Figure 5.1.1: Graph of feasible region showing lattice points

Therefore, the possible number of ads Cynthia Brown will purchase must be one of the lattice points shown in Figure 5.1.1.

5.1.1 The Effectiveness of Political Advertising

As mentioned earlier, Cynthia Brown is running for the office of governor in the next Michigan election. She wants to have an effective advertising strategy and has hired an advertising consultant to oversee this task. Cynthia Brown has a weekly budget of $650,000. Mr. Response is planning to develop a linear model that determines the most *effective* and *cost-efficient* advertising media.

In addition to TV and newspaper ads, the advertising media that can be used in the campaign are radio, the Internet, brochures and mailings, and billboards. Mr. Response has gathered some information related to these media for Michigan.

In a survey, probable voters were asked to evaluate each medium. From this survey, Mr. Response found the following results.
- 35% said political ads in newspapers are influential
- 25% said TV ads are influential
- 10% said radio ads are influential
- 5% said Internet ads are influential

Chapter 5 Optimize Effectiveness or Cost with Integer Programming

- 15% said brochures and mailings are influential
- 3% said billboard ads are influential

Mr. Response decides to purchase ads only from all local TV stations and newspapers, because these two media were the most influential, based on the survey results.

Listed below are some additional data that Mr. Response deemed important:
- 60% of Michigan's 10 million people are eligible to vote.
- Newspaper ads reach 70% of the electorate and cost of $80,000 per ad per day.
- TV ads reach 85% of the electorate and cost of $60,000 per ad per day.

Mr. Response knows that Cynthia Brown does not want to purchase more than one ad per day on any medium (i.e., no more than seven ads per week for any particular medium).

15. What is Mr. Response's objective?

16. What decision variables should Mr. Response use?

17. Who are the people being targeted for the advertisements?

18. Determine the number of people targeted in Michigan for the political advertisements.

19. Determine the number of these targeted people who are *reached* through newspaper advertisements.

20. How many eligible Michigan voters can be influenced by each newspaper advertisement?

The value you found in question 20 is the coefficient of a decision variable in the objective function. This coefficient represents the effectiveness rate for a newspaper ad. If one newspaper ad is purchased, it will reach and may influence the number of eligible Michigan voters that you found in question 20. Imagine that Cynthia Brown purchases two newspaper ads. She would likely reach about the same group of eligible Michigan voters who were reached with the first newspaper ad. These are the people who can be influenced by an ad in the newspaper. People are more likely to be influenced by repeated exposure to advertisements. Running multiple newspaper ads does benefit Cynthia Brown's campaign. The role played by repeated exposure to advertisements adds a level of complexity that is more subtle than our mathematical model can handle. For our purposes, we will make the assumption that repeated exposure to advertisements in the same medium has an additive effect. Three ads would produce three times as many influences, and so on. Therefore, it is possible to have over 10 million attempted influences, but we cannot know exactly how many eligible voters were actually influenced by our advertising campaign.

21. Use the same process as you did for newspaper ads to compute the effectiveness rate (coefficient) for the TV advertisement decision variable.

22. Use the decision variables and the coefficients from above to define the objective function.
23. Identify and represent any constraints Mr. Response faces in this problem.

24. Develop a spreadsheet representation of this LP formulation and solve it the way you normally solve a linear programming problem.

5.1.2 The Kernel of the Problem

The graph in Figure 5.1.2 shows the optimal solution to this LP formulation. The optimal solution is $x_1 = 2.875$ and $x_2 = 7$.

25. Is this optimal solution feasible? Explain.

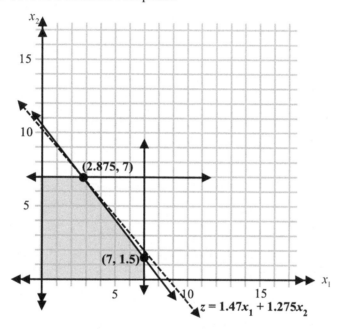

Figure 5.1.2: The political advertising solution

If Mr. Response considers an additional constraint—that the decision variables take on only integer values—the problem becomes an integer programming (IP) problem.

A common misconception is that the solution to an IP problem can be found simply by taking the LP solution and rounding the optimal solution when necessary. Consider the optimal solution for this political advertising problem, which is 2.875 newspaper ads per week and seven TV ads per week. Taking the rounding approach would give you an optimal solution of three newspaper ads and seven TV ads. Unfortunately, substituting $x_1 = 3$ and $x_2 = 7$ into the budget constraint violates shows a clear violation:
$$80(3) + 60(7) = 240 + 420 = 660 > 650$$

Thus, $x_1 = 3$ and $x_2 = 7$ is not feasible, much less optimal.

> **Rounding the Optimal Solution**
>
> In general, rounding the optimal solution from an LP problem will not produce an optimal solution from an IP problem. There are two difficulties with the rounding strategy.
> - The rounded solution may not be feasible
> - If it is feasible, it may not be optimal.
>
> Thus, the rounding strategy applied to an IP problem may not generate the optimal solution.

The set of feasible solutions for an IP problem contains only the lattice points within the constraint boundaries. In the political advertising problem, there are nearly 50 feasible lattice points, as shown in Figure 5.1.3.

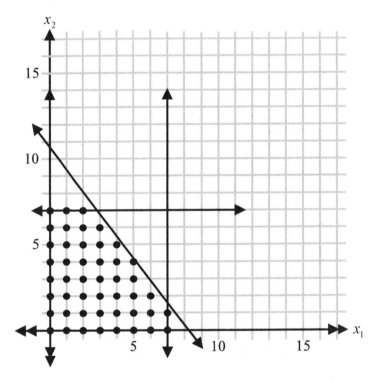

Figure 5.1.3: Graph showing feasible lattice points

Mr. Response can use brute force to check all of these points in order to find the optimal solution. For example, the top horizontal boundary line contains three lattice points: (0, 7), (1, 7) and (2, 7). If he checks these in the objective function, he finds that the lattice point closest to the LP optimal line of constant effectiveness will maximize the objective function. But that point is also the one that is farthest to the right of those three points. Since they are on the same horizontal line, all three points have the same value of x_2, namely seven. So it makes sense that the one that yields the largest value of the objective function is the one farthest to the right, because that point has the largest value of x_1. Putting this back in the context of the problem, if

Mr. Response is considering all the possible decisions that include purchasing seven TV ads, he knows the campaign will be more effective if it includes more newspaper ads rather than fewer.

In general, as Mr. Response considers lattice points closer to the optimal line of constant effectiveness, he will always get a better value for the objective function. Therefore, he can minimize the number of lattice points to check by only considering the lattice points closest to the boundary. This collection of lattice points is called the **kernel** of the feasible lattice points. Figure 5.1.4 shows the kernel for this problem.

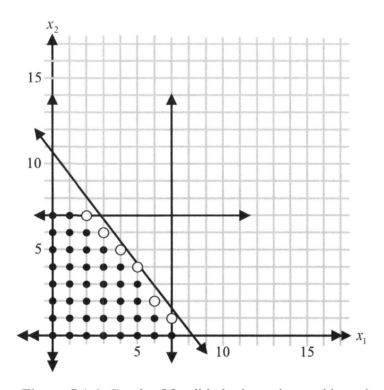

Figure 5.1.4: Graph of feasible lattice points and kernel

In this problem, the kernel consists of the following points: (2, 7), (3, 6), (4, 5), (5, 4), (6, 2), and (7, 1). Using brute force on these six points is significantly less work than applying brute force to the original 47 lattice points.

Table 5.1.1 displays the result of evaluating the objective function at these points. Notice that the optimal integer solution is (5, 4)—that is, five newspaper ads per week and four TV ads per week, with an advertising effectiveness of 12.45 million people per week.

Newspaper ads per week x_1	TV ads per week x_2	Advertising effectiveness $z = 1.47x_1 + 1.275x_2$
2	7	11.865
3	6	12.06
4	5	12.255
5	**4**	**12.45**
6	2	11.37
7	1	11.565

Table 5.1.1: Evaluating the objective function for the kernel points

26. What are the units of measure for advertising effectiveness?

27. What does an advertising effectiveness of 12.45 mean?

28. Is that answer reasonable? Why or why not?

29. Your response to question 28 relies on some assumptions that are built into the mathematical model Mr. Response created. Name some of these assumptions and discuss whether they are justifiable.

The IP optimal solution of (5, 4) is quite different from the LP optimal solution of (2.875, 7). Mr. Response tried rounding the LP solution up to (3, 7), and he found a solution that was not even feasible. If he had rounded down to two newspaper ads and seven TV ads, he would have a feasible solution, just not the optimal solution. The IP optimal solution, (5, 4), not only fails to be the point he would obtain by rounding the LP optimal solution, it is not even the feasible point closest to the LP optimal corner point.

Using Excel Solver, integer constraints are easily handled. In the Add Constraint dialog box, simply select the cells containing the decision variables for the Cell Reference box. Instead of an inequality, select "int," which is how Solver will know that the constraint is that these variables must be integers (see Figure 5.1.5).

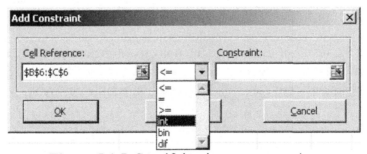

Figure 5.1.5: Specifying integer constraints

Depending on the version of Excel Solver you are using, you may need to instruct the program not to ignore the integer constraints. To do so, go into the Options menu and unclick "Ignore Integer Constraints," as shown in Figure 5.1.6.

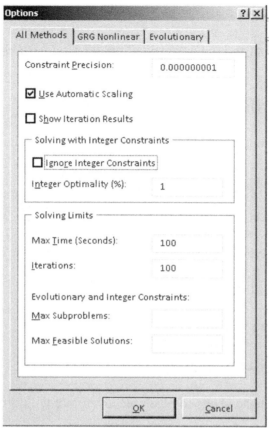

Figure 5.1.6: Do not ignore the integer constraints in the Options menu

30. Find the IP optimal solution using Excel Solver.

31. The optimal solution does not use the entire available weekly budget. What should Cynthia Brown do with the extra funds?

5.1.3 Gearing Up for the Election

After a very successful round of fundraising, Cynthia Brown asks Mr. Response to purchase ads using other possible media: all local radio stations, several Web sites, direct mailings to all residents, and billboards on several main roads. Cynthia Brown increases the weekly budget to $900,000. The information regarding these media is given in Table 5.1.2.

Medium	Influence	Achieved reach	Cost (per ad-day)
Radio (x_3)	10%	90%	$20K
Internet (x_4)	5%	60%	$10K
Direct mailings (x_5)	15%	95%	$180K
Billboards (x_6)	3%	40%	$15K

Table 5.1.2: Effectiveness, percentage of population reached, and cost of various advertising media

32. Define the decision variables (be sure to include the first two media).

33. Compute the coefficient for each decision variable.

34. Apply the coefficients and the corresponding decision variables to construct the objective function.

35. Represent the constraint based on the weekly budget.

36. Based on your new budget constraint, is it possible to spend all $900,000 this week?

37. Represent all the other constraints that must be considered.

38. Develop a spreadsheet representation of this IP formulation and solve it.

39. How many of each type of ad should Mr. Response purchase?

40. How many people will be influenced by the ads?

41. Which constraints are binding?

5.1.4 Adding another Requirement

Ms. Brown has decided that she wants to reach the electorate with direct mailings *at least once a week*.

42. What aspect(s) of the problem will be affected by this additional change? What aspect(s) will not be affected by this change?

43. Solve the problem again using Excel Solver with this additional constraint.

44. How does the requirement to have at least one direct mailing per week affect the optimal solution?

45. How many people will be influenced by the ads in this case?

46. Why might Ms. Brown want to stay with her decision regarding direct mailings?

Section 5.2: Opening and Operating the Pizza Palace

A new shopping district has recently been developed to revitalize downtown Burgville. Ms. Johnson is opening a restaurant that will be located in the heart of this district, on the corner of State and Main. She plans to open a pizzeria and call it Pizza Palace. As owner, Ms. Johnson wants this venture to be as successful and as profitable as possible. One of the decisions she has to make involves determining how many employees to schedule for each shift. The goal of this decision is to keep the cost of wages as low as possible.

There are three categories of employee for any given shift: crew members, shift supervisors, and assistant managers.
- Crew members take orders, prepare and cook the pizza, clean the facility, restock materials, and handle customer transactions.
- Shift supervisors perform all of the tasks of crew members. In addition, they supervise, train, and assist the crew members, and aid the assistant manager.
- Assistant managers perform all of the tasks of shift supervisors. In addition, they set the schedules, complete inventory, supervise all workers, manage productivity, make deposits, and settle customer issues. They also direct the operation of the restaurant during their shifts, ensure compliance with company standards, ensure compliance with health and safety codes, maintain fast and accurate service, interview job applicants, and motivate and discipline employees as necessary.

Assistant managers are the most productive employees. Shift supervisors are 80% as productive as assistant managers. Crew members are 75% as productive as shift supervisors.

Ms. Johnson decides on the following pay rates:
- Crew members: $8.00/hr
- Shift supervisors: $10.00/hr
- Assistant managers: $15.00/hr

After thorough research of similar restaurants, Ms. Johnson arrives at some significant conclusions. There are certain necessities for any given shift. In order to have a profitable, safe, and efficiently run eatery,
- There must be at least one assistant manager working each shift.
- There must be at least one assistant manager or shift supervisor for every five crew members.
- The makeup of any scheduled shift requires a level of productivity equivalent to at least ten assistant managers.
- Ms. Johnson has committed to schedule at least seven crew members to work each shift as a community jobs commitment to local teenagers.

1. Adhering to the required conditions, how many employees of each category do you think Ms. Johnson should hire for a given shift at Pizza Palace while keeping the wage cost as low as possible?

5.2.1 Problem Formulation

Ms. Johnson begins by defining the decision variables and the objective function.

Decision Variables
    Let: x_1 = number of crew members working per shift
                  x_2 = number of shift supervisors working per shift
                  x_3 = number of assistant managers working per shift

Objective Function
        Minimize: $z = 8x_1 + 10x_2 + 15x_3$, where z = the cost of wages

Two of the four constraints are straightforward: there must always be at least one assistant manager and at least seven crew members working. Thus, the next part of the formulation is as follows.

Constraints
    Subject to:
        Assistant Manager Minimum: $x_3 \geq 1$
        Crew Member Minimum: $x_1 \geq 7$

Formulating the supervision constraint takes some careful thought. There must be at least one shift supervisor or assistant manager for every five crew workers. This situation represents a ratio, which can be simplified:

$$\frac{\text{number of supervisors}}{\text{number of crew members}} \geq \frac{1}{5}$$

$$\frac{x_2 + x_3}{x_1} \geq \frac{1}{5}$$

$$5\left(\frac{x_2 + x_3}{x_1}\right) \geq 5\left(\frac{1}{5}\right)$$

$$\frac{5x_2 + 5x_3}{x_1} \geq 1$$

$$x_1\left(\frac{5x_2 + 5x_3}{x_1}\right) \geq x_1(1)$$

$$5x_2 + 5x_3 \geq x_1$$

The initial equation has x_1 in the denominator. This is a nonlinear equation. Solver requires a linear equation with a right hand side numeric value. That means all the variable terms must be on the left hand side, with only the constant on the right hand side. Ms. Johnson multiplied both sides of the equation by x_1. After subtracting x_1 from both sides of the equation, Ms. Johnson obtained the following constraint:

    Supervision Minimum $-x_1 + 5x_2 + 5x_3 \geq 0$

The productivity constraint is based on the productivity of each type of employee. Each shift requires productivity equivalent to at least ten assistant managers. Recall that shift supervisors are 80% as productive as assistant managers. Thus, one shift supervisor is as productive as 0.8 assistant managers. Finally, crew members are 75% as productive as shift supervisors, so their productivity is 75% of 80% of the productivity of an assistant manager. That is, one crew member is as productive as $(0.75)(0.8) = 0.6$ assistant managers. Based on this information, Ms. Johnson creates the following constraint:

Productivity Minimum: $0.6x_1 + 0.8x_2 + 1x_3 \geq 10$

5.2.2 Solution and Interpretation

Figure 5.2.1 contains the solved spreadsheet formulation and solution of the Pizza Palace problem. Figure 5.2.2 shows the Answer Report.

	A	B	C	D	E	F	G
1	5.2 Pizza Palace						
2	IP - Min						
3		Crew Members	Shift Supervisors	Assistant Managers			
4	Decision Variables	x1	x2	x3			
5	Decision Variable Values (# of workers)	7	6	1			
6					Total Cost		
7	Objective Function ($ of wages)	8	10	15	131		
8							
9	Constraints						
10	Assistant managers (#)			1	1	$\geq$	1
11	Crew members (#)	1			7	$\geq$	7
12	Supervision (Ratio - no units)	-1	5	5	28	$\geq$	0
13	Productivity (Assistant manager equivalents)	0.6	0.8	1	10	$\geq$	10

Figure 5.2.1: Formulation and solution for the Pizza Palace problem

	A	B	C	D	E	F	G
6	Target Cell (Min)						
7		Cell	Name	Original Value	Final Value		
8		$E$7	Objective Function ($ of wages) Total Cost	0	131		
9							
10							
11	Adjustable Cells						
12		Cell	Name	Original Value	Final Value		
13		$B$5	Decision Variable Values (# of workers) x1	0	7		
14		$C$5	Decision Variable Values (# of workers) x2	0	6		
15		$D$5	Decision Variable Values (# of workers) x3	0	1		
16							
17							
18	Constraints						
19		Cell	Name	Cell Value	Formula	Status	Slack
20		$E$10	Assistant managers (#)	1	$E$10>=$G$10	Binding	0
21		$E$11	Crew members (#)	7	$E$11>=$G$11	Binding	0
22		$E$12	Supervision (Ratio - no units)	28	$E$12>=$G$12	Not Binding	28
23		$E$13	Productivity (Assistant manager equivalents)	10	$E$13>=$G$13	Binding	0
24		$B$5	Decision Variable Values (# of workers) x1	7	$B$5=integer	Binding	0
25		$C$5	Decision Variable Values (# of workers) x2	6	$C$5=integer	Binding	0
26		$D$5	Decision Variable Values (# of workers) x3	1	$D$5=integer	Binding	0

Figure 5.2.2: Answer Report for the Pizza Palace problem

2. Identify the binding and nonbinding constraints, and interpret them in the context of the problem. How are your answers related to the slack for each variable?

Determine the sensitivity of the optimal solution to pay raises by considering the following two options.

> Option A: Offer a raise to the shift supervisors to $10.25, $10.50, $10.75, and so forth in $0.25 increments. This will naturally cause the value of the objective function to increase, but at what point will the optimal number of each type of employee change?

> Option B: Offer a raise to all employees, but make the increase proportional to the current rate; that is, increase each wage by 10%, 50%, and 100%. At what proportional wage increase for all employees, if any, do you think the optimal number of employees will change? Why do you think this is the case?

As Ms. Johnson's business grows, she finds that her productivity needs to change. Pizza Palace now requires that the makeup of any scheduled shift include a level of productivity equivalent to 12 assistant managers instead of 10.

3. Use Solver to determine how this affects the optimal solution.

Currently, the ratio of assistant mangers and shift supervisors to crew members must be no lower than 1:5. Over the first few months, her crew members have become more experienced and now need less supervision.

4. Write a new constraint if this ratio changes to 1:7; that is, there must be at least one assistant manager or shift supervisor for every *seven* crew members.

5. Use Solver to see the result of this new constraint (the productivity constraint should remain at the equivalent of 12 assistant managers, from question 3).

After three years of outstanding success, Ms. Johnson has decided to give back to the community by increasing her commitment. She will hire at least 9 inexperienced teenagers as crew members for each shift. At this point, the productivity constraint shows that the makeup of any scheduled shift includes a level of productivity equivalent to 12 (not 10), and the ratio of supervision is 1:7. That is, Ms. Johnson has kept the changes from question 3 and question 4.

6. In what ways will this new change affect the optimal solution?

7. How does the situation change if the she decides to schedule 10 crew members each shift?

Section 5.3: Schedule Tellers at Red River Teachers Bank

The previous section introduced a workforce planning decision. The decision optimized the number of workers of each type. The constraints related to meeting the workload demand and issues of supervision. The decision variables are readily defined as the number of workers of each type. There is another class of workforce planning problems that involves scheduling workers by hour of the day or by day of the week. In these contexts the key modeling challenge relates to defining the decision variables. There will be one decision variable for each possible work schedule. The challenge is to determine the optimum number of workers on each work schedule.

Robert Rich, VP of operations at Red River Teachers Bank is developing a new work schedule for better customer service on Mondays. Peak times are after school and during school lunch times. He is also planning for extended hours. The bank will be open from 8 a.m. until 6 p.m. The required number of tellers in each hour of a day is given in Table 5.3.1.

Time	Number of Tellers	Time	Number of Tellers
8 a.m. – 9 a.m.	2	1 p.m. – 2 p.m.	5
9 a.m. – 10 a.m.	4	2 p.m. – 3 p.m.	4
10 a.m. – 11 a.m.	3	3 p.m. – 4 p.m.	5
11 a.m. – Noon	4	4 p.m. – 5 p.m.	6
Noon – 1 p.m.	6	5 p.m. – 6 p.m.	3

Table 5.3.1: Hourly teller requirements

The bank employs both part-time and full-time tellers. Full-time tellers work an 8 hour shift. There is a half-hour lunch break for full-timers. The lunch hour begins after 4 consecutive hours of operation. After lunch they complete their shift. Bank management wants only full-timers on duty to start and end the day. They can start work at 8 a.m. and finish at 4 p.m. with a half-hour lunch break between noon and 1 p.m. Alternatively, they can start at 10 a.m. and finish at 6 p.m. with a half-hour break between 2 p.m. and 3 p.m.

A part-time teller works four consecutive hours. Starting times for these workers are limited to between 9 a.m. to 1 p.m. The bank pays a part-timer $10/hr. A full-timer costs the bank $15/hr. which covers both salary and fringe benefits.

There are two decision variables for full-timers. These correspond to the two possible starting times. There are five possible starting times for part-time workers. They can start at 9 a.m., 10 a.m., 11 a.m., noon, and 1 p.m. This group is represented by four separate decision variables. In total there are seven decision variables.

Let x_1 = number of full-time tellers starting work at 8 a.m. and ending at 4 p.m.
x_2 = number of full-time tellers starting work at 10 a.m. and ending at 6 p.m.
y_1 = number of part-time tellers starting work at 9 a.m. and ending at 1 p.m.
y_2 = number of part-time tellers starting work at 10 a.m. and ending at 2 p.m.
y_3 = number of part-time tellers starting work at 11 a.m. and ending at 3 p.m.
y_4 = number of part-time tellers starting work at noon and ending at 4 p.m.
y_5 = number of part-time tellers starting work at 1 p.m. and ending at 5 p.m.

1. What are the daily costs of a part-timer and a full-timer?

The constraints involve matching up the schedules with the hourly demands. The key to writing these constraints is identifying which schedules provide resources for each hour. For example, the constraint for the first hour is simple.

$x_1 \geq 2$ tellers

For the second hour between 9 a.m. and 10 a.m., there could be both full-time and part-time workers. This includes full-timers who started at 8 a.m. plus part-timers who started at 9 a.m. This constraint is

$x_1 + y_1 \geq 4$ tellers

Between 10 a.m. and 11 a.m., four groups of workers are available.

$x_1 + x_2 + y_1 + y_2 \geq 3$ tellers

2. Which scheduled full-timers and part-timers would be working between 11 a.m. and noon?

3. Represent mathematically the requirement that there are at least four workers on duty.

Between noon and 1 p.m. the early full-timers have a half-hour break. Thus, the inequality is

$0.5x_1 + x_2 + y_1 + y_2 + y_3 + y_4 \geq 6$ tellers

4. Write the inequality for the hour in which the workers who start at 10 a.m. take their half-hour break.

5. Who will be working between 4 p.m. and 5 p.m.? Do not include workers whose shifts have already ended.

6. Construct the entire formulation that minimizes the total cost of meeting the workload constraints.

The optimal schedule is presented in Figure 5.3.1. Two full-time workers start at 8 a.m. Two part-timers come on duty at 9 a.m. Another three full-timers start work at 10 a.m. Lastly, three part-timers come on duty at 1 p.m. The total labor cost for Monday is $800.

Chapter 5 — Optimize Effectiveness or Cost with Integer Programming

	A	B	C	D	E	F	G	H	I	J	K
1	Red River Bank tellers										
2		8 a.m.	10 a.m.	9 a.m.	10 a.m.	11 a.m.	Noon	1 p.m.	Cost		
3	Objective function	$120	$120	$40	$40	$40	$40	$40	800		
4		Full-timers		Part-timers							
5	Decision Variable	x1	x2	y1	y2	y3	y4	y5			
6	Tellers	2	3	2	0	0	0	3			
7		Starting Times									
8		8 a.m.	10 a.m.	9 a.m.	10 a.m.	11 a.m.	Noon	1 p.m.	Tellers		
9	Time Period	x1	x2	y1	y2	y3	y4	y5	On Duty		Needed
10	8-9	1							2	>=	2
11	9-10	1		1					4	>=	4
12	10-11	1	1	1	1				7	>=	3
13	11-12	1	1	1	1	1			7	>=	4
14	12-1	0.5	1	1	1	1	1		6	>=	6
15	1-2	1	1		1	1	1	1	8	>=	5
16	2-3	1	0.5			1	1	1	6.5	>=	4
17	3-4	1	1				1	1	8	>=	5
18	4-5		1					1	6	>=	6
19	5-6		1						3	>=	3

Figure 5.3.1: Red River Teachers Bank optimal teller schedule

Robert Rich noticed that there were no excess tellers at either the beginning or end of the day. However, between 10 a.m. and noon, there were many more tellers than were needed. The same thing happened between 1 p.m. and 4 p.m.

> 7. Explain how the current schedule results in many excess tellers during certain hours.

Yolanda Qwik, the director of analytics, carefully reviewed the plan. She had a suggestion. Consider allowing tellers to work a 10-hour shift and pay them for two hours overtime. Each hour of overtime would cost 50% more. The hourly cost for their final two hours would be $22.50 instead $15.

	A	B	C	D	E	F	G	H	I	J	K
1	Red River Bank Tellers										
2	IP Min	Full Time		Part Time							
3		8:00 AM	10:00 AM	9:00 AM	10:00 AM	11:00 AM	12:00 PM	1:00 PM			
4	Decision Variables	x1	x2	y1	y2	y3	y4	y5			
5	Decision Variable Values	2	1	4	0	0	0	3			
6									Cost		
7	Objective Function ($)	165	120	40	40	40	40	40	730		
8									Tellers		
9	Constraints								On Duty		Needed
10	8:00 - 9:00	1							2	>=	2
11	9:00 - 10:00	1		1					6	>=	4
12	10:00 - 11:00	1	1	1	1				7	>=	3
13	11:00 - 12:00	1	1	1	1	1			7	>=	4
14	12:00 - 1:00	0.5	1	1	1	1	1		6	>=	6
15	1:00 - 2:00	1	1		1	1	1	1	6	>=	5
16	2:00 - 3:00	1	0.5			1	1	1	5.5	>=	4
17	3:00 - 4:00	1	1				1	1	6	>=	5
18	4:00 - 5:00	1	1					1	6	>=	6
19	5:00 - 6:00	1	1						3	>=	3

Figure 5.3.2: Red River Teachers Bank schedule with overtime

8. What is the total cost for a teller who works a 10-hour day?

9. How was Column B changed in the spreadsheet to represent a teller working overtime?

The new schedule reduced the daily cost to $730, an almost 10% reduction.

10. Explain in detail how $70 is saved.

Ms. Qwik reviewed the plans. She noticed that the constraint for the hour starting at noon was a binding constraint. During this hour, tellers who start work at 8 a.m. take their half-hour break. In contrast, the hour between 11 a.m. and noon had excess tellers. Ms. Qwik wondered what would happen if their lunch breaks were moved to between 11 a.m. and noon.

11. How is the algebraic model modified in order to represent this change in lunch schedule?

With this option, the total cost decreases to $690. There will be one less part-timer starting at 9 a.m. (See Figure 5.3.3.)

	A	B	C	D	E	F	G	H	I	J	K
1	Red River Bank Tellers										
2	IP Min	Full Time		Part Time							
3		8:00 AM	10:00 AM	9:00 AM	10:00 AM	11:00 AM	12:00 PM	1:00 PM			
4	Decision Variables	x1	x2	y1	y2	y3	y4	y5			
5	Decision Variable Values	2	1	3	0	0	0	3			
6									Cost		
7	Objective Function ($)	165	120	40	40	40	40	40	690		
8										Tellers	
9	Constraints								On Duty		Needed
10	8:00 - 9:00	1							2	>=	2
11	9:00 - 10:00	1		1					5	>=	4
12	10:00 - 11:00	1	1	1	1				6	>=	3
13	11:00 - 12:00	0.5	1	1	1	1			5	>=	4
14	12:00 - 1:00	1	1	1	1	1	1		6	>=	6
15	1:00 - 2:00	1	1		1	1	1	1	6	>=	5
16	2:00 - 3:00	1	0.5			1	1	1	5.5	>=	4
17	3:00 - 4:00	1	1				1	1	6	>=	5
18	4:00 - 5:00	1	1					1	6	>=	6
19	5:00 - 6:00	1	1						3	>=	3

Figure 5.3.3: Half-hour break scheduled between 11 a.m. and 12 p.m.

VP Rich was worried that towards the end of the ten-hour day the workers would be tired and provide poor service. He asked Yolanda if it would be possible to offer these workers another half-hour break later in the day. Yolanda looked at the optimal schedule and said it would be easy to find additional break time.

12. How would Yolanda go about finding the right times to offer an additional half-hour break? Which times are available for this extra break?

Yolanda wondered if there were other work schedules that cost the same and satisfied all of the constraints. She knew that two part-timers had to start at 9 a.m. but the bank did not need three to start at that time. This would be good to know in case one of these part-timers would prefer to start his or her shift later in the morning.

>13. Identify alternative schedules for the six part-time tellers that satisfy all of the constraints.

There are a number of times with excess tellers. This problem cannot be avoided if the number of tellers needed varies by hour of day. The only way to match schedules to need would be for workers to come to work for just one hour.

>14. Why is scheduling workers to work one hour at a time not a good idea?

Section 5.4: Transporting Oranges to Midwest Markets

Efficient delivery of goods and services is a critical element of the U.S. economy. The large trucks that transport a wide range of goods on our highways are just one highly visible piece of a broader transportation network. This includes highways, railroads, airports, and pipelines.

Physical transport is part of a broader industry called logistics. **Logistics** is the process of planning, carrying out, and controlling the efficient, effective flow and storage of goods, services, and related information from point of origin to point of consumption to meet the needs of the final consumer.

Logistics starts with collecting all of the raw materials and transporting them to one or more facilities that converts them into products. This is called *inbound logistics.* These products or components are later assembled or mixed into the final product. The final product is then transported through a variety of means to a point where the final consumer can purchase it. These are called *outbound logistics.* The customer may bring the product home or have it delivered to his/her home. Logistics does not end there if the final product eventually gets recycled. The used-up product must then be picked up and transported to a recycling center.

Consider the logistics for sending a package overnight from a business to a customer 1,500 miles away.
- The sender may already have a contract with the delivery service. This includes a computer account, pricing, and packaging material.
- The sender inputs the relevant data into an information system that is used to track the package and also bill the sender. The sender attaches an information-rich label to the package.
- The package is deposited into a box for collection.
- A local truck collects the package from the collection box and brings it to a local processing center where the package is logged in.
- The package is shipped to a nearby airport to be transported to a major regional processing center.
- At the regional processing center, it is combined with other packages from all over the US that are ending up in the same local area.
- It is often placed on another plane and flown to an airport nearer the final destination.
- At the airport, the package is grouped with others into a specific route. It is assigned to a truck to be delivered along a specific route. All of the information is loaded into a portable device the driver carries with him/her.
- The driver organizes the packages in the truck so that he/she can reach each package in sequence as he/she travels the route.
- The driver travels the route in sequence and records into the computer when the package is delivered. Sometimes the driver also obtains a signature from the recipient.
- The sender is billed for the service.

Now imagine that Amazon starts this process millions of times each day. This is clearly a complicated process that few people think about.

5.4.1 Hermes Transportation: Moving Oranges across the Country

In the following example, just one piece of the logistics system is studied. Although this is not technically an integer programming problem, this example is included in the chapter on integer programming because only supply and demand constraints that are integer values for their right hand sides—representing full truckloads—are considered. Logistics managers often develop plans based on full truckloads. The cost of the driver, the equipment, and all of the information processing does not vary with the amount loaded in the truck. However, the cost of diesel fuel will depend on the overall weight of the truck and its shipment.

The Hermes Transportation Company has won a contract to deliver full truckloads of fresh oranges from California and Florida to six major markets in the Midwest throughout the prime harvest season. Now their logistics manager, Joshua Mercury, will form a plan for getting the oranges from their warehouses to their customers. Fresh oranges are shipped in refrigerated trucks. The total cost of a fully loaded truck on a long-distance trip is $1.40 per mile (this assumes the truck and its driver will be contracted to transport something on the return trip as well). Hermes Transportation has purchased contracts with several major orange groves in both states. The contracts allow for a maximum purchase of 180 truckloads from California and a maximum of 80 truckloads from Florida.

Typically, shipping is consolidated at a few central locations, as is receiving. This simplifies estimating shipping costs. Hermes Transportation has consolidation warehouses in Orange County, California and Orlando, Florida. Hermes Transportation will make deliveries to distribution centers located in large metropolitan areas within each state. These centers are usually located on a major interstate highway not far from a major city. For example, in Minnesota, the center is outside Minneapolis. The major metropolitan areas in this example are:
- Minneapolis (MN)
- Indianapolis (IN)
- Kansas City (MO)
- Oklahoma City (OK)
- Omaha (NE)
- Des Moines (IA)

Mr. Mercury estimates the miles between each origin and destination using the Internet. The distances between the two warehouse sources of oranges and the six markets are shown in Table 5.4.1. The last column and row includes the local supply and demand, respectively, for fresh oranges.

Distance (miles)		Market						Supply (truckloads)
		MN	IN	MO	OK	NE	IA	
Source	CA	1929	2076	1583	1329	1554	1686	180
	FL	1554	972	1241	1227	1428	1342	80
Demand (truckloads)		66	34	42	30	20	15	

Table 5.4.1: Travel distances between supply and demand locations

These distances are converted into costs by multiplying each mile by $1.40, which is the total cost per mile of a fully loaded truck on a long-distance trip. Mr. Mercury rounded values to the nearest dollar (see Table 5.4.2).

Cost ($/truckload)		Market						Supply (truckloads)
		MN	IN	MO	OK	NE	IA	
Source	CA	2,701	2,906	2,216	1,861	2,176	2,360	180
	FL	2,176	1,361	1,737	1,718	1,999	1,879	80
Demand (truckloads)		66	34	42	30	20	15	

Table 5.4.2: Transportation costs between supply and demand locations

The total available supply of 260 truckloads is more than enough to meet the total demand of 207 truckloads. Mr. Mercury's task is to meet demand without exceeding supply while minimizing Hermes Transportation's total cost of delivery.

5.4.2 Model Formulation

Mr. Mercury must decide on the number of truckloads of oranges to ship from each source (supply) to each market (demand) location. The goal of the decision is to minimize the total transportation cost. The decision is constrained by the available supply and the demand in each market. Mr. Mercury will use decision variables with double subscripts to formulate the decision. In general, the decision variable $x_{i,j}$ is the number of truckloads shipped from source i to market j. There are two sources and six markets for a total of 12 decision variables, as shown in Table 5.4.3. Whenever decision variables represent a quantity from one source to one destination, double-subscripted variables are helpful for making sense of what each variable represents. You also saw this in Chapter 3 in the Wisconsin watersheds problem.

Truckloads		Market (j)					
		MN	IN	MO	OK	NE	IA
Source (i)	CA	$x_{1,1}$	$x_{1,2}$	$x_{1,3}$	$x_{1,4}$	$x_{1,5}$	$x_{1,6}$
	FL	$x_{2,1}$	$x_{2,2}$	$x_{2,3}$	$x_{2,4}$	$x_{2,5}$	$x_{2,6}$

Table 5.4.3: Definition of decision variables

There are two constraints on the available supply (in truckloads):
California supply: $1x_{1,1} + 1x_{1,2} + 1x_{1,3} + 1x_{1,4} + 1x_{1,5} + 1x_{1,6} \leq 180$
Florida supply: $1x_{2,1} + 1x_{2,2} + 1x_{2,3} + 1x_{2,4} + 1x_{2,5} + 1x_{2,6} \leq 80$

There are six constraints that represent Hermes Transportation's contract to meet the demand (in truckloads) in each market:
Minnesota demand: $1x_{1,1} + 1x_{2,1} = 66$
Indiana demand: $1x_{1,2} + 1x_{2,2} = 34$
Missouri demand: $1x_{1,3} + 1x_{2,3} = 42$
Oklahoma demand: $1x_{1,4} + 1x_{2,4} = 30$
Nebraska demand: $1x_{1,5} + 1x_{2,5} = 20$

Iowa demand: $1x_{1,6} + 1x_{2,6} = 15$

The complete formulation is listed below. In setting up the formulation, Mr. Mercury has carefully aligned each column to represent one decision variable. Notice that each decision variable appears in exactly two constraints: one supply constraint and one demand constraint. In addition, the coefficient before each decision variable is always one. These observations apply to every transportation decision formulation.

Decision Variables
Let:
$x_{1,1}$ = the number of truckloads of oranges from CA to MN
$x_{1,2}$ = the number of truckloads of oranges from CA to IN
$x_{1,3}$ = the number of truckloads of oranges from CA to MO
$x_{1,4}$ = the number of truckloads of oranges from CA to OK
$x_{1,5}$ = the number of truckloads of oranges from CA to NE
$x_{1,6}$ = the number of truckloads of oranges from CA to IA
$x_{2,1}$ = the number of truckloads of oranges from FL to MN
$x_{2,2}$ = the number of truckloads of oranges from FL to IN
$x_{2,3}$ = the number of truckloads of oranges from FL to MO
$x_{2,4}$ = the number of truckloads of oranges from FL to OK
$x_{2,5}$ = the number of truckloads of oranges from FL to NE
$x_{2,6}$ = the number of truckloads of oranges from FL to IA

Objective Function
Minimize:
$z = 2701x_{1,1} + 2906x_{1,2} + 2216x_{1,3} + 1861x_{1,4} + 2176x_{1,5} + 2360x_{1,6} + 2176x_{2,1} + 1361x_{2,2} + 1737x_{2,3} + 1718x_{2,4} + 1999x_{2,5} + 1879x_{2,6}$,
where z = the cost of shipping

Subject to:
Constraints

CA supply: $1x_{1,1} + 1x_{1,2} + 1x_{1,3} + 1x_{1,4} + 1x_{1,5} + 1x_{1,6} \leq 180$
FL supply: $1x_{2,1} + 1x_{2,2} + 1x_{2,3} + 1x_{2,4} + 1x_{2,5} + 1x_{2,6} \leq 80$
MN demand: $1x_{1,1} + 1x_{2,1} = 66$
IN demand: $1x_{1,2} + 1x_{2,2} = 34$
MO demand: $1x_{1,3} + 1x_{2,3} = 42$
OK demand: $1x_{1,4} + 1x_{2,4} = 30$
NE demand: $1x_{1,5} + 1x_{2,5} = 20$
IA demand: $1x_{1,6} + 1x_{2,6} = 15$

All decision variables are non-negative.

Figure 5.4.1 shows the general pattern for any transportation problem. The example shown is for a problem with *n* supply sources and *m* demand markets, where *m* and *n* can be any non-negative integer. This structure produces an interesting result. Recall that all of the decision variable values must be integers to meet the requirement that only full truckloads are being modeled. In addition, the right hand side values for the supply and demand constraints are all integers. Due to the structure of this standard transportation model, the linear programming optimal solution of transportation problems will always be *integer* values!

Figure 5.4.1

$x_{1,1}$ $x_{1,2}$ ... $x_{1,m}$	$x_{2,1}$ $x_{2,2}$ ... $x_{2,m}$	...	$x_{n,1}$ $x_{n,2}$ ... $x_{n,m}$		Supply constraints
[1 1 ... 1]				≤	s_1
	[1 1 ... 1]			≤	s_2
			[1 1 ... 1]	≤	s_n
diagonal 1's	diagonal 1's		diagonal 1's	=	d_1
				=	d_2
				=	d_m

(Demand constraints on the right)

Figure 5.4.1: The general structure of a transportation problem

---

Integer Solutions in Transportation Problems

Only integer solutions will be given in a transportation problem if the following requirements are met:
- The structure of the formulation is as shown in Figure 5.4.1
- The right hand sides of the supply constraints are integer values
- The right hand sides of the demand constraints are integer values
- A feasible integer solution exists

This is important because it is not desirable to restrict the solution procedure to only integers. When using linear programming, Solver produces a Sensitivity Report that allows the user to easily explore the impact of changes in the parameters of the model. Solver cannot produce a Sensitivity Report for integer programming models. That is, in order to examine the Sensitivity Report, linear, not integer, programming must be used. Fortunately, the integer constraint does not need to be included in the formulation of a transportation problem. Integer solutions will be given due to the structure of the problem!

Although this section describes a transportation problem, any problem that meets the requirements described above will produce integer solutions. A problem does not necessarily need to have any "transporting" for it to be considered a transportation problem.

5.4.3 Optimal Solution and Sensitivity Analysis

Mr. Mercury seeks minimize the company's total cost of delivery while meeting all its supply and demand constraints. The optimal solution and total cost are shown in Figure 5.4.2.

	A	B	C	D	E	F	G	H	I	J	K	L	M	N	O	P
1	5.3 Hermes Transportion															
2	IP - Min															
3		CA-MN	CA-IN	CA-MO	CA-OK	CA-NE	CA-IA	FL-MN	FL-IN	FL-MO	FL-OK	FL-NE	FL-IA			
4	Decision Variables	x1,1	x1,2	x1,3	x1,4	x1,5	x1,6	x2,1	x2,2	x2,3	x2,4	x2,5	x2,6			
5	Decision Variable Values (truckloads)	20	0	42	30	20	15	46	34	0	0	0	0			
6														Total Cost		
7	Objective Function ($)	2701	2906	2216	1861	2176	2360	2176	1361	1737	1718	1999	1879	428,212		
8																
9	Constraints															
10	California supply (truckloads)	1	1	1	1	1	1							127	≤	180
11	Florida supply (truckloads)							1	1	1	1	1	1	80	≤	80
12	Minnesota demand (truckloads)	1						1						66	=	66
13	Indiana demand (truckloads)		1						1					34	=	34
14	Missouri demand (truckloads)			1						1				42	=	42
15	Oklahoma demand (truckloads)				1						1			30	=	30
16	Nebraska demand (truckloads)					1						1		20	=	20
17	Iowa demand (truckloads)						1						1	15	=	15

Figure 5.4.2: Spreadsheet formulation and optimal solution for orange transportation

In the optimal solution, all oranges shipped to Missouri, Oklahoma, Nebraska, and Iowa come from California. All of the oranges provided to Indiana come from Florida. Only Minnesota receives shipments from both California and Florida. All of the demands are met exactly since they were equality constraints. The optimal solution uses all of the available oranges in Florida but not from California.

1. Why is the available supply in California not used up?

Figure 5.4.3 shows the Sensitivity Report for the optimal solution. Recall that a Sensitivity Report can be generated for this problem because there were no integer constraints. The structure of the problem formulation forced the optimal solution to be integer values. Had Mr. Mercury included integer constraints when he set up Solver, he would have found the same optimal solution, but he would not have been able to generate the Sensitivity Report.

From the Sensitivity Report, Mr. Mercury considers the two supply constraints. The Shadow Price for California's supplies is zero because Hermes Transportation is not using up all of California's supply. However, the Shadow Price for Florida is -$525. This means that each additional truckload of oranges obtained in Florida would reduce the total transportation cost by $525. Because of this, the management at Hermes Transportation has begun exploring additional sources of supply in Florida.

	A	B	C	D	E	F	G	H
22	Constraints							
23				Final	Shadow	Constraint	Allowable	Allowable
24		Cell	Name	Value	Price	R.H. Side	Increase	Decrease
25		$N$10	California supply (truckloads)	127	0	180	1E+30	53
26		$N$11	Florida supply (truckloads)	80	-525	80	20	46
27		$N$12	Minnesota demand (truckloads)	66	2701	66	53	20
28		$N$13	Indiana demand (truckloads)	34	1886	34	46	20
29		$N$14	Missouri demand (truckloads)	42	2216	42	53	42
30		$N$15	Oklahoma demand (truckloads)	30	1861	30	53	30
31		$N$16	Nebraska demand (truckloads)	20	2176	20	53	20
32		$N$17	Iowa demand (truckloads)	15	2360	15	53	15

Figure 5.4.3: Constraints section of the Sensitivity Report for orange transportation

2. How much could be saved if the Florida supply increased by 15 truckloads?

3. The Shadow Price on the Minnesota demand constraint is $2,701. This is the exact cost of shipping from California to Minnesota, as reported in Table 5.4.2.
   a. Why does this make sense?
   b. Which of the other demand constraints have shadow prices that equal a specific transportation cost?
   c. Which of the other demand constraints have Shadow Prices that do not equal a specific transportation cost? Explain why this happened.

As mentioned above, the structure of the formulation of this problem forced the optimal solution to be all integer values. Suppose the formulation changed because each state slightly increased their demand or supply by non-integer amount. The revised supply and demand requirements

Cost ($/truckload)		Market						Supply (truckloads)
		MN	IN	MO	OK	NE	IA	
Source	CA	2,701	2,906	2,216	1,861	2,176	2,360	180.5
	FL	2,176	1,361	1,737	1,718	1,999	1,879	80.0
Demand (truckloads)		66.3	34.3	42.1	30.3	20.2	15.5	

Table 5.4.4: Transportation costs and non-integer supply and demand values

4. How is this change captured in the revised formulation?

Mr. Mercury uses Solver to find the new solution. He solves it as a linear programming problem. The new optimal solution and total cost are shown in Figure 5.4.4.

5.3 Hermes Transportation
IP - Min

	CA-MN	CA-IN	CA-MO	CA-OK	CA-NE	CA-IA	FL-MN	FL-IN	FL-MO	FL-OK	FL-NE	FL-IA			
Decision Variables	x1,1	x1,2	x1,3	x1,4	x1,5	x1,6	x2,1	x2,2	x2,3	x2,4	x2,5	x2,6			
Decision Variable Values (truckloads)	20.6	0	42.1	30.3	20.2	15.5	45.7	34.3	0	0	0	0			
													Total Cost		
Objective Function ($)	2701	2906	2216	1861	2176	2360	2176	1361	1737	1718	1999	1879	431,983.20		
Constraints															
California supply (truckloads)	1	1	1	1	1	1							128.7	≤	180.5
Florida supply (truckloads)							1	1	1	1	1	1	80.0	≤	80.0
Minnesota demand (truckloads)	1						1						66.3	=	66.3
Indiana demand (truckloads)		1						1					34.3	=	34.3
Missouri demand (truckloads)			1						1				42.1	=	42.1
Oklahoma demand (truckloads)				1						1			30.3	=	30.3
Nebraska demand (truckloads)					1						1		20.2	=	20.2
Iowa demand (truckloads)						1						1	15.5	=	15.5

Figure 5.4.4: Formulation and optimal solution with new constraint right hand sides

Mr. Mercury notices that the optimal solution no longer contains integer values. He considers this a problem because all of the costs were based on shipping completely full truckloads. A half a truckload does not cost half as much to transport. The driver is paid the same amount whether the truck is completely or partially full.

5. Why will the decision variables in the optimal solution have non-integer values in this context?

Chapter 5 (Integer Programming) Homework Questions

1. Giovanni Carducci is an expert tailor. He makes wool sport coats and pants. It is two weeks before a major holiday. The customers he will serve want to be absolutely sure that they will have their garments in time for the holiday. If Giovanni cannot complete the tailoring in time, they will go elsewhere. Mr. Carducci was able to purchase at a discount 100 square yards of high quality wool material to make the sport coats and pants. He is willing to work long hours to meet this seasonal demand. Giovanni estimates that under time pressure he can work 120 hours during these two weeks and still deliver a quality product. A sport coat brings $100 in profit. It uses 3 square yards of material and takes 8 hours to make. A pair of pants brings $75 in profit. It uses 5 square yards of material and takes 4 hours to make. Giovanni would like to maximize his profit.

   a. Define the decision variables for an integer programming model of the problem.

   b. Use the decision variables you defined in (a) to define the objective function.

   c. Use the decision variables to formulate any constraints in the problem.

   d. Find the optimal mix of sport coats and pants the tailor should in the next two weeks.

   e. What is the maximum profit he can earn?

   f. If Giovanni were able to obtain 5 more square yards of wool, what would be the effect, if any, on the optimal solution?

   g. Giovanni is considering finding time to work a few more hours. What is the value of 2, 4, 6, or 8 more hours? Explain why the impact is not a linear function of the extra hours. (Note: Assume the original value of 100 square yards of fabric.)

   h. If this were a linear programming problem, what information could you have used to answer f and g without rerunning Solver?

2. Bob and Sue Penny produce handmade wooden furniture in a workshop on their farm during the winter months. In late December, they purchased 900 board feet of solid cherry wood, and they plan to make kitchen tables, chairs and stools during January, February and March. Once April starts, farm work takes up all of their time. Thus, everything must be completed and sold by the end of March. Each table requires 25 hours to make, each chair requires 15 hours, and each stool requires 8 hours. A table uses 36 board feet of wood, a chair uses 14 board feet, and a stool uses 5 board feet. Between them, Bob and Sue have 650 hours that they can devote to this project. A table earns $650 profit, a chair earns $175, and a stool earns $100. Most people who buy a table also want four chairs, so they decide to make enough chairs so that there are at least four chairs for every table. How many of each piece of furniture should they make in order to maximize their profit?

3. Natty likes making and wearing jewelry. One day she saw an article on the Internet talking about the best ways of earning money. After reading it, she decided to turn her jewelry-making skills into moneymaking. Natty read about the local art fair to be held the following weekend. She decided to use this fair as an opportunity to see if she can make jewelry that will sell in the $20 to $40 price range. Natty has already bought some tools and supplies for her home business.

She is going to make and sell bracelets, earrings, and rings. She will use different sizes of beads, clasps, crimps, thread, etc. She wants to work at least 30 hours and no more than 35 hours between now and the fair. The amounts of each resource needed and how much time it will take to make each type of jewelry are given in Table 1. The labor hours are at the bottom of the table.

	Bracelet	Earring	Ring
6 mm faceted iridescent bead	4	2	1
4 mm faceted iridescent bead	2	2	2
4 mm silver beads	4	2	2
silver clasp	1	2	-
crimp	2	2	-
labor hours	3.5	1.5	1

Table 1: Usage amounts from each resources

Natty found a jewelry supply shop that sells all of the materials she needs. Table 2 shows the prices of the supplies, and Table 3 shows the amounts that Natty purchased.

Item	Cost
6 mm faceted iridescent bead	0.95
4 mm faceted iridescent bead	0.75
4 mm silver beads	1.75
silver clasp	0.60
crimp	0.52

Table 2: Expenses ($)

Item	Available
6 mm faceted iridescent bead	40
4 mm faceted iridescent bead	40
4 mm silver beads	50
silver clasp	20
crimp	26

Table 3: Available Supplies

Natty is planning to sell a bracelet for $36.50, an earring for $28.95 and a ring for $23.50.

a. How many of each item should she produce to make the most profit? How much would she earn per hour of work?

b. With integer programming there is no sensitivity analysis report. Sensitivity analysis involves rerunning the model. Determine whether or not a 20% increase in any of the three bead types would change the optimal solution.

4. Products that we purchase from a local store have to be transported from an original location where the product is grown or manufactured to the retail outlet. Some examples are:
    - Fruits and vegetables: fresh or processed
    - Meat: fresh or processed
    - Fish: fresh or processed
    - Milk
    - Toys
    - Clothing
    - Cars
    - Bottled soda

    a. Which of the above products are transported in specialized vehicles at some point in the transportation network?

    b. Which of the above products are made in a few centralized locations with finished products often transported a thousand miles or more?

    c. Which of these products are likely to be transported across international borders?

    d. Which products are usually produced within 1,000 miles of the final customer? Might it vary by time of year?

    e. Construct a logistics chain for one of the above products from production to customers.

5. Recall that the transportation problem format always generates an integer optimal solution even without adding a specific integer constraint. Thus, a sensitivity analysis report can be provided. Regarding the Hermes Transport problem in the text,

    a. The Allowable Increase for the Florida constraint is 20 truckloads. What would be the total reduction in cost if Florida's supply were increased by 20 truckloads?

    b. What would be the total reduction in cost if Florida's supply were increased by 30 truckloads? What is the new shadow price?

6. ShiftyGears, Inc. manufactures three types of bicycle, a simple 10-speed, a slightly more elaborate 12-speed, and a deluxe 24-speed mountain bike. The production manager at ShiftyGears, Joe Sprocket, does his production schedule daily, and he wants to complete any bike the company begins to manufacture on the same day when its production began. Table 4 shows the relevant manufacturing details for each of the three bikes that ShiftyGears produces.

    There are 30 production workers available to work on the bikes, and the ShiftyGears factory works an 8-hour workday. Therefore, there are 240 production hours available to build the bikes. How many of each type of bike should Mr. Sprocket schedule for production in order to maximize the company's profit?

	SG-10	SG-12	MB-24
Market estimate of maximum to produce	20	15	6
Number of hours to build one bike	6	8	12
Net profit per bike	$85	$110	$150

Table 4: ShiftyGears manufacturing data

a. Define the decision variables for this problem. Are those decision variables non-negative? Are they integer?

b. Using the decision variables you have defined, define the objective function.

c. How many constraints are there? Define all of the constraints.

d. Use your problem formulation from parts a, b, and d to develop a spreadsheet representation of the problem.

e. Use a spreadsheet solver to find the optimal solution.

f. ShiftyGears, Inc. is considering expanding production by adding one or even two workers. It costs approximately $3,200 to recruit and train each new worker. What would you recommend? How many days of work will it take to recover the cost of this investment?

7. In your original optimal solution to problem 6, the value of one of the decision variables should be 0.

   a. What does that mean in the context of the problem?

   b. Why do you think this happened?

   c. Joe Sprocket believes there is an opportunity to increase the price and profit of their fanciest bike, MB-24. In steps of $10, increase the profitability of that bicycle model until its value in the optimal solution is greater than zero. For what amount of profit does it first happen?

   d. What is its new production plan and its profitability?

   e. Refer back to the original profit values in problem 6. The industrial engineering department has some ideas for reducing the time required to manufacture MB-24. If they, if they can reduce production time by 5%, how would this impact the optimal plan? If they could reduce production time by 10%, how would this impact the optimal plan?

Personnel Staffing

8. The chief of the Arlington University Public Safety Department has determined the number of police officers he would like to have on duty during different two-hour time periods. During each of these time periods, the number of officers required is constant. Police officers work eight-hour shifts. The chief is considering only using starting times: from 6 a.m., until midnight. Nobody wants to start work at 2 a.m. or 4 a.m. The chief recruited an Industrial Engineering student to create a schedule that meets the hourly requirements and uses the fewest number of officers. The minimum personnel required for each time period is given in Table 5.

Time	Number of Officers Required
Midnight – 2 a.m.	15
2 a.m. – 4 a.m.	7
4 a.m. – 6 a.m.	6
6 a.m. – 8 a.m.	13
8 a.m. – 10 a.m.	16
10 a.m. – Noon	22
Noon – 2 p.m.	24
2 p.m. – 4 p.m.	17
4 p.m. – 6 p.m.	15
6 p.m. – 8 p.m.	17
8 p.m. – 10 p.m.	18
10 p.m. – Midnight	20

Table 5: Number of Police Officers Required for Each Period

   a. Define the decision variables for the problem and write the objective function. Explain why you would not include shifts that start at 7 a.m. or 9 a.m. or any odd-numbered hour.

   b. Which police officers will be working in the 4 a.m. – 6 a.m. time interval? Use your answer to write the minimum police requirement constraint for the 4 a.m.–6 a.m. time interval.

   c. Formulate the integer programming model to determine the optimal schedule.

   d. Determine the optimal schedule.

   e. What constraints are not binding? What does it mean to the Chief that a constraint is not binding? How might he use this information?

9. After reviewing the results and talking to his personnel, he found that none of the officers really want to be on a shift that ends at 2 a.m. or 4 a.m. He is concerned that officers who are forced to work these shifts will be unhappy and look for police jobs elsewhere.

a. Describe two different ways the spreadsheet could be modified to deal with this new limitation. Which do you believe is the better approach? What is the new optimal solution?

b. The chief was very surprised by number of extra officers required. He estimates it costs the department approximately $400 in salary and benefits for each 8 hour shift. How much would this new policy increase his daily costs? How much would it increase his annual budget? He also noticed a large excess of officers at certain times in the day. Could these extra officers be used effectively?

c. The industrial engineering student suggested offering $80 premium pay for officers working a shift that ends at 2 a.m. or 4 a.m. He believes this premium will be able to attract as many as 10 officers to volunteer for these shifts. Rewrite the objective function to minimize daily cost. How does this solution compare to the solutions found in part d) of the previous question? What do you think the chief should do?

d. The police officers have requested that the chief consider using 10 hour shifts. With this schedule, officers would work only four days per week. This schedule has the advantage that officers will frequently have three-day weekends off. If they were allowed to work 10 hour shifts, the union agreed to ensure that officers would accept without complaint shift assignments that end in the early morning hours. The chief is evaluating the impact of having 10-hour shifts on the total number of officers required to meet the workload requirements. What problems do you see with this optimal schedule? Why do 10 hour shifts create these types of problems?

e. How does the cost of this schedule compare to earlier plans?

10. In the text, the management of Red River Teachers Bank determined an optimal schedule that involved three types of schedules. Some workers worked ten hours including overtime. Others worked eight hours and part-timers worked four hours. In the optimal schedule there were many hours with excess tellers, Vice President Rich asked Yolanda Qwik to evaluate several new possibilities.

a) What if the bank could find people willing to work two hour shifts? To attract them to work, the hourly salary would be increased to $12.50 for a total of $25. Would this option save money?

b) What if some of the full-timers were paid an extra dollar or two per hour to work only a six hour shift on Monday. This shift would start at noon and end at 6 p.m. Would this plan save money, if these workers cost $16 per hour or $17 per hour?

11. The emergency room of MedCare Hospital operates 24 hours a day, 7 days a week. Nurses work 5 consecutive days a week. The daily nurse requirements for the daytime hours are as given in Table 7.

Day	Mon	Tue	Wed	Thurs	Fri	Sat	Sun
Nurse	6	5	7	5	5	10	11

Table 7: Daily Nurse Requirements

    a. Why do you think a hospital might need more nurses on Saturday and Sunday?

    b. What are the decision variables of the problem?

    c. Use these decision variables to write the equation that there be a minimum of 5 nurses on duty on Friday. Use these decision variables to write the equation that there be a minimum of 7 nurses on duty on Wednesday.

    d. Develop a weekly staffing schedule that can handle the daily demand of the MedCare Emergency Room with the minimum number of nurses.

    e. They have found that they often do not need as many nurses as they thought. If you could reduce the requirements by 1 nurse on one specific day, which day would you choose and why? Reduce that day's total by 1 and demonstrate that the total number required goes down. Try any other day with no slack and see if the same thing happens?

12. Kars Department Store located in a business area has three types of employees for any given day: sales associate, sales representative, and supervisor. Sales associates handle sales transactions and keep the shelves organized. Sales representatives can handle all of the tasks of sales associates. They also deal with customer problems, product changes and product returns. Supervisors can perform the tasks of sales representatives, as well as, supervise and train all employees. Sales associates are 75% productive as Supervisors. Sales representatives are 90% productive as Supervisors. Pay rates for the sales associate, sales representative, and supervisor are $500, $700, and $1000, respectively per 40 hour week.

    The store is open only on weekdays. The workers work 4 consecutive days for ten hours. There has to be 2 sales representatives each day to handle all of the customer problems. The company's policy is to have at least 1 supervisor for every 6 sales staff (associates plus representatives) each day. The daily staff requirements are based on the workload demand which varies by the day of the week. The workload has been converted into equivalent supervisors. It is reported in Table 8.

Day	Mon	Tue	Wed	Thurs	Fri
Staff	9	8	7	6	12

Table 8: Supervisor Equivalent Number of Staff Requirement

    a. Formulate the equation of the requirement that there be at least 12 equivalent supervisors on Friday. Formulate the equation of the requirement that there be at least 8 equivalent supervisors on Tuesday.

    b. How many decision variables and constraints are required to formulate the decision?

Chapter 5   Optimize Effectiveness or Cost with Integer Programming

c. Develop a work schedule that creates the minimum weekly cost for Kars Department Store. (Important: Set the tolerance level at 1% in the Options section of the Solver window.) There can be multiple optimal solutions.

d. How many workers of each type are needed? How many workers of each type will be on duty each day? How many workers are needed in total? Fill in the Table 9 below.

	Employee Type			
Day of Week	Associate	Representative	Supervisor	Totals
Mon				
Tue				
Wed				
Thurs				
Fri				
Total Workdays				
Total Workers				

Table 9: Number of workers of each type on duty each day

e. Which day of the week has the most extra workers? Any suggestions as to how to deal with this overstaffing?

Transportation Problems

Recall that the transportation problem format always generates an integer optimal solution even without adding a specific integer constraint. Thus, a sensitivity analysis report can be provided.

13. Spuds Inc, manages four suppliers of potatoes that are disbursed geographically. Every month it ships truckloads of potatoes to regional distributors in the suburbs of six big cities. These distributors supply restaurants within a 100 mile radius. These four suppliers are located in Idaho, Washington (State), Wisconsin, and North Dakota. Their main customers are in New York City, Washington DC, Los Angles, Chicago, Dallas, and Denver. Spuds, Inc. is committed to meeting all of the regional demands. Tables 10 and 11 contain the demand in truckloads for each city and supply in truckloads available each producer. Table 12 reports the distances between supply and demand locations.

	Idaho	Washington(State)	Wisconsin	North Dakota
Supply	13,000	9,000	7,000	5000

Table 10: Supply of Potatoes – Truckloads

	NYC	Wash. DC	LA	Chicago	Dallas	Denver
Demand	7000	3000	6000	2500	5000	7000

Table 11: Demand for Potatoes – Truckloads

	NYC	Wash. DC	LA	Chicago	Dallas	Denver

Lead Authors: Thomas Edwards and Kenneth Chelst   Page 243

Idaho	2,471	2,381	893	1,693	1,626	836
Washington(State)	2,992	2,902	1,212	1,979	2,266	1508
Wisconsin	961	871	2,048	172	1,065	1034
North Dakota	1,662	1,572	1,628	873	1,225	735

Table 12: The distances (miles) between supply and demand locations

a. The cost per mile of shipping from the Idaho and North Dakota producers to each city is $2.73 per truckload. Due to higher salaries, the cost per mile is $3.20 for shipping potatoes from Washington (State) and Wisconsin. Complete a table of shipment costs by using the given information.

b. Formulate the problem in order to decide on the number of truckloads to ship from each producer to each city in order to minimize cost. Write the objective function and all constraints.

c. Use Excel solver to find the optimal shipment plan.

d. Why is shadow price of the Washington (State) supply constraint zero?

e. Why are shadow prices of Idaho, Wisconsin, and North Dakota supply constraints negative?

f. Washington producers have approached Spuds Inc. with a proposal to increase their supply. Do you think Spuds should pursue this offer?

g. Some restaurants in Los Angeles have informed Spuds Inc, that they are planning to expand their business and will need more potatoes each month. How much does each extra truckload add to the total cost?

h. Due to the reputation of Idaho potatoes, each of regional markets wants at least 100 truckloads of Idaho potatoes. Are you able to predict whether this requirement will cause the total cost to increase or decrease?

i. .

j. Reformulate the model and use Solver to determine the impact of this new requirement on the optimal plan. How much does this change the total cost?

14. Pod Inc. produces canned vegetables at plants located in Michigan (E. Lansing), Indiana (Fort Wayne), and California (Truckee). They ship their products to five main distribution centers across the U. S. Tables 13 and 14 contain the demand and supply (in truckloads) for each producer and each distribution center. The demand must be met.

	Michigan	Indiana	California
Supply	1500	700	900

Table 13: Supply of canned vegetables – Truckloads

	Dallas	Kansas City	Denver	NYC	Atlanta

Chapter 5   Optimize Effectiveness or Cost with Integer Programming

Demand	1000	300	500	250	350

Table 14: Demand for canned vegetables – Truckloads

    a. Use the Internet to complete the following table. Find the distances between suppliers and distribution centers:

	Dallas	Kansas City	Denver	NYC	Atlanta
East Lansing (MI)					
Fort Wayne (IN)					
Truckee (CA)					

    b. If the shipping cost from each producer to each distribution center is $2.55 per mile, what are the shipment costs per truckload between each supplier and each distributor? Complete a table for the shipment costs:

    c. Formulate the problem in order to decide on the number of truckloads to ship from each producer to each distribution center. Write the objective function and all constraints.

    d. Use Excel solver to find the optimal shipments. Generate a sensitivity report for the constraints.

    e. East Lansing producer has found an investor who is willing to invest in their production facility in order to increase the production. Do you think that this investment has any effect on Pod Inc. shipment costs? Explain.

    f. Pod Inc. is considering expanding capacity at Fort Wayne by 200 truckloads. The expansion will cost $40,000. Should they expand?

    g. Due to the bankruptcy of several competitors who also service Kansas City, the Kansas City distribution center may request 200 more truckloads from Pod Inc. Which producer will be used to satisfy this demand? How much will it cost Pod Inc. for each extra truckload?

    h. The Dallas distribution center has started a marketing campaign in order to increase the demand in their region. If they would be able to increase demand to 1200 truckloads. How much will transportation costs increase? Where will the truckloads come from?

15. Lactose Inc. buys milk from seven large milk producers in Wisconsin. As a result they can ship milk from seven different locations. They have contracts to supply each month the local distributors in four cities: Milwaukee, Appleton, Eau Claire, and Shawano. Table 15 and 16 contain the supply (in gallons) from each producer and the demand (also in gallons) for each city. Table 17 has the distances between supply and demand locations.

(gal)	Milwaukee	Kenosha	Madison	Green Bay

Demand	1,500,000	600,000	1,000,000	500,000

Table 15: Supply of milk – Gallons

(gal)	La Crosse	Wausau	Brookfield	Washburn	Merrill	Antigo	Bloomer
Supply	500,000	650,000	700,000	900,000	400,000	650,000	600,000

Table 16: Supply of milk – Gallons

The distances between milk producers and demand centers are presented in the following table:

(miles)	La Crosse	Wausau	Brookfield	Washburn	Merrill	Antigo	Bloomer
Milwaukee	211	189	12	359	205	184	267
Kenosha	246	224	48	394	240	219	302
Madison	144	143	71	313	159	168	199
Green Bay	210	97	130	266	113	89	200

Table 17: The distances between supply and demand locations

a. If the shipment cost for each gallon is $0.05 per mile, complete the following shipment cost table

b. Formulate the problem in order to decide on the number of truckloads to ship from each milk producer to each market so as to minimize the cost of shipping. Write the objective function and all constraints.

c. Use Excel solver to find the optimal shipment plan.

d. Kenosha, is seeking to buy an additional 50,000 gallons per month from one of the producers. Which producer should they target first and why?

e. Explain why Washburn is not a good candidate.

16. Companies that market cell phones usually only design their products and outsource production. The production companies are called contract manufacturers (CM). CMs do not have design teams and do not sell their products to retailers. Now, assume that the Zykes Company sells four different cell phones. There are five certified CMs in the market that Zykes can outsource its manufacturing to. The demand for each cell phone and production capacity of each CM is given in Table 18.

Product	Demand
Cell phone 1	700,000
Cell phone 2	1,100,000
Cell phone 3	800,000
Cell phone 4	900,000

Manufacturer	Capacity
CM 1	800,000
CM 2	600,000
CM 3	700,000
CM 4	800,000
CM 5	1,000,000

Table 18: Demand for cell phones and manufacturing capacity

The company asked each CM to submit their manufacturing price for each cell phone. Table 19 presents the manufacturing costs of each cell phone in the different CMs.

($)	CM 1	CM 2	CM 3	CM 4	CM 5
Cell 1	15.5	14.5	16	15	14.5
Cell 2	23	20.5	22	21	21.5
Cell 3	19	19.5	20	21	17.5
Cell 4	17	16	17.5	15.5	16

Table 19: Manufacturing costs for each cell phone supplier

Zykes will allocate the cell phones to the CMs so as to minimize total manufacturing costs.

   a. There is no mention of transportation or shipping in this example. Explain why a transportation model can be used to solve this problem?

   b. What are the decision variables? Formulate the model using decision variables with double subscripts. Explain each subscript.

   c. What is the objective function?

   d. How many constraints does the model have? What are they?

   e. What is the optimal contracting plan and total cost?

   f. Zykes is considering sharing the cost of increasing capacity by 200,000 units at one of the CMs. Zykes share of the cost will be $50,000. Which CM should the operations research team pick for expansion?

   g. How much would this expansion decrease production costs?

   h. How would the expansion affect the optimal production plan?

17. In a transportation problem, why are supply constraints less-than-or-equal-to inequalities while demand constraints are equations?

Chapter 5 Summary

What have we learned?

As in the previous two chapters on mathematical programming (LP Max and LP Min), **Integer Programming** (IP) is a method of modeling a situation in which a decision has to be made to optimize some objective while being constrained by limited resources. The process for solving an integer programming problem is the same as other linear programming problems.

1. Formulate the problem.
    - Identify and define the decision variables.
    - Write the objective function.
    - Identify and write the functional constraints.

2. Enter the problem formulation into a spreadsheet.
    - Enter decision variables, objective function, and constraint coefficients.
    - Create formulas for objective function values and constraints' RHS.
    - Set up Solver Parameters and Options.
    - Add integer constraint for decision variable values.
      Note: Some integer programming problems are maximization and others are minimization problems.
    - Solve and generate Answer Report.

3. Interpret the results.
    - Answer Report shows status and amount of slack for constraints
    - Solver cannot create a Sensitivity Report for integer programming problems.

Terms

Feasible Region In integer programming problems, the feasible region is the set of lattice points that satisfy all the constraints.

Integer Programming Integer programming is another example of mathematical programming. It has the added constraint that all decision variables can only take on integer values.

Kernel The kernel is the set of lattice points in the feasible region of an integer programming problem that are candidates for the optimal solution, because they are closest to the boundary of the feasible region. For example, in a two-dimensional maximization problem, the kernel is the set of lattice points that have the largest x_1-coordinate for each x_2-value and the largest x_2-coordinate for each x_1-value.

Lattice Point On a two-dimensional plane, a lattice point is any point whose x- and y-coordinates are both integers.

Logistics The process of planning, carrying out, and controlling the efficient, effective flow and storage of goods, services, and related information from point of origin to point of consumption to meet the needs of the final consumer.

Chapter 5 (Integer Programming) Objectives

You should be able to:

- Identify the objective of the problem

- Identify and define the decision variables

- Write the objective function

- Identify the limited resources involved in the problem

- Write the functional constraints as inequalities

- Enter the problem formulation into Excel

- Set up Solver Parameters and Options

- Interpret the optimal solution in the context of the problem

- Analyze the Answer Report

Chapter 5 Study Guide

1. What is the objective function?

2. What are decision variables?

3. What is different about the decision variables in an integer programming (IP) problem compared to a linear programming (LP) problem?

4. In an LP problem with just two decision variables, the feasible region is part of a plane bounded by lines representing the constraints. Each point in that region satisfies all the constraints and is thus a feasible solution. Compare and contrast the notion of the *feasible region* in an LP problem with that of an IP problem.

5. What are functional constraints?

6. Besides functional constraints, describe two other types of constraints found in IP problems.

7. What information is found in the Answer Report for an IP problem?

8. In an IP problem, is it possible for an integer constraint to be nonbinding? Explain why or why not.

9. To what can the "Final Value" on the Answer Report refer?

10. Define *slack*.

11. In the Answer Report, what information is in the "Cell Value" column?

CHAPTER 6:

Optimize Selection with Binary Programming

Chapter 6 Optimize Selection with Binary Programming

Section 6.0: Introduction

Binary programming is a form of integer programming. The word "binary" refers to the decision variables. When the decision variables are binary, this means that they can only take on the values of either zero or one. That might seem overly restrictive, but there are many situations that can easily be modeled using binary decision variables. For example, the following decisions could be modeled with binary decision variables:
- Should we locate a new automobile dealership at this location?
- Should I choose to apply to this college?
- Should I invest in this stock?

1. What are the possible answers to each of these questions?

2. How could 0 and 1 be used to model those answers?

To explore the idea of binary decision variables further, a brief introductory example is given. In this example, Jarvis is choosing whether to work on two projects.

Section 6.0.1: Jarvis Selects Projects

Jarvis is considering working on two short-term projects, one with a $5,000 profit and the other with a $7,000 profit. Project 1 requires seven days, and Project 2 requires 11 days.

To formulate this problem, Jarvis needs to define some decision variables. A binary decision variable is an integer decision variable that has been further restricted to just two values, zero and one. Jarvis will use two **binary decision variables**, one for each project. A value of one means he decides to work on the project; a value of zero means he decides not to.

Jarvis develops the following problem formulation:

Decision Variables
  Let: x_1 = the binary decision variable for Project 1
        x_2 = the binary decision variable for Project 2

Objective Function
  Maximize: $z = 5{,}000x_1 + 7{,}000x_2$, where z = the amount of profit

Constraints
  Subject to:
    Time Available: $7x_1 + 11x_2 \leq A$, where A is the number of days Jarvis is available to work on the projects
    Binary Decision Variables: x_1 and x_2 equal 0 or 1

3. What are the four possible feasible solutions for this problem?

Lead Authors: Thomas Edwards and Kenneth Chelst

4. What is the value of the objective function for each of these possible solutions?

In order to determine the optimal solution, Jarvis needs to know how many days he has available to work on the projects. He considers several possible values for A. Suppose $A = 5$ days. In this case, there is not enough time to complete either project, so the optimal solution is $x_1 = 0$ and $x_2 = 0$.

5. For what other values of A is $x_1 = 0$ and $x_2 = 0$ the optimal solution?

6. What is the smallest value of A that changes the optimal solution?
    a. What is the new optimal solution?
    b. What other values of A generate the same optimal solution?

7. What is the smallest value of A that causes the optimal solution to change again?
    a. What is the optimal solution for that value of A?
    b. What other values of A produce that same optimal solution?

8. What is the smallest value of A that changes the optimal solution a third time?
    a. What is that new optimal solution?
    b. What other values of A produce that same optimal solution?

9. Explain why continuing to increase A cannot change the optimal solution again.

As you can see, sometimes increasing the value of A does not change the optimal solution. Other times, increasing the value of A does change the optimal solution. In this case, the constraint is non-binding.

For example, consider $A = 10$ days. That is, Jarvis has 10 days available to work on the projects. In a *linear programming* example, there would be no value in increasing this resource above 10 days (assuming it stays within the allowable increase for the constraint). Furthermore, reducing this resource by up to 3 days would also not impact the optimal solution. As a result, a $0 shadow price would be reported because this constraint is non-binding.

In *linear programming*, the decision variables are continuous. Thus, if there's slack, getting more of the resources does not increase the value.

In *binary programming*, if there's slack, getting more resources may or may not increase the optimal value of the objective function. That is, if Jarvis has 11 days—rather than 10 days—available, the optimal solution changes. However, if he has 10 days—rather than 9 days—available, the optimal solution does not change.

In other words, if A were 10, the optimal solution clearly would be to do Project 1, as there is not enough time to do Project 2. The constraint would not be binding because it will have three days of slack. However, there would be value in increasing the number of days available from 10 days to 11 days. If that were done, Jarvis could now complete Project 2 and earn $7,000 instead of only $5,000, and the constraint would then be binding. For this reason, Solver does not generate Sensitivity Reports for binary programming problems, like integer programming problems.

Since one extra day could potentially earn him $7,000 rather than $5,000, Jarvis may try to reschedule his time to provide an extra day to work on the project.

10. What is the marginal value of the extra day of work?

Section 6.1: Flipping Houses

Dream Homes, Inc. is a company that buys houses that are in need of repair. The houses are renovated, updated, and then sold for a profit. This process is referred to as *house flipping*.

Mr. Dale, a retired realtor, owns Dream Homes, Inc. and wants to provide a summer job for each of his five grandchildren, all of whom are in college. Mr. Dale would pay each grandchild by dividing the profits evenly amongst them.

Each of Mr. Dale's five grandchildren—Ani, Benita, Cameron, Dante, and Edwin—has specific skills appropriate for house restoration. The skills for each grandchild are shown in Table 6.1.1. Notice that all the grandchildren are capable of cleaning.

Grandchild	Skills				
	Plumbing	Painting	Landscaping	Carpentry	Cleaning
Ani	✓			✓	✓
Benita	✓		✓		✓
Cameron		✓		✓	✓
Dante		✓	✓		✓
Edwin			✓	✓	✓

Table 6.1.1: List of each grandchild and their skills

Mr. Dale has identified ten available houses on the market. Each house has an estimated profit margin. This estimate is based on the difference between the total cost and the estimated resale value of the house. The total cost includes the purchase price, the cost of necessary materials, and the expense of hiring any outside contractor assistance. Then, this total cost is subtracted from the estimated resale value of the house to find the estimated total profit. Mr. Dale's goal is to earn the highest possible total profit.

The grandchildren are all college students, so they have just 12 weeks in which to work. They are focused on the goal of acquiring the most lucrative houses so they can earn the most profit. The grandchildren first need to select which houses to flip in order to yield the highest profit. They have all agreed to work 8 hours a day, 6 days a week.

Each house requires the appropriate skill set of the grandchildren (i.e., plumbing, carpentry, landscaping, painting, and cleaning). Mr. Dale needs to determine which house, or set of houses, will allow the grandchildren to earn the largest profit over the course of the summer.

6.1.1: Problem Formulation

Mr. Dale decides the maximum amount that he can invest in this summer venture for his grandchildren's benefit is $500,000. This includes the costs to purchase the homes and the materials, and to pay any contractors' fees. Since he plans on investing in more than one house, the question becomes: in which of the ten available houses should Mr. Dale invest to maximize the profit for his grandchildren?

Chapter 6 — Optimize Selection with Binary Programming

The formulation of this problem utilizes **binary decision variables**. A binary decision variable is defined for each house under consideration:
- A value of 1 indicates that a particular house will be purchased
- A value of 0 indicates that particular house will not be purchased.

Mr. Dale lets the decision variable x_1 represent whether or not House 1 is purchased; he lets the decision variable x_2 represent whether or not House 2 is purchased, and so on. For example, if the value of x_1 is 1, then Mr. Dale will purchase House 1; if the value of x_1 is 0, then Mr. Dale will not purchase House 1. In general, he will let x_i represent the binary decision variable for House i. Here is another way of stating this definition for the decision variables:

$$\text{Let } x_i = \begin{cases} 1 & \text{if Mr. Dale purchases House } i \\ 0 & \text{if Mr. Dale does not purchases House } i \end{cases}$$

Furthermore, Mr. Dale lets the constant P_1 represent the estimated profit of House 1; he lets the constant P_2 represent the estimated profit of House 2, and so on. Thus, the objective function is:

$$z = P_1 \cdot x_1 + P_2 \cdot x_2 + P_3 \cdot x_3 + P_4 \cdot x_4 + P_5 \cdot x_5 + P_6 \cdot x_6 + P_7 \cdot x_7 + P_8 \cdot x_8 + P_9 \cdot x_9 + P_{10} \cdot x_{10}.$$

1. Explain what the value of z represents in this problem context. How do you know?

A more compact way to write this objective function uses summation notation: $z = \sum_{i=1}^{10} P_i \cdot x_i$.

(See Appendix A for an explanation of summation notation.)

Mr. Dale seeks to maximize z subject to constraints related to total cost and available labor.

2. What constraints must Mr. Dale consider?

Mr. Dale is constrained by the total cost. The total cost restriction takes into consideration the cost to purchase the home, the costs for necessary material, and the costs to any outside contractors. This total cannot exceed $500,000.

Mr. Dale lets:
V_i = the expected resale value of House i,
H_i = the cost to purchase House i,
M_i = the cost of materials for House i, and
C_i = any contractors' fees for House i.

Then, the *estimated* total cost for House i, is $H_i + M_i + C_i$. "House i" is a general label for any of the houses. The subscript i could be any number between 1 and 10, inclusive.

Thus, the estimated net profit for House i, is $P_i = V_i - (H_i + M_i + C_i)$.

The actual total cost for that particular house depends on whether they decide to purchase it. The binary decision variable x_i equals 1 if the house is purchased and 0 if not. Therefore, multiplying the potential total cost by x_i gives that actual total cost for a particular house: $(H_i + M_i + C_i) \cdot x_i$. For example, if Mr. Dale purchase House 2, the total cost for this house is $(H_2 + M_2 + C_2) \cdot 1 = H_2 + M_2 + C_2$. On the other hand, if Mr. Dale does not purchase House 2, the total cost is: $(H_2 + M_2 + C_2) \cdot 0 = 0$

That is, it will not cost Mr. Dale anything if he does not purchase the house.

Finally, using summation notation, the total cost constraint is:

$$\sum_{i=1}^{10}(H_i + M_i + C_i) \cdot x_i \leq 500,000.$$

3. In your own words, explain the meaning of this constraint.

Next, Mr. Dale writes the constraints on the number of hours of labor available for each type of work (i.e., plumbing, carpentry, landscaping, painting, and cleaning). To do so, he must analyze the availability and skills of each grandchild. Each grandchild has committed to working a maximum of 8 hours a day, 6 days a week, for 12 weeks during summer vacation. This amounts to each grandchild being able to work a maximum of 576 hours.

Table 6.1.1 showed the skills of each grandchild. There are only two grandchildren skilled at plumbing (Ani and Benita) and each is available for a maximum of 576 hours. Mr. Dale must ensure that the required plumbing hours are less than or equal to $2 \cdot 576 = 1,152$ hours.

Similarly, the total painting hours cannot exceed 1,152 hours because only two grandchildren are skilled at painting (Cameron and Dante).

There are three grandchildren who are skilled at landscaping as well as carpentry, so he can assign a maximum of $3 \cdot 576 = 1,728$ hours to each of those trades.

Finally, all the grandchildren are able to clean. The hours for plumbing, painting, landscaping, and carpentry have been accounted for. Therefore, Mr. Dale only needs to ensure that the total number of hours does not exceed $5 \cdot 576 = 2,880$, the number of hours his grandchildren are available to work.

Given all of these factors, Mr. Dale lets:
B_i = the number of plumbing hours required for House i,
N_i = the number of painting hours required for House i,
L_i = the number of landscaping hours required for House i,
R_i = the number of carpentry hours required for House i,
Z_i = the number of cleaning hours required for House i,
T_i = the total number of labor hours required for House i.

Then, he has the following labor constraints:

plumbing labor constraint $\sum_{i=1}^{10} B_i \cdot x_i \leq 1,152$ hours,

painting labor constraint $\sum_{i=1}^{10} N_i \cdot x_i \leq 1,152$ hours,

landscaping labor constraint $\sum_{i=1}^{10} L_i \cdot x_i \leq 1,728$ hours, and

carpentry labor constraint $\sum_{i=1}^{10} R_i \cdot x_i \leq 1,728$ hours.

There is also a constraint on the total amount of time available:

total labor time available $\sum_{i=1}^{10} T_i \cdot x_i \leq 2,880$ hours.

Notice that the total cleaning hours is not really a constraint by itself, because it is included in the constraint that restricts the total hours worked. Each of the five grandchildren can work a maximum of 576 hours, and thus the total amount of labor spread over all areas cannot exceed 5 · 576 = 2,880 hours.

4. In your own words, explain why there is no need for a constraint on cleaning hours.

In order to complete this formulation, Mr. Dale adds the information about the houses he may purchase. First, Table 6.1.2 shows the estimates of costs and total potential value for each house.

($)	House 1	House 2	House 3	House 4	House 5	House 6	House 7	House 8	House 9	House 10
Expected Value (V_i)	125,000	135,000	178,000	110,000	108,000	124,000	244,000	192,000	130,000	275,000
Purchase Price (H_i)	65,000	100,000	125,000	70,000	35,000	99,000	140,000	115,000	88,000	129,000
Materials Cost (M_i)	5,000	3,000	3,750	5,500	7,500	2,000	8,000	6,000	7,000	16,000
Contractors' Fees (C_i)	6,000	3,000	8,000	1,000	17,500	0	10,000	9,300	0	22,000

Table 6.1.2: Value and cost of each house

5. Calculate the net profit for each house by adding up all the costs and subtracting this value from the potential value of the house.

6. Based on this information, predict which houses Mr. Dale will purchase. Explain your reasoning.

Next, Mr. Dale collects estimates about the labor requirements for each house. This information is shown in Table 6.1.3.

(hours)	House 1	House 2	House 3	House 4	House 5	House 6	House 7	House 8	House 9	House 10
Plumbing (B_i)	238	211	264	145	211	100	400	422	185	304
Painting (N_i)	150	125	115	80	130	50	250	160	100	200
Landscaping (L_i)	210	175	161	112	182	70	350	224	140	280
Carpentry (R_i)	264	330	143	242	165	180	220	385	330	396
Cleaning (Z_i)	300	280	310	410	335	200	350	325	390	325
Total (T_i)	1,162	1,121	993	989	1,023	600	1,570	1,516	1,145	1,505

Table 6.1.3: Labor hours for each house

7. Write the complete problem formulation.

6.1.2: Neighborhood Constraints

Mr. Dale decided to map the location of each of these houses. Houses 1 through 6 were in the Downriver neighborhood of the city. The neighborhood was stable but not especially attractive to home buyers. He was concerned about selling all the homes he fixed up if he bought too many Downriver houses. He therefore added the restriction that he would purchase no more than two Downriver houses. All of the decision variables are binary. As a result it is relatively easy to represent this restriction algebraically.

$$x_1 + x_2 + x_3 + x_4 + x_5 + x_6 \leq 2$$

The other four houses were all in the Newling neighborhood. This neighborhood was attracting a great deal of interest from first time home buyers. He, therefore, decided that he wanted to purchase at least one home from this neighborhood. This requirement is captured by the following constraint.

$$x_7 + x_8 + x_9 + x_{10} \geq 1$$

6.1.3: Solving the Problem

Next, Mr. Dale creates a spreadsheet with the problem formulation from the previous section. The spreadsheet is set up in a way similar to the previous chapters. However, Mr. Dale needs to add an additional constraint to tell Solver that the decision variables are binary. To do so, he chooses "bin" from the drop-down menu, as shown in Figure 6.1.1.

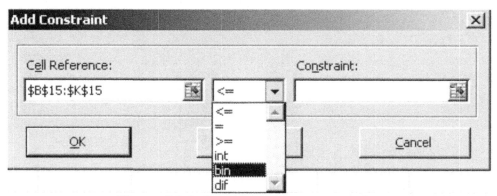

Figure 6.1.1: Specifying binary constraints

Mr. Dale's solution is shown in Figure 6.1.2. Use this spreadsheet to answer the following questions.

8. What is the largest profit the grandchildren can make during the summer flipping houses?

9. Which houses will be purchased to produce this profit?

10. How much money will each person receive if the profit is divided equally among the grandchildren?

11. Are any of the neighborhood restrictions binding?

The optimal solution identifies two houses to be purchased and flipped to maximize profits. One of these houses has the largest margin for profit. But the second home in the optimal solution is fourth out of the possible ten homes when ranked based on profit margin.

12. Why do you suspect this has happened?

In binary programming, decision variables change in units of one. That means they consume resources in discrete increments. To explore this phenomenon, Mr. Dale changes the values of two decision variables, without resolving the problem. He changes the binary decision value for house 1 from a 1 to a 0. Then he changes the binary decision value for house 7—the second most profitable house—from a 0 to a 1. These changes are shown in Figure 6.1.3. (The neighborhood restrictions had no effect on any of the analyses that follow. They were left out of the figures to save space.)

13. Looking at the spreadsheet in Figure 6.1.3, explain why Mr. Dale cannot purchase House 10 and House 7.

6.1 House Flipping
BIP - Max

	Variable		House 1	House 2	House 3	House 4	House 5	House 6	House 7	House 8	House 9	House 10			
5	V_i	Expected Value ($)	125,000	135,000	178,000	110,000	108,000	124,000	244,000	192,000	130,000	275,000			
6	H_i	Purchase Price ($)	65,000	100,000	125,000	70,000	35,000	99,000	140,000	115,000	88,000	129,000			
7	M_i	Materials Cost ($)	5,000	3,000	3,750	5,500	7,500	2,000	8,000	6,000	7,000	16,000			
8	C_i	Contractors Cost ($)	6,000	3,000	8,000	1,000	17,500	0	10,000	9,300	0	22,000			
9		Total Cost ($)	76,000	106,000	136,750	76,500	60,000	101,000	158,000	130,300	95,000	167,000			
10	P_i	Net Profit ($)	49,000	29,000	41,250	33,500	48,000	23,000	86,000	61,700	35,000	108,000			
13	x_j	Decision Variables	x1	x2	x3	x4	x5	x6	x7	x8	x9	x10			
14		Dec. Var. Values	1	0	0	0	0	0	0	0	0	1		Total Profit ($)	
16	P_i	Objective Function ($)	49,000	29,000	41,250	33,500	48,000	23,000	86,000	61,700	35,000	108,000		157,000	
18		Constraints													
19		Budget ($)	76,000	106,000	136,750	76,500	60,000	101,000	158,000	130,300	95,000	167,000		243,000	≤	500,000
20	B_i	Plumbing (hrs)	238	211	264	145	211	100	400	422	185	304		541	≤	1,152
21	N_i	Painting (hrs)	150	125	115	80	130	50	250	160	100	200		350	≤	1,152
22	L_i	Landscaping (hrs)	210	175	161	112	182	70	350	224	140	280		490	≤	1,728
23	R_i	Carpentry (hrs)	264	330	143	242	165	180	220	385	330	396		660	≤	1,728
24	Z_i	Cleaning (hrs)	300	280	310	410	335	200	350	325	390	325				
25	T_i	Total Labor (hrs)	1,162	1,121	993	989	1,023	600	1,570	1,516	1,145	1,505		2,666	≤	2,880

Figure 6.1.2: Flipping houses spreadsheet with the optimal solution

6.1 House Flipping
BIP - Max

Variable			House 1	House 2	House 3	House 4	House 5	House 6	House 7	House 8	House 9	House 10		
5	V_i	Expected Value ($)	125,000	135,000	178,000	110,000	108,000	124,000	244,000	192,000	130,000	275,000		
6	H_i	Purchase Price ($)	65,000	100,000	125,000	70,000	35,000	99,000	140,000	115,000	88,000	129,000		
7	M_i	Materials Cost ($)	5,000	3,000	3,750	5,500	7,500	2,000	8,000	6,000	7,000	16,000		
8	C_i	Contractors Cost ($)	6,000	3,000	8,000	1,000	17,500	0	10,000	9,300	0	22,000		
9		Total Cost ($)	76,000	106,000	136,750	76,500	60,000	101,000	158,000	130,300	95,000	167,000		
10	P_i	Net Profit ($)	49,000	29,000	41,250	33,500	48,000	23,000	86,000	61,700	35,000	108,000		
13	x_i	Decision Variables	x1	x2	x3	x4	x5	x6	x7	x8	x9	x10		
14		Dec. Var. Values	0	0	0	0	0	0	1	0	0	1		
													Total Profit ($)	
16	P_i	Objective Function ($)	49,000	29,000	41,250	33,500	48,000	23,000	86,000	61,700	35,000	108,000	194,000	
Constraints														
19		Budget ($)	76,000	106,000	136,750	76,500	60,000	101,000	158,000	130,300	95,000	167,000	325,000	≤ 500,000
20	B_i	Plumbing (hrs)	238	211	264	145	211	100	400	422	185	304	704	≤ 1,152
21	N_i	Painting (hrs)	150	125	115	80	130	50	250	160	100	200	450	≤ 1,152
22	L_i	Landscaping (hrs)	210	175	161	112	182	70	350	224	140	280	630	≤ 1,728
23	R_i	Carpentry (hrs)	264	330	143	242	165	180	220	385	330	396	616	≤ 1,728
24	Z_i	Cleaning (hrs)	300	280	310	410	335	200	350	325	390	325		
25	T_i	Total Labor (hrs)	1,162	1,121	993	989	1,023	600	1,570	1,516	1,145	1,505	3,075	≤ 2,880

Figure 6.1.3: Flipping houses spreadsheet with Houses 7 and 10 chosen

As a result, Mr. Dale has identified the fact that the group's ability to make a profit is restricted by the total available hours, even though the constraint is not binding. He explores this restriction to see whether it is advantageous to hire a cleaning crew.

To explore this issue, Mr. Dale increases the available hours in 100-hour increments until he sees a change in the optimal solution. The first time a change occurs is when the total labor hours is 3,080 (a 200-hour increase). The new optimal solution is shown in Figure 6.1.4.

14. What is the new largest profit the grandchildren can make during the summer flipping houses?

15. How much money will each person receive if the profit is divided equally among the grandchildren?

16. Which houses will be purchased to produce this profit?

17. Is this new optimal solution what you expected? Why or why not?

18. Would you recommend that Mr. Dale hires a cleaning crew so that he has an extra 200 labor hours available?

19. Suppose Mr. Dale decides to hire a cleaning crew. How much per hour should he pay the cleaning crew? Explain your reasoning.

Chapter 6
Finding Optimal Solutions—Binary Programming

	A	B	C	D	E	F	G	H	I	J	K	L	M	N	O
1		BIP - Max													
2	Variable														
3															
4			House 1	House 2	House 3	House 4	House 5	House 6	House 7	House 8	House 9	House 10			
5	V_i	Expected Value ($)	125,000	135,000	178,000	110,000	108,000	124,000	244,000	192,000	130,000	275,000			
6	H_i	Purchase Price ($)	65,000	100,000	125,000	70,000	35,000	99,000	140,000	115,000	88,000	129,000			
7	M_i	Materials Cost ($)	5,000	3,000	3,750	5,500	7,500	2,000	8,000	6,000	7,000	16,000			
8	C_i	Contractors Cost ($)	6,000	3,000	8,000	1,000	17,500	0	10,000	9,300	0	22,000			
9		Total Cost ($)	76,000	106,000	136,750	76,500	60,000	101,000	158,000	130,300	95,000	167,000			
10	P_i	Net Profit ($)	49,000	29,000	41,250	33,500	48,000	23,000	86,000	61,700	35,000	108,000			
11															
12															
13	x_i	Decision Variables	x1	x2	x3	x4	x5	x6	x7	x8	x9	x10			
14		Dec. Var. Values	0	0	0	0	0	0	1	0	0	1			
15													Total Profit ($)		
16	P_i	Objective Function ($)	49,000	29,000	41,250	33,500	48,000	23,000	86,000	61,700	35,000	108,000	194,000		
17															
18		Constraints													
19		Budget ($)	76,000	106,000	136,750	76,500	60,000	101,000	158,000	130,300	95,000	167,000	325,000	≤	500,000
20	B_i	Plumbing (hrs)	238	211	264	145	211	100	400	422	185	304	704	≤	1,152
21	N_i	Painting (hrs)	150	125	115	80	130	50	250	160	100	200	450	≤	1,152
22	L_i	Landscaping (hrs)	210	175	161	112	182	70	350	224	140	280	630	≤	1,728
23	R_i	Carpentry (hrs)	264	330	143	242	165	180	220	385	330	396	616	≤	1,728
24	Z_i	Cleaning (hrs)	300	280	310	410	335	200	350	325	390	325			
25	T_i	Total Labor (hrs)	1,162	1,121	993	989	1,023	600	1,570	1,516	1,145	1,505	3,075	≤	3,080

Figure 6.1.4: Optimal solution when right hand side of Total Labor constraint is increased by 200 hours

Section 6.2: Sam Johnson Makes a Hard Decision

Sam Johnson lives in Brooklyn, New York. He is deciding which colleges to apply to this fall. He took the SAT, and his total score was 1650. He wants to find colleges that are more than 50 miles from home but no more than 300 miles from home. He also wants to find colleges with acceptance rates of at least 50%. Sam has lived in a large city all of his life and enjoys life in the city, so he would like to find a college in or near an urban setting. He also prefers a medium-sized college.

Sam decides to do some Internet searches to learn more about potential colleges that satisfy his criteria. Along with the web sites for individual colleges, he finds these resources useful:
- http://www.uscollegesearch.org/
- http://www.collegeview.com/collegesearch/index.jsp
- http://cnsearch.collegenet.com/cgi-bin/CN/index
- https://bigfuture.collegeboard.org/college-search

Sam decides that the most important criteria to him are distance from home, tuition, average debt at graduation, acceptance rate, and size. Table 6.2.1 contains Sam's short list of 11 colleges that fit his criteria. His problem is that his mom and dad will pay the application fee for only five colleges. Sam needs to decide where he should apply.

College	Location	Distance from Home	Tuition	Average Debt at Graduation	Acceptance Rate	Student Population
Western Connecticut State University	Danbury, CT	55 miles	$15,344	No data available	57%	6,001
Rutgers: Camden Regional Campus	Camden, NJ	80 miles	$18,263	$18,645	53%	49,760
Drexel University	Philadelphia, PA	97 miles	$28,780	No data available	76%	20,821
College of Saint Rose	Albany, NY	141 miles	$20,620	$24,732	68%	5,000
Johnson & Wales University	Providence, RI	155 miles	$21,717	$19,890	81%	16,095
Worcester State College	Worcester, MA	157 miles	$11,619	$13,742	56%	5,470
Loyola College in Maryland	Baltimore, MD	169 miles	$34,250	$16,073	64%	6,131
University of Mass. Boston	Boston, MA	190 miles	$19,977	$14,805	63%	13,433
Suffolk University	Boston, MA	190 miles	$24,250	No data available	69%	5,196
University of Mass. Lowell	Lowell, MA	194 miles	$19,714	$14,833	70%	11,635
Syracuse University	Syracuse, NY	197 miles	$31,686	$24,000	51%	19,082

Table 6.2.1: Sam's list of 11 colleges meeting his criteria with their data

6.2.1: Developing the Constraints

As seen in Table 6.2.1, Sam collected data on the criteria most important to him: distance from home, tuition, average debt at graduation, acceptance rate, and size. However, he is not able to learn the average debt at graduation for several of the colleges on his list, so he decides not to use that criterion in his selection process.

Now, Sam wants to decide which colleges to apply to. He bases this decision on the colleges' acceptance rates and on attractiveness scores that he will determine. The attractiveness scores will be based on each college's distance from his home, its tuition, and some personal preferences.

To calculate the attractiveness score for each college, Sam relies on some techniques used in the Multi-Criteria Decision Making process from Chapter 1. Specifically, he performs the following steps:
1. Rescale scores by assigning a score between zero (worst) and one (best) for each criterion.
2. Weight scores by multiplying the criterion scores by their assigned weight.
3. Add these three weighted scores to obtain the attractiveness score for each college.

First, Sam rescales the scores for each criterion. Table 6.2.2 shows how he will assign the scores for distance, tuition, and size.

Distance (weight = 0.2)		Tuition (weight = 0.2)		Size (weight = 0.2)	
Range (miles)	Score	Range ($)	Score	Range (students)	Score
50 – 90	0	< 10,000	1	< 5,000	0
91 – 130	0.33	10,000 – 14,999	0.67	5,000 – 9,999	0.8
131 – 170	0.67	15,000 – 19,999	0.33	10,000 – 19,999	1
≥ 171	1	≥ 20,000	0	≥ 20,000	0.2

Table 6.2.2: Sam's scheme for assigning scores

Sam also creates personal preference scores based on what he learned from his Internet searches about the quality of life on each campus. These scores are entirely subjective. They reflect Sam's personal opinions and values that will affect where he will attend college. They are shown in Table 6.2.3.

College	Personal Preference Score
Western Connecticut State University	0.14
Rutgers: Camden Regional Campus	0.71
Drexel University	0.61
College of Saint Rose	0.63
Johnson & Wales University	0.33
Worcester State College	0.44
Loyola College in Maryland	0.40
University of Massachusetts Boston	0.81
Suffolk University	0.60
University of Massachusetts Lowell	0.72
Syracuse University	0.53

Table 6.2.3: Sam's personal preference scores for each college

Second, Sam wants to give equal weight to the objective criteria (distance from home, tuition, and size) and a higher weight to his personal preferences. He decides to weight distance, tuition, and size at 20% each and his personal preferences at 40%.

Third, Sam computes the attractiveness score for each college as follows:
(0.2)(distance score) + (0.2)(tuition score) + (0.2)(size score) + (0.4)(personal preference score).

Table 6.3.4 shows the original data Sam collected, the rescaled scores, and the total attractiveness scores for each of the 11 colleges he is considering. Rescaling scores to a common unit takes the data from Table 6.2.1 and the ranges from Table 6.2.2 into account. The personal preference scores were given in Table 6.2.3. The three steps described in the preceding paragraphs outline the process of transforming raw data from a variety of measures into a single attractiveness scores for each college.

College	Raw Data			Rescaled Scores				Weighted Sum
	Distance from home (miles)	Tuition ($)	Enrollment (students)	Distance score (20%)	Tuition score (20%)	Size score (20%)	Personal preference score (40%)	Attractiveness score
Western Connecticut State University	58	15,344	6,001	0	0.33	0.8	0.14	0.28
Rutgers: Camden Regional Campus	78	18,263	49,760	0	0.33	0.2	0.61	0.35
Drexel University	97	28,780	20,821	0.33	0	0.2	0.71	0.39
College of Saint Rose	141	20,620	5,000	0.67	0	0.8	0.63	0.55
Johnson & Wales University	155	21,717	16,095	0.67	0	1	0.33	0.46
Worcester State College	157	11,619	5,470	0.67	0.67	0.8	0.44	0.60
Loyola College in Maryland	169	34,250	6,131	0.67	0	0.8	0.40	0.45
University of Massachusetts Boston	189	19,977	13,433	1	0.33	1	0.81	0.79
Suffolk University	191	24,250	5,196	1	0	0.8	0.72	0.65
University of Massachusetts Lowell	194	19,714	11,635	1	0.33	1	0.53	0.68
Syracuse University	197	31,686	19,082	1	0	0.2	0.60	0.48

Table 6.2.4: Data, rescaled scores, and weighted sum to determine attractiveness scores

Consider the row of information for Western Connecticut State University (WCSU) as an example. WCSU is 58 miles from Brooklyn, this falls in the range 50-90 miles, which was Sam's least preferred range. The rescaled score for WCSU's distance from home is therefore zero. Its tuition of $15,344 per year is in the $15,000-$19,999 range, and Sam assigns any value in that range a score of 0.33. The enrollment at WCSU is 6,001 students. Sam likes medium-sized schools, so he gives any school whose enrollment is between 5,000 and 9,999 students the relatively high score of 0.8. Sam gave the personal preference score of 0.14 to WCSU based on the impression he formed of the University while conducting his research. The weights for each of these measures, which were determined based on Sam's values and priorities, are multiplied by their respective rescaled scores and then added. The weighted sum of the rescaled scores is calculated as follows to determine the attractiveness score for WCSU.

$$(0.2 \cdot 0) + (0.2 \cdot 0.33) + (0.2 \cdot 0.8) + (0.4 \cdot 0.14) = 0.28$$

Sam will determine which colleges to apply to based on these attractiveness scores as well as on the acceptance rates of the colleges. Sam wants to be fairly certain that he gets accepted into at least two of the colleges to which he is applying. To ensure this happens, he creates a constraint that will force at least two of the colleges in his portfolio to have acceptance rates of at least 70%. Furthermore, Sam is seeking balance in acceptance rates. He wants to apply to colleges where he will be challenged, so he creates a constraint requiring the average of all the acceptance rates at the colleges where he will apply to be less than 65%. Taken together, these two constraints will ensure Sam applies to a portfolio of colleges that is balanced in their academic rigor.

In a similar way, Sam could balance his preferences for distance from home, size, and cost of the colleges. He notes that he could set up additional constraints to ensure that he is applying to a wide range of colleges when considering any criterion, but he decides not to do so at this time.

Below are the constraints on Sam's portfolio of colleges that he decides to use:
- At least *two* colleges with acceptance rates of at least 70% must be selected.
- The average acceptance rate of all the selected colleges must be less than 65%.
- He will apply to exactly *five* colleges.

6.2.2: Formulating the Problem

To formulate this problem, Sam must define some variables. The first variables he defines are **binary decision variables.** He uses binary decision variables to determine which colleges to apply to. As shown in the previous sections, a binary decision variable is an integer decision variable that has been further restricted to just two values: 0 or 1. Sam will use 11 binary decision variables, one for each college he is considering. Sam's problem formulation will maximize his objective function, subject to constraints. When the problem is solved, the binary decision variables for those colleges that will appear in Sam's portfolio (i.e., the colleges Sam *will* apply to) will be assigned a value of one. The binary decision variables for those colleges that will not appear in Sam's portfolio (i.e., the colleges Sam *will not* apply to) will be assigned a value of zero.

Therefore, Sam defines the following decision variables:
Let:
x_1 = the binary decision variable for College 1 (Western Connecticut State U)
x_2 = the binary decision variable for College 2 (Rutgers: Camden Rgnl. Campus)
x_3 = the binary decision variable for College 3 (Drexel University)
x_4 = the binary decision variable for College 4 (College of Saint Rose)
x_5 = the binary decision variable for College 5 (Johnson & Wales University)
x_6 = the binary decision variable for College 6 (Worcester State College)
x_7 = the binary decision variable for College 7 (Loyola College in Maryland)

x_8 = the binary decision variable for College 8 (University of Mass. Boston)
x_9 = the binary decision variable for College 9 (Suffolk University)
x_{10} = the binary decision variable for College 10 (University of Mass. Lowell)
x_{11} = the binary decision variable for College 11 (Syracuse University)

More generally, we could say,

let $x_i = \begin{cases} 1 \text{ if Sam will apply to College } i \\ 0 \text{ if Sam does not apply to College } i \end{cases}$

To organize his data, Sam also defines some constants for acceptance rate, size, tuition, distance from home, and attractiveness. These constants will eventually be used as coefficients in the objective function and constraints. Table 6.2.5 lists these constants.

Constant	Description
R_i	Acceptance rate (%) for College i
S_i	Size (# of students) of College i
T_i	Tuition ($) for College i
H_i	Distance (miles) from home to College i
A_i	Attractiveness score for College i

Table 6.2.5: The constants Sam will use

Finally, Sam defines a **binary indicator coefficient**. Like binary decision variables, a binary indicator coefficient is restricted to two values: 0 or 1. To develop a binary indicator coefficient, Sam thinks about checking each of the 11 colleges in his list against the list of the three constraints appearing at the end of the previous section. For each college, its data must be checked to see if it meets each of the constraints.

For example, the first constraint is that at least two of the colleges selected have acceptance rates greater than or equal to 70%. So, for each college, he must ask the question, "Is the acceptance rate greater than or equal to 70%?" This is a yes-or-no question, and this is where binary indicator coefficients come into play. If the answer is yes, the binary indicator will be assigned a value of 1, but if the answer is no, the binary indicator will be assigned a value of 0.

Sam defines the following binary indicator coefficient:

Let: $_bR_i = \begin{cases} 1 \text{ if College } i \text{ has an acceptance rate of at least 70\%} \\ 0 \text{ if College } i \text{ does not have an acceptance rate of at least 70\%} \end{cases}$

He uses the b subscript to the left of the R_i to show it is a binary indicator coefficient.

These binary indicators will be used to count the number of colleges in the portfolio that satisfy the constraint. Referring to Table 6.2.1, the first college on the list is Western Connecticut State University, and its acceptance rate (R_1) is 57%. So it does not meet the first constraint. Therefore, the binary indicator is assigned the value zero: $_bR_1 = 0$. Similarly, $_bR_2 = 0$ because Rutgers's acceptance rate is also less than 70%. However, the third college in the list, Drexel University, has an acceptance rate of 76%. Therefore, $_bR_3 = 1$. The values of $_bR_i$ are shown in Table 6.2.6.

College	Acceptance Rate (%)	Binary Indicator Coefficient
Western Connecticut State University	$R_1 = 57$	$_bR_1 = 0$
Rutgers: Camden Regional Campus	$R_2 = 53$	$_bR_2 = 0$
Drexel University	$R_3 = 76$	$_bR_3 = 1$
College of Saint Rose	$R_4 = 68$	$_bR_4 = 0$
Johnson & Wales University	$R_5 = 81$	$_bR_5 = 1$
Worcester State College	$R_6 = 56$	$_bR_6 = 0$
Loyola College in Maryland	$R_7 = 64$	$_bR_7 = 0$
University of Massachusetts Boston	$R_8 = 63$	$_bR_8 = 0$
Suffolk University	$R_9 = 69$	$_bR_9 = 0$
University of Massachusetts Lowell	$R_{10} = 70$	$_bR_{10} = 1$
Syracuse University	$R_{11} = 51$	$_bR_{11} = 0$

Table 6.2.6: Binary indicator values for acceptance rate constraint

To complete the formulation, Sam defines an objective function that takes into account the acceptance rates (R_i) and attractiveness scores (A_i) of the selected colleges. The acceptance rates express the likelihood of Sam actually being admitted to the college. The attractiveness scores were based on tuition, distance from home, size, and personal preferences. If Sam just wanted to apply to the most attractive group of colleges he possibly could, he would simply maximize attractiveness. However, Sam wants to adjust what he maximizes to reflect the probability that he will actually be admitted to the schools where he applies. Multiplying the attractiveness score and the acceptance rate to be objective function coefficients will accomplish this. For example, Syracuse only accepts 51% of the students who apply. This is the lowest acceptance rate of any college Sam is considering. If Syracuse is included in Sam's portfolio, it will not add much to the objective function. This is because its attractiveness score is being "discounted" by the relatively low likelihood that Sam will be admitted there. Sam, therefore, writes his objective function as follows.

Maximize: $z = (R_1 \cdot A_1) \cdot x_1 + (R_2 \cdot A_2) \cdot x_2 + (R_3 \cdot A_3) \cdot x_3 + (R_4 \cdot A_4) \cdot x_4 + (R_5 \cdot A_5) \cdot x_5 + (R_6 \cdot A_6) \cdot x_6 + (R_7 \cdot A_7) \cdot x_7 + (R_8 \cdot A_8) \cdot x_8 + (R_9 \cdot A_9) \cdot x_9 + (R_{10} \cdot A_{10}) \cdot x_{10} + (R_{11} \cdot A_{11}) \cdot x_{11}$

As in the previous section, this sum can be written using summation notation:

$$z = \sum_{i=1}^{11} (R_i \cdot A_i) \cdot x_i.$$

This objective function sums 11 products, one for each college. Each product consists of three factors:
- the binary decision variable x_i representing whether a particular college is selected,
- that college's acceptance rate R_i, and
- that college's attractiveness score A_i.

The summation of these products represents the total value of Sam's portfolio. Sam is referring to the value of his portfolio as "desirability." Since exactly five colleges will be selected, six of the values of x_i will be 0. Thus, those resulting products will also be 0. The sum of the products $R_i \cdot A_i$ must be maximized to determine which colleges will be selected.

Finally, Sam's three constraints will be:
- The average acceptance rate of all the selected colleges must be at least 65%:

$$\frac{\sum_{i=1}^{11} R_i \cdot x_i}{5} \leq 0.65$$

- At least *two* colleges with acceptance rates greater than or equal to 70% must be selected:

$$\sum_{i=1}^{11} {}_bR_i \cdot x_i \geq 2$$

- Sam will apply to exactly five colleges:

$$\sum_{i=1}^{11} x_i = 5$$

In the first constraint, the average acceptance rate is the sum of the acceptance rates at the selected colleges $\left(\sum_{i=1}^{11} R_i \cdot x_i\right)$ divided by the number of colleges selected for the portfolio. In this case it is five.

In the second constraint, the binary indicator coefficient (${}_bR_i$) for each college is multiplied by the binary decision variable x_i for that college. Since both factors in this constraint are binary, every one of the products also is binary. If the particular college does not meet the constraint, then the left hand side of the constraint is zero. If the college does meet the constraint, then it will be one. Therefore, the sum of those 11 products is the number of colleges in the portfolio that meet the particular constraint (e.g., acceptance rate greater than or equal to 70%). Notice that the binary indicator ${}_bR_i$ is crucial to the formulation of this constraint.

In the last constraint, the sum of all of the binary decision variables x_i is just the number of colleges in the portfolio. This value must be five because Sam will apply to exactly five colleges.

Now, Sam has the complete problem formulation:

Decision Variables

Let: x_1 = the binary decision variable for College 1 (Western Conn. State University)
x_2 = the binary decision variable for College 2 (Rutgers: Camden Rgnl. Campus)
x_3 = the binary decision variable for College 3 (Drexel University)
x_4 = the binary decision variable for College 4 (College of Saint Rose)
x_5 = the binary decision variable for College 5 (Johnson & Wales University)
x_6 = the binary decision variable for College 6 (Worcester State College)
x_7 = the binary decision variable for College 7 (Loyola College in Maryland)
x_8 = the binary decision variable for College 8 (University of Mass. Boston)
x_9 = the binary decision variable for College 9 (Suffolk University)
x_{10} = the binary decision variable for College 10 (University of Mass. Lowell)
x_{11} = the binary decision variable for College 11 (Syracuse University)

Objective Function

Maximize: $z = \sum_{i=1}^{11}(R_i \cdot A_i)x_i$, where z = the total value of the portfolio

Constraints

Subject to:

The average acceptance rate of all the selected colleges should be at least 65%:

$$\frac{\sum_{i=1}^{11} R_i \cdot x_i}{5} \leq 0.65$$

At least two colleges with acceptance rates greater than or equal to 70% should be selected:

$$\sum_{i=1}^{11} {}_bR_i \cdot x_i \geq 2$$

Sam will apply to exactly five colleges:

$$\sum_{i=1}^{11} x_i = 5$$

6.2.3: Solving the Problem

The process of checking data against constraints and assigning values to the binary variables would be very tedious and time-consuming if done by hand. Fortunately spreadsheet solvers allow the user to define variables as binary, just as they allow the definition of integer variables. Figure 6.2.1 shows Sam's spreadsheet formulation after he has set up and solved the problem.

1. A coefficient for the objective function was computed for each college (see Row 16 in Figure 6.2.1) to show its desirability. Which rows in the spreadsheet were used to do this?

2. Which colleges are included in Sam's portfolio?

3. Did the five selected colleges have the five largest objective function coefficients? Why or why not?

4. Which of the selected colleges met the acceptance rate constraint?

5. What is the average (mean) acceptance rate of the five selected colleges?

6. What is the range of the distance from home of the five selected colleges?

7. What is the median tuition of the five selected colleges?

8. What is the average (mean) size of the five selected colleges?

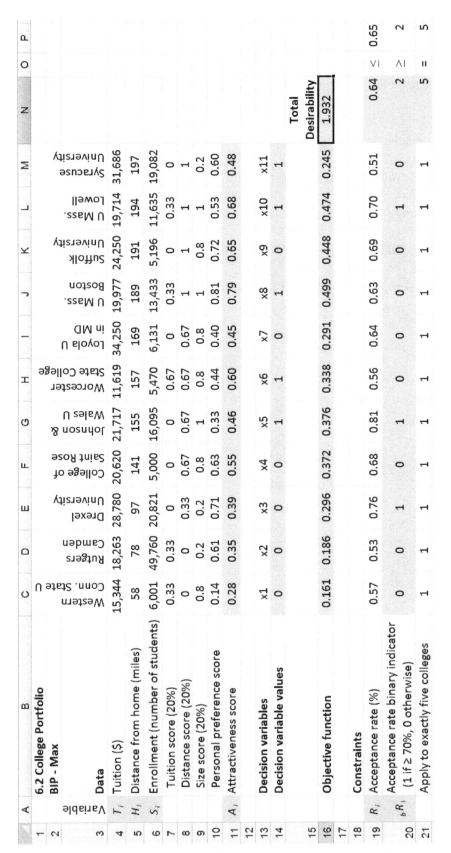

Figure 6.2.1: Sam's spreadsheet showing the optimal portfolio of colleges to which to apply

Section 6.3: Coach Bass's Problem

6.3.0 An Application of Binary Programming: Assignment Problems

Assignment problems are a special case of binary programming. An assignment problem arises whenever one type of thing, such as a person, must be assigned to another type of thing, such as a task. To solve an assignment problem, a **cost matrix** is usually defined based on the parameters of the given problem. The following is the general form of a cost matrix:

$$C = \begin{bmatrix} c_{1,1} & c_{1,2} & c_{1,3} & & c_{1,n} \\ c_{2,1} & c_{2,2} & c_{2,3} & & c_{2,n} \\ c_{3,1} & c_{3,2} & c_{3,3} & & c_{3,n} \\ & & & & \\ c_{m,1} & c_{m,2} & c_{m,3} & & c_{m,n} \end{bmatrix}$$

Suppose m workers must be assigned to n tasks. Then, each entry of the cost matrix, $c_{i,j}$ could represent the time needed by Worker i to perform Task j.

For example, suppose five workers need to be assigned to four tasks. The goal is to finish the tasks as quickly as possible. Then, the following cost matrix can be used, where the entries represent the time, in minutes, it takes each worker to complete each task.

	shaping	assembling	painting	inspecting
Worker 1	10	8	14	17
Worker 2	13	9	17	16
C = Worker 3	14	11	20	18
Worker 4	9	10	15	14
Worker 5	14	12	16	16

1. Based on this cost matrix, which workers do you think should perform which tasks? Explain your reasoning.

To explore assignment problems further, two problem contexts are discussed. In the first context, a coach of a girls' swimming team must decide which swimmer to assign to each part of a 400-yard medley relay. In the second context, a teacher wants to determine how to assign her students into groups for a class project. In the first context, a cost matrix is utilized, but in the second, a special set of matrices is developed. Then, binary programming is used to find the optimal solution.

6.3.1 The Medley Relay Team

Bill Bass, coach of the Jefferson High School girls' swimming team, has four swimmers he always assigns to compete in the 400-yard medley relay. In this event, there are four 100-yard legs that each must be swum by a different competitor using a different stroke. The four legs in the relay are butterfly, backstroke, breaststroke, and freestyle.

Coach Bass knows the best times for each of his swimmers for each leg of the relay. He wonders how to use their best times to develop the optimal medley relay team..

2. Other than using best times, what other criteria could Coach Bass use to assign the swimmers?

3. Formulate a general cost matrix for Coach Bass's problem, with entries represented by $c_{i,j}$.

4. In the case of the medley relay team, what would $c_{1,2}$ represent? What would $c_{i,j}$ represent?

5. What is Coach Bass's objective in the medley relay problem?

6. What could possibly complicate this problem?

6.3.2 Formulating the Problem

Coach Bass must assign each of the four swimmers to one of the four 100-yard legs in the 400-yard medley relay: butterfly, backstroke, breaststroke, and freestyle. Coach Bass has decided to use the best times for each of his swimmers for each of the legs of the relay to determine which swimmer to use for which leg of the relay. He wants to assign the four swimmers so as to complete the four legs of the relay in the minimum total time. He wonders what assignment would be the best, based on their best times. Table 6.3.1 contains those best times for the four swimmers.

(seconds)		Leg			
		100-yd butterfly	100-yd backstroke	100-yd breaststroke	100-yd freestyle
Swimmer	Schmidt	59.59	59.83	66.83	52.61
	Reid	60.45	59.56	68.14	53.31
	Sanchez	61.84	64.63	67.69	53.70
	Lamartina	62.37	59.13	68.36	54.77

Table 6.3.1: Best times (in seconds) for four swimmers in the medley relay legs

7. Coach Bass first considered using the best swimmer in each event. Why does this strategy not work?

To solve this problem, Coach Bass decides to use Excel Solver. First, he needs a formulation of the problem. To do that, he will use a binary decision variable to represent the possibility that each of the four swimmers could be assigned to any one of the four legs. He defines the following binary decision variables:

Chapter 6   Finding Optimal Solutions—Binary Programming

Let: $x_{1,1}$ = the binary decision variable for Swimmer 1 (Schmidt) in Leg 1 (butterfly)
         $x_{1,2}$ = the binary decision variable for Swimmer 1 (Schmidt) in Leg 2 (backstroke)
         $x_{1,3}$ = the binary decision variable for Swimmer 1 (Schmidt) in Leg 3 (breaststroke)
         $x_{1,4}$ = the binary decision variable for Swimmer 1 (Schmidt) in Leg 4 (freestyle)
         $x_{2,1}$ = the binary decision variable for Swimmer 2 (Reid) in Leg 1 (butterfly)
         $x_{2,2}$ = the binary decision variable for Swimmer 2 (Reid) in Leg 2 (backstroke)
         $x_{2,3}$ = the binary decision variable for Swimmer 2 (Reid) in Leg 3 (breaststroke)
         $x_{2,4}$ = the binary decision variable for Swimmer 2 (Reid) in Leg 4 (freestyle)

8.   Write the remaining binary decision variables.

Writing out these binary decision values is very long and repetitive. Therefore, Coach Bass decides to use a shortcut, using generic letter subscripts:

        For $1 \leq i \leq 4$ and $1 \leq j \leq 4$,
Let $x_{i,j}$ = the binary decision variable representing whether Swimmer i is assigned to Leg j.

9.   How many binary decision variables are there in this problem formulation?

Next, Coach Bass needs to define an objective function in terms of the decision variables. His objective is to minimize the time for the relay team to swim the event. Thus, for each of the four swimmers, he will need her best time in each of the four legs. So, for $1 \leq i \leq 4$ and $1 \leq j \leq 4$, let $T_{i,j}$ = the best time for Swimmer i in Leg j.

10.   How are all of the $T_{i,j}$ values different from all of the $x_{i,j}$ values?

Next, using $T_{i,j}$ and $x_{i,j}$ Coach Bass will represent the total time for the medley relay event. For each possible pair of values for i and j, he will multiply $T_{i,j}$ by $x_{i,j}$. Most of the time, $x_{i,j}$ will equal 0 and the corresponding product will also be 0. There are 16 binary decision variables, but only four of them are going to equal 1. These represent the four swimmers who are assigned to swim the four legs of the relay. In those cases, the product $T_{i,j} \cdot x_{i,j}$ will be the best time of the swimmer for the leg she is assigned to swim. Finally, adding all of the products, $T_{i,j} \cdot x_{i,j}$, gives the total of those four best times. Thus, the objective function that Coach Bass would like to minimize is:

$$z = \sum_{i=1}^{4} \left(\sum_{j=1}^{4} T_{i,j} \cdot x_{i,j} \right).$$

What does this complicated-looking double-summation mean? Looking first at the summation outside the parentheses, it is changing the values of i. In other words, it is changing the rows in a matrix such as this:

$$\begin{bmatrix} T_{1,1} \cdot x_{1,1} & T_{1,2} \cdot x_{1,2} & T_{1,3} \cdot x_{1,3} & T_{1,4} \cdot x_{1,4} \\ T_{2,1} \cdot x_{2,1} & T_{2,2} \cdot x_{2,2} & T_{2,3} \cdot x_{2,3} & T_{2,4} \cdot x_{2,4} \\ T_{3,1} \cdot x_{3,1} & T_{3,2} \cdot x_{3,2} & T_{3,3} \cdot x_{3,3} & T_{3,4} \cdot x_{3,4} \\ T_{4,1} \cdot x_{4,1} & T_{4,2} \cdot x_{4,2} & T_{4,3} \cdot x_{4,3} & T_{4,4} \cdot x_{4,4} \end{bmatrix}$$

So, when $i = 1$, the summation inside the parentheses is computed. Looking at the inside summation, it changes the values of j, which are the columns. So, row-by-row, this double summation computes the sum of the products for each row and then adds those row sums to get the sum of all 16 of those products.

The sum can be written out as follows:

$$z = \sum_{i=1}^{4}\left(\sum_{j=1}^{4} T_{i,j} \cdot x_{i,j}\right) = \underbrace{T_{1,1} \cdot x_{1,1} + T_{1,2} \cdot x_{1,2} + T_{1,3} \cdot x_{1,3} + T_{1,4} \cdot x_{1,4}}_{i=1} + \underbrace{T_{2,1} \cdot x_{2,1} + T_{2,2} \cdot x_{2,2} +}_{i=2} \cdots \underbrace{+ T_{4,3} \cdot x_{4,3} + T_{4,4} \cdot x_{4,4}}_{i=4}$$

But that's the sum of a bunch of zeros plus the best times of the four swimmers who are actually assigned to swim the legs of the medley relay. In other words, that double-summation computes exactly the quantity that Coach Bass wants to minimize!

11. Arbitrarily assign each swimmer to a leg of the race.
    a. Calculate the value of z for this assignment.
    b. Interpret the meaning of z in terms of the problem context.

Finally, Coach Bass needs to add the constraints to the formulation. The constraints apply to the assignment of swimmers and can be stated simply as:
- Every leg of the relay must have exactly one swimmer assigned to it.
- Every swimmer must be assigned to exactly one leg of the relay.

Because each leg of the relay must have exactly one swimmer assigned to it, there are four different leg constraints. Since the legs of the relay are represented by the subscript j, the four leg constraints can be written as follows:

$$x_{1,1} + x_{2,1} + x_{3,1} + x_{4,1} = 1$$
$$x_{1,2} + x_{2,2} + x_{3,2} + x_{4,2} = 1$$
$$x_{1,3} + x_{2,3} + x_{3,3} + x_{4,3} = 1$$
$$x_{1,4} + x_{2,4} + x_{3,4} + x_{4,4} = 1$$

Notice that in each of the four leg constraints, the second subscript, representing the leg, has a fixed value. Then the first subscript, representing the swimmers, takes on each value from 1 to 4. Finally, the value of the second subscript is changed and the process is repeated until the second subscript has also taken on each value from 1 to 4. In a more compact notation, these four constraints can be written as follows:

$$\text{For } j = 1, 2, 3, \text{ and } 4, \sum_{i=1}^{4} x_{i,j} = 1.$$

Similarly, because every swimmer must be assigned to exactly one event, there are also four different swimmer constraints. Since the four swimmers are represented by the subscript i, the four swimmer constraints can be written:

$$x_{1,1} + x_{1,2} + x_{1,3} + x_{1,4} = 1$$
$$x_{2,1} + x_{2,2} + x_{2,3} + x_{2,4} = 1$$
$$x_{3,1} + x_{3,2} + x_{3,3} + x_{3,4} = 1$$

$$x_{4,1} + x_{4,2} + x_{4,3} + x_{4,4} = 1$$

Once again, a more compact notation is as follows:

For $i = 1, 2, 3,$ and 4, $\sum_{j=1}^{4} x_{i,j} = 1.$

Putting the various parts of the formulation all together, Coach Bass has the following.

<u>Decision Variables</u>
  Let
  $x_{i,j}$ = the binary decision value representing whether Swimmer i is assigned to Leg j, for $1 \leq i \leq 4$ and $1 \leq j \leq 4$

<u>Objective Function</u>
  Minimize: $z = \sum_{i=1}^{4}\left(\sum_{j=1}^{4} T_{i,j} \cdot x_{i,j}\right)$, where z = the total of the best times of the four swimmers assigned to swim the relay and $T_{i,j}$ = the best time for Swimmer i in Leg j

<u>Constraints</u>
  Subject to:

  for $j = 1, 2, 3,$ and 4, $\qquad \sum_{i=1}^{4} x_{i,j} = 1$

  for $i = 1, 2, 3,$ and 4, $\qquad \sum_{j=1}^{4} x_{i,j} = 1$

6.3.4 Putting the Formulation into a Spreadsheet

Coach Bass creates a spreadsheet formulation of the medley relay team problem, as shown in Figure 6.3.1. In this spreadsheet, he creates two matrices. One matrix has the best times in each leg for each swimmer ($T_{i,j}$). The other matrix—called the *assignment matrix*—contains the decision variables ($x_{i,j}$). The assignment of swimmers to the legs of the relay will be recorded in this assignment matrix.

At the beginning of the solution process, no swimmer has yet been assigned to any leg of the relay. Thus, each cell in this assignment matrix appears empty (i.e., the value of each of the decision variables it contains is 0).

	A	B	C	D	E	F	G	H	I	J
3				100-yard Legs				Obj. Function		
4			Butterfly	Backstroke	Breaststroke	Freestyle		Total Time		
5		Schmidt	59.59	59.83	66.83	52.61		0		
6	Swimmer	Reid	60.45	59.56	68.14	53.31				
7		Sanchez	61.84	64.63	67.69	53.7				
8		Lamartina	62.37	59.13	68.36	54.77				
9										
10										
11				100-yard Legs						
12			Butterfly	Backstroke	Breaststroke	Freestyle	Assigned		Swimmer	
13		Schmidt	0	0	0	0	0	=	1	
14	Swimmer	Reid	0	0	0	0	0	=	1	
15		Sanchez	0	0	0	0	0	=	1	
16		Lamartina	0	0	0	0	0	=	1	
17		Assigned	0	0	0	0				
18			=	=	=	=				
19		Must swim	1	1	1	1				
20										

Figure 6.3.1: Spreadsheet formulation of the Medley Relay Problem

The numbers to the right of the assignment matrix are the row sums. They represent the four swimmer constraints. Since each swimmer must be assigned to swim exactly one leg, each swimmer's row of decision variables must contain one 1 and three 0s. Therefore, in a valid assignment, each row sum must equal 1.

Similarly, the numbers below the assignment matrix are the column sums. They represent the four leg constraints. Each leg of the relay must have exactly one swimmer assigned to it. As a result, in a valid assignment, each column will contain one 1 and three 0s. Thus, each column sum must also equal 1.

To tell Excel to add up values, the SUM function is used. For example, in cell G13, Coach Bass writes the formula:
=SUM(C13:F13).

This formula tells Excel to add the values in cells C13, D13, E13, and F13.

The only other constraint is that each of the decision variables must be binary. This constraint is added in the Solver window, so Coach Bass does not need to write this information into the spreadsheet at this time.

The objective function is the sum of the best times of the four swimmers who are assigned to the medley relay. It is computed in cell H10 by multiplying each cell in the assignment matrix by the corresponding cell of the swimmers' times matrix and adding all of the products. For example, the cell containing Schmidt's best time in the Butterfly (59.59) is multiplied by the cell in the assignment matrix opposite Schmidt's name and below "Butterfly". In Figure 6.3.1, this means that the value in cell C6 is multiplied by the value in cell C13. Then, each of these products is added together.

This can easily be done using the SUMPRODUCT function of Excel. That is, in cell H11, Coach Bass types

=SUMPRODUCT(C6:F9,C13:F16).

This function will allow him to find the total of the minimum times of the four swimmers assigned to the medley relay. We usually use the SUMPRODUCT formula to multiply a single line of data by the decision variable values. In this case the same formula is used to multiply multiple rows of data by an equally sized array of decision variable values. All of the individual products will then be summed, as usual. Looking back to Figure 6.3.1, the cells in the rectangular array whose top left corner is C6 and whose bottom right corner is F9 are where the data for the swimmers' best times in each leg are stored. The second rectangular array, from cell C13 diagonally to cell F16, is where Solver will place the optimal values of the decision variables.

The current value of the objective function is 0 because no swimmer has yet been assigned to swim any leg of the medley relay. At this point, Coach Bass does not have a valid assignment. None of the rows or columns sum to 1. He must use Solver to find the optimal assignment.

6.3.5 Solving the Problem

Coach Bass sets up Solver as in previous chapters. He makes sure to include the binary constraints on the decision variables. Figure 6.3.2 shows the spreadsheet after the Solver has found a solution.

	A	B	C	D	E	F	G	H	I
3					100-yard Legs			Obj. Function	
4			Butterfly	Backstroke	Breaststroke	Freestyle		Total Time	
5	Swimmer	Schmidt	59.59	59.83	66.83	52.61		239.72	
6		Reid	60.45	59.56	68.14	53.31			
7		Sanchez	61.84	64.63	67.69	53.7			
8		Lamartina	62.37	59.13	68.36	54.77			
9									
10									
11					100-yard Legs				
12			Butterfly	Backstroke	Breaststroke	Freestyle	Assigned		Swimmer
13	Swimmer	Schmidt	1	0	0	0	1	=	1
14		Reid	0	0	0	1	1	=	1
15		Sanchez	0	0	1	0	1	=	1
16		Lamartina	0	1	0	0	1	=	1
17		Assigned	1	1	1	1			
18			=	=	=	=			
19		Must swim	1	1	1	1			

Figure 6.3.2: Spreadsheet showing a solution to Coach Bass's Problem

12. What is the optimal assignment of swimmers to legs of the medley relay?

13. In this optimal assignment, which swimmers were the fastest for the swim strokes they were assigned? Which swimmers were not the fastest for the strokes assigned to them? Explain why

14. What is the minimum total of the best times for each of the swimmers in the legs to which they have been assigned?

6.3.6 A Complication: More Swimmers

In the previous section, Coach Bass had to solve a *balanced assignment problem*. He had the same number of swimmers to assign as he had legs of the relay. What if the problem were out of balance? For example, suppose that Coach Bass has a large and very competitive swim team. Suppose there are eight girls in competition for the four legs of the medley relay. Table 6.3.2 contains the best times for each of the eight girls in each of the four events.

(seconds)		Leg			
		100-yd butterfly	100-yd backstroke	100-yd breaststroke	100-yd freestyle
Swimmer	Schmidt	59.59	59.83	66.83	52.61
	Reid	60.45	59.56	68.14	53.31
	Sanchez	61.84	64.63	67.69	53.70
	Lamartina	62.37	59.13	68.36	54.77
	Wu	60.33	64.30	66.74	54.05
	Greene	62.41	59.03	66.19	56.61
	Kleinfeld	62.43	67.63	68.05	55.55
	Lapinski	59.44	65.06	66.74	52.49

Table 6.3.2: Best times for eight swimmers in the medley relay legs

Much as he did in the first problem, Coach Bass sets up a spreadsheet solver to tackle the problem. First, he needs a problem formulation. He will use a binary decision variable to represent whether each of the eight swimmers is assigned to any one of the four legs. So, for $1 \leq i \leq 8$ and $1 \leq j \leq 4$, let $x_{i,j}$ = the binary decision variable representing whether Swimmer i is assigned to Leg j. Another way of putting this is:

$$\text{Let } x_{i,j} = \begin{cases} 1 & \text{if Swimmer } i \text{ is assigned to Leg } j \\ 0 & \text{if Swimmer } i \text{ is not assigned to Leg } j \end{cases}, \text{ for } i = 1, 2, \ldots, 8 \text{ and } j = 1, 2, 3, 4.$$

15. How many binary decision variables are there in this problem formulation?

Next, Coach Bass defines an objective function in terms of the decision variables. His objective is still to minimize the time for his relay team to swim the event. Thus, for each of the eight swimmers, he will need her best time in each of the four legs. So, he

let $T_{i,j}$ = the best time for Swimmer i in Leg j, for $1 \leq i \leq 8$ and $1 \leq j \leq 4$.

Next, using $x_{i,j}$ and $T_{i,j}$, Coach Bass represents the total time for the medley relay event. For each pair of values for i and j, he multiplies $T_{i,j}$ by $x_{i,j}$. Finally, adding all of the products, $T_{i,j} \cdot x_{i,j}$, gives him the total of those four best times. Therefore, the objective function is:

$$\text{Minimize} \quad z = \sum_{i=1}^{8} \left(\sum_{j=1}^{4} T_{i,j} \cdot x_{i,j} \right).$$

Finally, Coach Bass needs to add the constraints to the formulation. One set of constraints is almost the same as before: Every leg must have exactly one swimmer assigned to it. The only difference this time is that there are eight possible swimmers.

However, the other set of constraints is a bit different. It is no longer true that every swimmer must be assigned to exactly one leg because four of the swimmers will not be assigned to any leg. Thus, the second set of constraints is: No swimmer can be assigned to more than one leg.

Because each leg must have exactly one of the eight swimmers assigned to it, the four leg constraints are:

For $j = 1, 2, 3$, and 4: $\quad \sum_{i=1}^{8} x_{i,j} = 1.$

However, because no swimmer can be assigned to more than one event, there are now eight swimmer constraints:

For $i = 1, 2, 3, 4, 5, 6, 7$, and 8: $\quad \sum_{j=1}^{4} x_{i,j} \leq 1.$

Putting the various parts of the formulation all together, Coach Bass has the following:

Decision Variables

Let $x_{i,j} = \begin{cases} 1 & \text{if Swimmer } i \text{ is assigned to Leg } j \\ 0 & \text{if Swimmer } i \text{ is not assigned to Leg } j \end{cases}$, for $i = 1, 2, \ldots, 8$ and $j = 1, 2, 3, 4.$

Objective Function

Minimize: $\quad z = \sum_{i=1}^{8} \left(\sum_{j=1}^{4} T_{i,j} \cdot x_{i,j} \right)$, where z = the total of the best times of the four swimmers assigned to swim the relay and $T_{i,j}$ = the best time for Swimmer i in Leg j

Constraints
Subject to:

for $j = 1, 2, 3$, and 4: $\quad \sum_{i=1}^{8} x_{i,j} = 1$

for $i = 1, 2, 3, 4, 5, 6, 7$, and 8: $\quad \sum_{j=1}^{4} x_{i,j} \leq 1$

	A	B	C	D	E	F	G	H	I
3				100-yard Legs				Obj. Function	
4			Butterfly	Backstroke	Breaststroke	Freestyle		Total Time	
5		Schmidt	59.59	59.83	66.83	52.61		0	
6		Reid	60.45	59.56	68.14	53.31			
7	Swimmer	Sanchez	61.84	64.63	67.69	53.70			
8		Lamartina	62.37	59.13	68.36	54.77			
9		Wu	60.33	64.3	66.74	54.05			
10		Greene	62.41	59.03	66.19	56.61			
11		Kleinfeld	62.43	67.63	68.05	55.55			
12		Lapinski	59.44	65.06	66.74	52.49			
13									
14				100-yard Legs					
15			Butterfly	Backstroke	Breaststroke	Freestyle	Assigned	Not Required	Swimmer
16		Schmidt	0	0	0	0	0	<=	1
17		Reid	0	0	0	0	0	<=	1
18	Swimmer	Sanchez	0	0	0	0	0	<=	1
19		Lamartina	0	0	0	0	0	<=	1
20		Wu	0	0	0	0	0	<=	1
21		Greene	0	0	0	0	0	<=	1
22		Kleinfeld	0	0	0	0	0	<=	1
23		Lapinski	0	0	0	0	0	<=	1
24		Assigned	0	0	0	0			
25			=	=	=	=			
26		Must swim	1	1	1	1			

Figure 6.3.3: Spreadsheet formulation of the medley relay team with eight swimmers.

Figure 6.3.3 shows a spreadsheet formulation of the medley relay team problem with eight swimmers. Notice that, just as before, the spreadsheet contains two matrices. One matrix has the best times for each event for each swimmer ($T_{i,j}$). Another matrix shows the decision variable values, the assignment of swimmers to events ($x_{i,j}$). Each cell in the assignment matrix contains a zero because at the beginning of the solution process, no swimmer has been assigned to any event.

16. If Sanchez is assigned to swim the freestyle leg, what value will appear in cell F19?

17. What value will appear in every other cell in that row of the assignment matrix?

18. What value will appear in every other cell in that column of the assignment matrix?

The objective is to minimize the total of the best times of the four swimmers who will be assigned to the medley relay. The objective function has been stored under the cell labeled "Total Time (sec)."

19. Why is there a zero under the cell labeled "Total Time (sec)" in the spreadsheet depicted in Figure 6.3.3?

Notice that the cell directly below each leg column of the assignment matrix contains a 0. The cell below the 0 contains an equals sign, and the cell below that contains a 1. Notice also that the row of zeros is labeled "Assigned" and the row of ones is labeled "Required."

Chapter 6 Finding Optimal Solutions—Binary Programming

20. What does that mean in the context of the problem?

Similarly, notice that the cell directly to the right of each row of the assignment matrix contains a 0. The cell to the right of the 0 contains a less-than-or-equal-to sign. Lastly, the cell to the right of the inequality sign contains a 1. Also notice that the column of zeros is labeled "Assigned" and the column of ones is labeled "Capacity."

21. What does that mean in the context of the problem?

22. Why are these eight cells labeled "≤", while the other four were labeled "="?

Figure 6.3.4 shows the previous spreadsheet after Excel Solver has found a solution for the problem of assigning eight swimmers to the four legs of the medley relay.

	A	B	C	D	E	F	G	H	I
3				100-yard Legs				Obj. Function	
4			Butterfly	Backstroke	Breaststroke	Freestyle		Total Time	
5		Schmidt	59.59	59.83	66.83	52.61		237.37	
6		Reid	60.45	59.56	68.14	53.31			
7	Swimmer	Sanchez	61.84	64.63	67.69	53.70			
8		Lamartina	62.37	59.13	68.36	54.77			
9		Wu	60.33	64.3	66.74	54.05			
10		Greene	62.41	59.03	66.19	56.61			
11		Kleinfeld	62.43	67.63	68.05	55.55			
12		Lapinski	59.44	65.06	66.74	52.49			
13									
14				100-yard Legs					
15			Butterfly	Backstroke	Breaststroke	Freestyle	Assigned	Not Required	Swimmer
16		Schmidt	0	0	0	1	1	<=	1
17		Reid	0	0	0	0	0	<=	1
18	Swimmer	Sanchez	0	0	0	0	0	<=	1
19		Lamartina	0	1	0	0	1	<=	1
20		Wu	0	0	0	0	0	<=	1
21		Greene	0	0	1	0	1	<=	1
22		Kleinfeld	0	0	0	0	0	<=	1
23		Lapinski	1	0	0	0	1	<=	1
24		Assigned	1	1	1	1			
25			=	=	=	=			
26		Must swim	1	1	1	1			

Figure 6.3.4: Optimal assignment of swimmers to the medley relay

23. What is the optimal assignment of swimmers to legs of the relay?

24. Are there any swimmers in the optimal solution who are not the fastest swimmer for a specific leg? Are they all at least the second fastest swimmer in the event they have been assigned?

25. What is the minimum total time?

26. What is that time when converted to minutes and seconds?

Section 6.4: Ms. Newman Assigns Students to Teams

Ms. Cynthia Newman wants to assign her 20 eighth-grade students to four teams of five students so that they can work on a science project. She wants to be fair about the team assignments, so that no team has an advantage over any other team. She knows that each team must have at least one effective leader to keep the team organized and on-task. She is going to require a written report of the project, so she decides to require at least two students with good writing skills on each team. The project will require some analysis of data that will have to be collected, so she decides to require that each team have two analysts and two data collectors. Finally, she is also going to require each team to make an oral presentation to the rest of the class, so each team is going to need at least two students who are good presenters.

Ms. Newman identifies the roles that she believes each of her students might satisfy based on their skills. For example, she thinks that Don McAllister, one of her students, could play the role of leader, analyst, or data collector.

She identifies at least two possible roles for each of her students. Table 6.4.1 shows the possible roles for each of her students. In the table, Ms. Newman uses a "1" to indicate that she believes that a particular student can fill a particular role. Otherwise, she uses a "0". So, in this sense, these cell values are binary indicators.

		Skill				
		Leader	Writer	Analyst	Presenter	Data Collector
Student	1	1	0	1	0	1
	2	0	1	0	1	0
	3	0	1	0	0	1
	4	1	0	1	1	0
	5	0	1	0	0	1
	6	0	1	0	1	1
	7	1	0	1	0	1
	8	0	1	0	1	1
	9	0	1	0	0	1
	10	1	0	1	1	0
	11	0	1	0	1	1
	12	0	0	1	0	1
	13	0	1	0	1	0
	14	1	0	1	0	1
	15	0	1	0	1	0
	16	0	0	1	0	1
	17	1	1	0	1	0
	18	0	1	0	1	0
	19	0	1	0	0	1
	20	1	0	1	1	1

Table 6.4.1: Roles Ms. Newman has identified for her students

1. Which roles does Ms. Newman believe that Student 17 could fill?

2. Which students does Ms. Newman believe could fill the role of "Leader"?

3. For each role except for "Leader", Ms. Newman is going to require that each team have at least two students who can fill that role. Why might it make sense to only require one leader on each team?

Mrs. Newman also does not want to create a team with a weak academic track record. She has the grade point averages (GPAs) for each of her students, but she wonders what the best way to use that information would be. She knows that the average GPA of each of the four teams is certainly going to be different. She also knows that the average GPA for the entire class is 2.94. Table 6.4.2 contains the GPA for each of her students.

Student	GPA
1	3.13
2	3.43
3	3.70
4	2.84
5	3.81
6	2.70
7	3.75
8	2.35
9	3.44
10	2.42
11	2.04
12	2.46
13	2.47
14	2.39
15	2.92
16	2.24
17	2.14
18	3.82
19	3.20
20	3.64

Table 6.4.2: Class GPAs

Ms. Newman would like for the teams to be balanced academically and is especially concerned that there be no academically weak team. Ms. Newman's younger brother, Fred, is a graduate student in operations research. He suggests that she might use a technique called *maximizing the minimum* in order to solve her problem. Fred explains that she could form teams in such a way that the minimum average of each team's GPA would be as large as possible. The teams would also have to meet all of the constraints.

Assume, for example, she creates four teams, and the average GPAs of these teams are 2.71, 2.99, 3.00, and 3.06. The minimum average GPA for these four teams is 2.71. This GPA is substantially lower than the others. Maximizing the minimum team GPA is one way to make the weakest academic team as strong as possible. Ms. Newman decides to use Fred's suggestion.

6.4.1 Formulating the Problem

Fred helps his sister formulate her team problem. He recognizes it as an assignment problem with many constraints and an unusual type of objective function. The decision variables are all binary. A zero represents the decision not to assign a particular student to a particular team, and a one represents assigning that particular student to that team. Therefore, each student will have four different decision variables, one for each team. When a solution is found, for each student, three out of the four associated decision variables will equal 0; the other will equal 1.

4. Suppose Student 1 ends up with the four decision values found in Table 6.4.3. What does this mean in the context of the problem?

	Team 1	Team 2	Team 3	Team 4
Student 1	0	1	0	0

Table 6.4.3: Example decision values for student 1

Since there are 20 students in Ms. Newman's class and four possible teams, there are a total of $20 \cdot 4 = 80$ decision variables. Now Fred is ready to begin the formulation. First, he defines those decision variables:

For $1 \leq i \leq 20$ and $1 \leq j \leq 4$, let $x_{i,j} = \begin{cases} 1 & \text{if student } i \text{ is assigned to team } j \\ 0 & \text{if student } i \text{ is } not \text{ assigned to team } j \end{cases}$.

There are four categories of constraints
- Each and every student assigned to one team
- Each team has five students
- Set of team skills must meet or exceed team role requirements
- Minimum and maximum GPA for each team

With the help of Fred, Ms. Newman establishes the basic assignment constraints for students and teams. First, every student must be assigned to exactly one team. So, the first constraint is:

For $1 \leq i \leq 20$, $\sum_{j=1}^{4} x_{i,j} = 1$.

Next, every team must consist of exactly five students. So, the second constraint is:

For $1 \leq j \leq 4$, $\sum_{i=1}^{20} x_{i,j} = 5$.

In addition, each team must be made up of the following:
- At least one leader
- At least two writers
- At least two analysts

- At least two presenters
- At least two data collectors.

In order to formulate these constraints, Fred creates a matrix, R, using the information in Table 6.4.1. If Student i has been identified as capable of handling Role k, then the value of $R_{i,k}$ is 1. Otherwise, the value of $R_{i,k}$ is 0. Now for each team, there is a constraint for each possible role.

There are five distinct roles: leader, writer, analyst, presenter, and data collector for each of the four teams. This is a total of 20 constraints.

For example, for the role of leader, Ms. Newman must check to see if, among the five students assigned to Team 1, there is at least one whom she has identified as a possible leader. Then she must do the same for each of the other four roles. Finally, she must repeat the entire process for each of the other three teams.

To formulate all of this, Fred defines the following constraints:

Leader:
$$\text{For } 1 \leq j \leq 4, \text{ for } k = 1, \sum_{i=1}^{20} R_{i,1} \cdot x_{i,j} \geq 1,$$

Writer:
$$\text{For } 1 \leq j \leq 4, \text{ for } k = 2, \sum_{i=1}^{20} R_{i,2} \cdot x_{i,j} \geq 2,$$

Analyst:
$$\text{For } 1 \leq j \leq 4, \text{ for } k = 3, \sum_{i=1}^{20} R_{i,3} \cdot x_{i,j} \geq 2,$$

Presenter:
$$\text{For } 1 \leq j \leq 4, \text{ for } k = 4, \sum_{i=1}^{20} R_{i,4} \cdot x_{i,j} \geq 2, \text{ and}$$

Data Collector:
$$\text{For } 1 \leq j \leq 4, \text{ for } k = 5, \sum_{i=1}^{20} R_{i,5} \cdot x_{i,j} \geq 2.$$

Next, Fred defines a constant, G_i, to represent the value of each student's GPA:

For $1 \leq i \leq 20$, let G_i = the GPA of Student i.

With this and the binary decision variables, he can write an expression to calculate a team's average GPA

$$\text{Average GPA for Team } i = \frac{\sum_{i=1}^{20} G_i \cdot x_{i,j}}{5}.$$

Recall that the objective Fred has suggested is to maximize the minimum average GPA of the four teams. However, there is no simple way to write this in a linear equation. Fred learned an algebraic trick in one of his courses that is designed to handle this situation.

Let z represent the objective to be maximized. Since z is supposed to be equal to the minimum GPA, it must be less than or equal to every team's GPA. This idea is represented with the following set of four constraint equations.

$$\text{For } 1 \leq j \leq 4, \quad \frac{\sum_{i=1}^{20} G_i \cdot x_{i,j}}{5} \geq z.$$

By telling Solver to maximize z, it will naturally force the value of z to be equal to the lowest of the team GPAs. This last set of equations is rearranged to standard form with a numeric right hand side value.

$$\text{For } 1 \leq j \leq 4, \quad \frac{\sum_{i=1}^{20} G_i \cdot x_{i,j}}{5} - z \geq 0.$$

So, the objective function is simply: Maximize z.

Chapter 6 Finding Optimal Solutions—Binary Programming

6.4.2 Interpreting the Spreadsheet Solution

	A	B	C	D	E	F	G	H	I	J	K	L	M	N	O	P	Q	R
1			**Advisory - Run Time:** The theoretical best is all teams have exactly 2.944 GPA. Under options, you should specify how good a solution you are seeking. If you specify within 2% of optimality, my computer finds an answer in less than a minute with the lowest GPA 2.892. If it is set it at 1%, it takes less than 2.5 minutes to find solution with GPA 2.915. At 0.1% it can take 10 minutes to find solution with GPA 2.920. At 0.01% it was still running after 20 minutes but found a solution with GPA 2.942 which is better than textbook solution.														
3			**Ms. Newman Creates Balanced (GPA) Student Teams**															
4										**Decision variables**								
5					Student Skills						Team Number							
6			Leader	Writer	Analyst	Presenter	Data Collector	GPA	1	2	3	4						Student Number
7		1	1		1		1	3.13	0	1	0	0	1	=	1			
8		2		1		1		3.43	0	0	0	1	1	=	1			
9		3		1			1	3.70	1	0	0	0	1	=	1		Team One	
10		4	1		1	1		2.84	0	0	1	0	1	=	1			
11		5		1			1	3.81	0	0	1	0	1	=	1			
12		6		1		1		2.70	0	0	0	1	1	=	1			
13		7	1		1		1	3.75	0	0	0	1	1	=	1			
14		8			1		1	2.35	0	0	0	1	1	=	1		Team Two	
15		9		1			1	3.44	0	1	0	0	1	=	1			
16		10	1		1			2.42	0	0	1	0	1	=	1			
17		11		1		1	1	2.04	1	0	0	0	1	=	1			
18		12		1			1	2.46	0	0	0	1	1	=	1			
19		13		1		1		2.47	0	0	1	0	1	=	1		Team Three	
20		14	1		1	1	1	2.39	1	0	0	0	1	=	1			
21		15		1				2.92	1	0	0	0	1	=	1			
22		16			1		1	2.24	0	1	0	0	1	=	1			
23		17	1	1		1		2.14	0	1	0	0	1	=	1			
24		18		1		1		3.82	0	1	0	0	1	=	1		Team Four	
25		19		1			1	3.20	0	0	1	0	1	=	1			
26		20	1		1	1	1	3.64	1	0	0	0	1	=	1			
27							Average GPA	2.944	2.938	2.953	2.949	2.938					Optimal	Cell is BOTH changing cell and Optimal cell in Solver
28					The difference between the minimum GPA and the Team GPA =>				0.000	0.015	0.011	0.000	>=		0		2.938	
29					Number of Skill Types on Each Team												This cell is referenced in cells I27, J27, K27, and L27. The Q28 cell value is subtracted from each team's average GPA and recorded in row 28. This difference is constrained to always be greater than 0. By maximizing Q28 cell value, Solver will force it to be as large as the smallest average GPA. This is equivalent to maximizing the minimum average GPA.
30			Leaders	Writers	Analysts	Presenters	Data Collectors											
31		1	2	3	2	3	4		5	5	5	5						
32	Team	2	2	3	2	2	3		=	=	=	=						
33		3	2	3	2	3	2		5	5	5	5						
34		4	1	3	2	3	4											
35			>=	>=	>=	>=	>=											
36			1	2	2	2	2											

Figure 6.4.1: Spreadsheet formulation of Ms. Newman's Team Assignment Problem

Figure 6.4.2: Solver Parameters window of Ms. Newman's Team Assignment Problem

Figure 6.4.1 shows this formulation in a spreadsheet format. Figure 6.4.2 shows the Solver Parameters Window. It is a complicated but efficiently laid out spreadsheet.

5. Which cells in the spreadsheet contain the values of the decision variables?

6. Cell I27 contains the average GPA of Team 1. Write the Excel command used to calculate that value.

7. Column N and P are used to specify the constraint that every student must be assigned to exactly one team. Write an expression for cell N7 to be used to determine the left hand side of the constraint.

8. Row 36 contains the minimum requirements for each role. Write a formula for cell C31 that ensures that Team 1 meets the Leader requirement.

9. Which role constraints does Team 1 meet exactly and which does it exceed?

10. Cell H27 contains the value 2.944. What does it represent and how was that value obtained?

11. The values in the cells of the spreadsheet in Figure 6.4.1 represent an optimal solution. What is that optimal solution? What is the smallest team average GPA that satisfies all the constraints?

6.4.3 A Complication: Students Who Cannot Work Together on a Team

After looking at the spreadsheet solution to her team assignment problem, Ms. Newman realizes that she did not account for some important information. She knows from past experience that Students 7 and 8, who were assigned to the same team, cannot work with each other. She also knows that one other pair of students, numbered 16 and 17, is also incapable of working together. She tells her brother Fred about this, to see if there is any way to account for this additional complication. Fred assures her that this conflict can be avoided. We can eliminate the possibility of Students 7 and 8 being on the same team by adding the following constraint for each of the four teams.

For $1 \leq j \leq 4$, $x_{7,j} + x_{8,j} \leq 1$.

Restricting the sum to less than or equal to 1 ensures that Students 7 and 8 cannot be on the same team j. A similar constraint is created for students 16 and 17.

For $1 \leq j \leq 4$, $x_{16,j} + x_{17,j} \leq 1$.

Although this will add constraints to the formulation, the problem will still be solvable. These eight additional constraints were added to the spreadsheet formulation. Table 6.4.4 contains the new set of teams.

		Team Number			
		1	2	3	4
Student	1	0	1	0	0
	2	0	0	1	0
	3	1	0	0	0
	4	0	0	1	1
	5	0	0	1	0
	6	0	0	0	1
	7	0	1	0	0
	8	0	0	1	0
	9	0	0	0	1
	10	0	0	0	0
	11	1	0	0	1
	12	1	0	0	0
	13	0	1	0	0
	14	0	0	0	0
	15	1	0	0	0
	16	0	0	1	0
	17	0	1	0	0
	18	0	0	0	0
	19	0	1	0	0
	20	1	0	0	1
Average GPA		2.953	2.950	2.937	2.936

Table 6.4.4: New team assignments without student conflicts

12. What teams were Students 7 and 8 on in the original formulation? What teams are they on in the new solution?

13. What teams were Students 16 and 17 on in the original formulation? What teams are they on in the new solution?

14. What effect did the conflict avoidance constraints have on the minimum team GPA and range of team GPAs?

Chapter 6 (Binary Programming) Homework Questions

1. The City Council in Monroe, Michigan is considering four proposed new recreational facilities: a swimming pool, a tennis center, athletic fields (football/soccer, baseball/softball, and track), and a gymnasium. The Council wants to construct the facilities that will maximize the expected daily use, but there are budgetary and land restrictions. The expected daily use, cost, and land requirements for each of the proposed facilities are given in Table 1.

Facility	Expected Use (people/day)	Cost ($ million)	Land Required (acres)
Swimming pool and fitness center	550	5.2	1.2
Tennis center	150	1.2	1.5
Athletic fields	325	1.7	2.5
Gymnasium	400	4.3	1

   Table 1: Information on proposed recreational facilities

   The Council has planned on offering a capital funding bond for up to $10 million to construct new recreational facilities. There are 4.7 acres of land available. However, the section available for an enclosed building, indoor pool or gymnasium, is not large enough to construct both. Thus, only one of these two facilities can be built.

   a. Define a set of binary decision variables for this problem.

   b. Use the decision variables you defined to define the objective function.

   c. Formulate the constraints in terms of the decision variables.

   d. Use solver to obtain the optimal solution.

   e. Which of the proposed facilities should the Council build? What is the average daily usage?

   f. There is a one acre swamp adjacent to the 4.7 acres. The council is considering reclaiming this piece of property at a cost of $1 million. Would this change the optimal solution?

2. The Research Triangle Electronics Company is considering eight new research and development projects. The company cannot conduct all eight projects due to limitations on their R & D budget and the number of research scientists available. Table 2 contains the resource requirements and estimated profit for each of the projects. In addition, not more than two of projects 4, 5, and 6 can be undertaken, because they require many of the same research scientists. The budget for the research is $3.5 million. There are 30 scientists in the research group.

Project	Cost ($1,000s)	Research Scientists Required	Estimated Profit (Millions of $)
1	650	7	8.2
2	1,200	6	9.5
3	350	8	3.7
4	450	9	4.1
5	800	10	5.3
6	850	8	5.2
7	750	7	8.2
8	700	4	5.8

Table 2: Information about eight possible research projects

a. Formulate the decision as a binary programming model.

b. Which projects should be selected in order to maximize estimated profit?

c. The company is considering increasing its R&D budget by $1 million. What impact would this have on the optimal solution?

3. TopTen Recording Studios is considering funding recording projects with four different artists over the next three years. The studio has allocated $12 million per year to cover the expenses of the new projects. The artists, necessary expenditures per year, and the expected profit from their projects are given in Table 3. Which projects should be selected to maximize the expected total profit?

Artist	Expenditures (millions of $/year)			Profit (millions of $)
	Year 1	Year 2	Year 3	
Rambling Lou	3	2	6	32
Rita Rivera	5	5	2	19
Nightrider	4	3	7	45
SoozieQT	3	5	3	23
Available funds/year	12	12	12	

Table 3: Expenditures per year and expected profit from 4 recording projects

a. Define a set of binary decision variables to fit this problem.

b. Use your decision variables to define the objective function.

c. Formulate the annual budget constraints.

d. Which artists' projects should Top Ten Recording select and how much would they earn?

4. Karen Studio has an extra credit assignment with problems numbered 1 through 7. Each problem Karen works on earns the points listed in Table 4. The time required to complete

each problem is also listed in the table. There will be no partial credit. Therefore, Karen cannot allocate a fraction of the time required to complete a problem.

	Problem Number						
	1	2	3	4	5	6	7
Points	7	9	5	12	14	6	12
Time (hours)	3	4	2	6	7	3	5

Table 4: Extra credit problems with no partial credit

a. If Karen allocates a total of 15 hours, which parts should she do? How many points would she earn? How many points on average did she earn per hour of study?

b. Karen is thinking of increasing her hours of study. How much is an extra hour worth? How does the optimal set of problems change?

c. What about 2 or 3 more hours? Are there any problems that are consistently part of the optimal study plan? Are there any that are never in the study plan? Provide some justification for this pattern.

d. Karen has a lot of other tasks to complete over the next week. Karen has decided she only needs 25 extra credit points. She wants to minimize the time required to earn these points. Formulate the problem and solve it. How much more time would it take to earn a total of 26 points? Explain the differences.

5. Donald Karr specializes in repairing and upgrading Chevrolet Corvettes built in the 1950s. Most models in mint condition sell for more than $50,000. Some cars that are in limited supply sell for more than $100,000. He buys cars that need extensive rework for less than $25,000. He just bought one of these classic cars. There is a gathering of Corvette enthusiasts in 5 weeks. He wants to have his car ready for resale at that time.

In Table 5 below he estimates the cost of parts to repair or upgrade each system. The table also includes his estimate of the number of hours it will take to make those repairs or upgrades. He asked his friend Nina Ferucchi, a renowned car auctioneer, to estimate the added value resulting from improving the appearance or performance of each system. Don has a budget of $5,000 for parts. He is willing to work long hours over the next few weeks. Don estimates he can work 280 hours without burning out on this short-term project.

His goal is to maximize the potential net profit of his work. This is measured by the difference between the added value and the cost of parts. The decision to work on any system is an all or nothing decision. Don asked his former high school math teacher, Janet Cosby, to quickly develop a recommendation.

	Systems					
	Body Work	Interior	Engine	Transmission	Suspension, tires, and brakes	Air intake and exhaust
Cost	$2,500	$1,200	$1,800	$2,400	$1,100	$600
Hours	100	40	130	90	50	30
Added Value	$7,300	$3,800	$6,500	$6,500	$4,000	$2,800

Table 5: System data

    a. Formulate the decision.

    b. What is the optimal work plan?

6. His sister, an amateur race car driver, said she was not racing next week. She estimated she could work with him for 40 hours.

    a. How much more money would he make?

    b. Don was thinking of asking his mom to loan him $1,000 to buy parts. Should he ask?

    c. Don wondered about the added value in seeking help from both his sister and mother. Could he effectively use some or all of their help? What would you recommend?

7. Ed Thomas is the basketball coach at Lincoln H. S. He is nearly ready to select his team following two weeks of try-outs. He is down to 20 candidates, and league rules limit his roster to 12 players. Coach Thomas believes that there are three keys to winning basketball: good rebounding, good defense, and making your free throws. Therefore, during try-outs, he scored each player's rebounding and defensive abilities on a scale of 1 to 10 with 10 being the highest. He also had each player shoot 25 free throws each day, and recorded their results. Table 6 contains the year in school and the positions each candidate can play (1 = point guard, 2 = shooting guard, 3 = small forward, 4 = power forward, 5 = center). For example player 1 is a 10th grader. He can play positions 2 or 3, shooting guard or small forward. The table also includes each student's rating and free throw percentage.

    a. Explain how Coach Thomas's selection process could be modeled using binary decision variables. How many binary decision variables would there be? What would their values represent?

Coach Thomas decides to use his data in the following ways. First, he wants the average rebounding score for the 12 players he selects to be at least 8. Likewise, he wants their average defensive score to be at least 8. He decides to use the free throw percentages to define his objective: He wants to maximize the average free throw percentage of the 12 players he selects.

    b. Use binary decision variables in a spreadsheet formulation to model Coach Thomas's objective function.

Player	Year	Positions	Rebound score	Defensive score	Free throw Pct.
1	10	2, 3	6	6	57
2	10	4, 5	6	8	40
3	10	4, 5	5	8	73
4	11	4, 5	10	6	41
5	11	4, 5	6	9	65
6	11	4, 5	7	8	45
7	11	3, 4	10	9	71
8	11	2, 3	5	7	40
9	11	1, 2	5	9	53
10	11	4, 5	9	9	51
11	10	1, 2	6	9	81
12	11	1, 2, 3	7	6	70
13	12	1, 2, 3	7	9	75
14	12	1, 2	6	8	59
15	12	2, 3	8	9	76
16	12	2, 3	7	9	67
17	12	4, 5	9	8	67
18	12	4, 5	9	7	39
19	12	4, 5	5	6	74
20	12	4, 5	7	8	51

Table 6: Coach Thomas's data on 20 basketball candidates

c. Construct Coach Thomas's rebounding and defensive constraints.

Coach Thomas also has a few other constraints that he wants to use in his selection process. First, because he always looks to the future, he wants at least 5 of the 12 players he selects to be sophomores or juniors. He is also concerned about having enough players who can play each of the five positions. He wants at least two players who can play point guard, at least four who can play shooting guard and at least four who can play small forward. He wants at least three players who can play strong forward and at least three players who can play center.

d. Model each of these new constraints.

e. Find the optimal solution to Coach Thomas's problem, if one exists and what is the average free throw percentage?

Chapter 6   Finding Optimal Solutions—Binary Programming

8. Regarding Coach Thomas's problem,

   a. What would be the effect on the optimal solution if Coach Thomas decided he wanted the average rebounding score to be 8.5 instead of 8?

   b. What would be the effect on the optimal solution if he decided he wanted the average defensive score to be 8.5 instead of 8?

   c. Three of the players who were not selected in the optimal solution had free throw percentages that are higher than the average of the optimal selection. Why do you think those three were not selected?

   d. Suppose Coach Thomas relaxes the average rebounding constraint from 8 to 7.5. What is the effect on the optimal solution? How does the average free throw percentage change?

   e. Suppose Coach Thomas relaxes the average defensive constraint from 8 to 7.5. What is the effect on the optimal solution? How does the average free throw percentage change?

   f. Suppose he relaxes both of these constraints? What is the effect on the optimal solution? What does that tell you about the two constraints?

9. Smith & Jones Holdings is an investment company. They have $15-million available to invest this month and are considering ten different companies from four different sectors of the economy. For each of these potential investments, Smith & Jones must invest a specific amount, and will receive a guaranteed rate of return on the investment. Of course, they would like to maximize the annual yield on their investments.

   Table 7 contains the name of each company Smith & Jones is considering, It includes company's economic sector, the fixed amount of the investment, and the percent and amount of annual yield. The investment experts at Smith & Jones also do not want to invest more than one-half the funds they have available in any one of the four economic sectors. They would also like to invest at most 40% of the available funds in transportation and manufacturing sectors combined. They want to invest at least 1/3 of the available funds in technology companies.

   Use a binary integer programming formulation and a spreadsheet solver to make a recommendation to Smith & Jones about which companies they should invest in. How much would they earn? What is the average yield? (We are assuming that all ten investments are relatively risk-free.)

Company	Economic Sector	Amount Required to Invest ($ millions)	Annual Yield (%)
Soft Solutions	Technology	3.5	3.7
ABM	Technology	5.6	6.8
Innovators	Technology	4.3	5.1
Unified	Transportation	2.7	6.9
Long Haul	Transportation	3.6	6.0
Cord	Manufacturing	6.0	4.3
AmCo	Manufacturing	5.9	4.1
Trot	Communications	4.1	6.5
ust&t	Communications	2.7	4.6
Horizon	Communications	3.9	3.4

Table 7: Ten investments under consideration by Smith & Jones Holdings

10. Your recommendation in the previous question should have included three of the top four percentage annual yield investments. However, it should not have included the investment with the highest percentage annual yield.

    a. Why do you think that happened?

    b. In increments of $200,000 increase the specific amount of investment in Unified until it enters the optimal solution. At what point does that happen? Are there any other changes to the optimal solution when it does?

Your recommendation to Smith & Jones in answer to question 9 should not have included either of the manufacturing investments.

    c. For either of the manufacturing investments, increase the percent annual yield in increments of 0.001 (one-tenth of a per cent) until it enters the optimal solution. At what per cent annual yield does that first happen? Are there any other changes to the optimal solution when it does?

Assignment Problems

11. **Housework for Kids.** Joanne Mankowski has three sons: Brad, Mike, and Paul. She is getting tired of doing the housework by herself. She wants her sons' cooperation in keeping the house clean. She offered them payment if they share the housework on the weekend. She determined three types of tasks that are doable for her sons: washing the dishes after dinner, vacuuming the family room, and dusting the furniture in the living room. The kids told Joanne their preferred payment amount for each task secretly. Those amounts are represented in Table 8. Assign each son to a task so that the assignment generates the least cost to their mom. Describe in words the two sets of constraints needed to formulate this decision.

($)		Bid		
		Dish Washing	Vacuuming	Dusting
Son	Brad	6	11	7
	Mike	9	13	9
	Paul	7	14	10

Table 8: Preferred payment amount of each son

12. **Flight Attendants.** Triangle Airlines is assigning six new flight attendants to fly on the six types of aircraft flown by the airline. Each of the new attendants has been trained on each type of aircraft, but the number of training hours the new attendants have on the different aircraft varies. The airline wants to assign the attendants based on their number of hours of training on each aircraft type. Table 9 provides the number of training hours on each aircraft type for each attendant. How should the airline assign the attendants if it wants to assign them based on their training experiences? (Hint: Is this a maximization or minimization problem?)

(hours of training)		Aircraft					
		CRJ	DC-9	A320	747	757	767
Attendant	Albert	4	4	2	4	2	8
	Jack	4	4	4	4	4	4
	Mary	4	2	2	4	8	4
	Katie	2	2	4	4	4	8
	Dave	2	2	4	6	6	4
	Matthew	4	2	2	6	6	4

Table 9: Training experience of the new attendants

13. **Leaders for Projects.** Mr. Summit has four projects, one each in marketing, product development, logistics, and finance. He has chosen four employees with good leadership skills, Ahmad, John, Julia, and Subhash. Now, it is time to assign the right person to the right project. First, he developed a specific test of 20 questions for each project and asked the four employees to take all four tests. He wants to assign the leaders to the tasks so that their total combined number of mistakes for the project they are assigned will be as small as possible. The number of mistakes each employee made on each test is displayed in Table 10. Determine an optimal strategy for assigning employees to projects. What is the objective function?

		Project			
		Marketing	Product Development	Logistics	Finance
Employee	Ahmad	1	2	3	3
	John	4	4	3	4
	Julia	1	2	1	3
	Subhash	4	2	2	2

Table 10: Number of mistakes made in the test

14. **Industrial Training.** Industrial Training Consultants is offering four types of courses in August. There are five instructors who have experience teaching all of the subjects. The assignment will be based on past student evaluations of the five instructors. The student evaluation scores appear in Table 11. How should instructors be assigned to courses so that the total of the student evaluation scores is maximized? Which instructor is assigned no course?

	(% positive)	Course			
		Lean Manufacturing	Six Sigma	Logistic Management	Simulation
Student	Randolph	93	96	86	87
	Angela	90	94	92	89
	Anthony	91	87	84	88
	Deborah	92	88	90	85
	Myles	95	97	94	88

Table 11: Student evaluations

15. **Renovación Home Improvement Store.** The Renovación Home Improvement Store will assign an employee to each of the five departments: Appliances, Flooring, Outdoor Living, Kitchen, and Tools. There are seven employees available who have past experience in all of these five departments. The company collected sales performance information for each worker on each day he or she was assigned to a department. The average daily sales of each employee are shown in Table 12.

   a. Assign employees to departments so that the average daily sales of the five employees assigned are as large as possible. What would the daily average sales be in an optimal assignment?

		Department Sales ($)				
		Appliances	Flooring	Outdoor Living	Kitchen	Tools
Employee	Joshua	1,555	525	370	275	560
	Adan	1,250	450	285	250	540
	Ha	850	500	320	330	550
	Tyson	1,675	490	375	350	580
	Valley	1,125	510	365	345	190
	Lacole	950	500	195	335	350
	Haemon	1,050	300	345	200	545

Table 12: Average daily sales of employees by department

b. In the optimal solution, which individuals were the best in their selected category? Which individuals were not the best? Explain why the optimal solution did not pick the best in each.

16. **VogueTech Computer.** The VogueTech Customer Service provides 24-hour online technical support. There are three 8-hour shifts and thirteen representatives. Five representatives will be assigned to the morning shift, five will be assigned to the afternoon shift, and 3 will be assigned to the night shift. The manager wants to assign them according to their preferences. Table 13 shows the shift preferences of each representative. "3" indicates the most preferred and "1" indicates the least preferred shift.

		Shift		
		6 a.m. – 2 p.m.	2 p.m. – 10 p.m.	10 p.m. – 6 a.m.
Representative	1	3	2	1
	2	2	3	1
	3	3	1	2
	4	2	3	1
	5	3	1	2
	6	3	2	1
	7	1	3	2
	8	3	2	1
	9	3	1	2
	10	1	2	3
	11	1	3	2
	12	3	2	1
	13	3	1	2

Table 13: The preferences of each representative

a. What is the optimal assignment plan? How many representatives are assigned their lowest preference? To their second lowest preference?

b. Experience shows that representatives 5 and 6 do not work well together. Therefore, management decides not to assign them to the same shift. How does this affect the optimal solution?

17. **Disaster Kits.** Counselor Cynthia Walker at Foster High School assigns students to community service work as a part of their graduation requirements. She recently received a notice about a project from the Community Help organization to pack kits for disaster relief. A section of the country needs assistance after a hurricane struck the area. Six types of kit are needed: Emergency Food Packs, Children's, Personal Care, Food Support, Layette, and Household. Only one student can be assigned to pack each of the six different types of kit, but ten students have signed up. All of these students have previously packed relief kits for this organization. The organization would like to pack 60 of each type of kit.

In order to assign the most efficient volunteer to the right task, the Community Help organization calculates and records the packing rate for each volunteer. These data are shown in Table 14. The packing rate is the number of kits packed per hour. The organization would like to assign the volunteers so that 60 of each type kit are packed in the least time.

(kits per hour)		Kit					
		Emergency Food Packs	Children's	Personal Care	Food Support	Layette	Household
Student	Abdullah	6	11	6	9	4	5
	Susan	12	18	7	10	4	9
	Jeff	9	12	6	10	5	7
	Briana	8	13	6	13	4	6
	Naomi	7	14	7	10	5	6
	Brenden	12	15	7	12	6	9
	Carlos	11	17	7	11	5	7
	LaQuita	7	16	8	11	6	6
	Matthew	9	14	9	10	7	8
	Erika	8	17	7	14	4	6

Table 14: Packing times for different kinds of kits

a. How can the Community Help organization determine which student to assign to each task?

b. Which student should be assigned to each task in order to pack the kits in the least amount of time?

c. What if any kits are NOT assigned to the fastest student?

d. With the optimal assignment, which set of sixty kits will take the shortest time to pack? How long will it take?

e. With the optimal assignment, which set of sixty kits will take the longest time to pack? How long will it take? What would you suggest to speed up completion of this task?

18. After reviewing the plan, Counselor Cynthia Walker was concerned that it would take too many hours to complete the 60 sets of some of the kits. Cynthia feels she made a mistake by creating the rule that only student could be assigned to each kit. She now believes that the kits that take the longest to assemble should have two or even three students assigned. According to the original policy only six of the 10 students were assigned. That leaves four more students to be assigned to the slowest tasks.

   a. How many students should be assigned to each kit type in order to ensure that each set of 60 kits is completed in less than 6 hours?

   b. There are ten students in total. How many students should be allocated to each kit type? Be sure to allocate a total of ten students. This involves assigning totals and not specific students to tasks.

   c. Use the assignment you created to change the constraints in the assignment formulation. Determine an optimal solution with these new constraints.

   d. Which set of 60 kits will now take the longest to assemble? How long will it take?

19. **School Bus Route Assignment.** The school district in Mayonia, Michigan makes annual contracts with school bus companies. There are three companies who are bidding for nine routes in the Mayonia School District. First, the companies simultaneously submit their sealed bids for the routes they are interested in. Then, the school district decides which company to assign to each route. The submitted bids of each company are shown in Table 15. A blank cell in the table indicates that the company did not bid for the route. None of the companies that are bidding can be assigned more than three routes. Help the Mayonia Schools assign companies to all routes with a minimum total cost.

($)		Route								
		1	2	3	4	5	6	7	8	9
Company	1	$22,000		$12,400		$16,400	$19,700	$14,000	$18,500	$9,800
	2	$22,300	$17,400		$25,900	$17,600		$13,200	$16,800	$9,900
	3		$16,600	$12,500	$25,200	$16,300	$19,500	$13,000		$8,600

Table 15: Bids on routes of school bus companies

   a. Identify the lowest and the second lowest bids on each route.

   b. None of the companies can be assigned to more than three routes. Assign the companies to the routes manually without violating the three-route restriction.

   c. In what way is this problem different than the typical assignment problem?

d. Formulate the decision model.

e. Solve the problem using a spreadsheet solver. Make sure the changing cells include only those routes for which a company actually bid. What is the total cost of the contracts?
When you list changing cells that are not just one range of continuous cells, you separate the groups of cells with a comma. When you list changing cells that are not just one range of continuous cells, you separate the groups of cells with a comma. For example, the following includes two "changing cells" in column B (B8 and B9), two in column C (C9 and C10), and three in column D (D8, D9, and D10):
$B$8:$B$9,$C$9:$C$10,$D$8,$D$10
However, you do not need to make any adjustments to the SUMPRODUCT formulae. The cells not included in the "changing cells" will always stay empty with an assumed value of 0.

20. As you discovered in the previous problem, identifying the changing cells and writing the equations can be tedious. An easier way to prevent assigning a route to a non-bidder is to place a very large number such as 999,999 into the blank cell as shown in Table 16.

($)		Route								
		1	2	3	4	5	6	7	8	9
Company	1	$22,000	999,999	$12,400	999,999	$16,400	$19,700	$14,000	$18,500	$9,800
	2	$22,300	$17,400	999,999	$25,900	$17,600	999,999	$13,200	$16,800	$9,900
	3	999,999	$16,600	$12,500	$25,200	$16,300	$19,500	$13,000	999,999	$8,600

Table 16: Modified bids of school bus companies

a. Why do you think you can find the same solution using the values above?

b. Formulate the problem.

c. Solve it again using Solver. You can use the SUMPRODUCT function to write the objective function.

21. Four more school bus companies are added to the bidding. Now there are seven companies that are bidding for nine routes in the Mayonia School District. The announced bids are shown in Table 17. Now each company can be assigned at most two routes. Use the idea from the previous problem to simplify the spreadsheet modeling.

($)		Route								
		1	2	3	4	5	6	7	8	9
Company	1	$22,000		$12,400		$16,400	$19,700	$14,000	$18,500	$9,800
	2	$22,300	$17,400		$25,900	$17,600		$13,200	$16,800	$9,800
	3		$16,600	$12,500	$25,200	$16,300	$19,500	$13,000		$8,600
	4	$20,600	$16,800	$13,550	$26,000	$16,700	$19,700	$13,300	$15,800	
	5	$20,000	$17,600	$13,800	$24,000		$18,000		$14,550	$9,200
	6	$20,800	$17,000	$13,900			$19,000	$12,700	$16,400	$9,000
	7			$12,900	$25,200	$17,500	$19,300	$12,800	$15,400	$8,800

Table 17: Bids of School Bus Companies

a. What will the new assignment be? How many companies are assigned zero, one, and two routes?

b. The city wants to see the effect of allowing each company at most three routes instead of only two. What will the impact be? Is the difference large enough to justify changing the policy?

Chapter 6 Summary

What have we learned?

This is the last in a series of four chapters on mathematical programming. As in the previous three chapters (LP Max, LP Min, and IP), Binary Programming (BP) is method of modeling a situation in which a decision has to be made to optimize some objective while being constrained by limited resources.

The process for solving a binary integer programming problem is the same as other linear programming problems.

1. Formulate the problem.
   - Identify and define the decision variables.
   - Write the objective function.
   - Identify and write the functional constraints.

2. Enter the problem formulation into a spreadsheet.
   - Enter decision variables, objective function, and constraint coefficients.
   - Create formulas for objective function values and constraints' RHS.
   - Set up Solver Parameters and Options.
   - Add binary constraint for decision variable values.
     Note: Some integer programming problems are maximization and others are minimization problems.
   - Solve and generate Answer Report.

3. Interpret the results.
   - Answer Report shows status and amount of slack for constraints
   - Solver cannot create a Sensitivity Report for binary integer programming problems.

We have also learned that assignment problems are special cases of binary programming problems. The mathematical formulation of these problems has many requirements.
- Matrices are used extensively.
- The binary decision variables are arranged in a compact matrix X. If agent i is assigned to perform task j, then x_{ij} will equal one, otherwise it will equal zero.
- The "cost" of agent i performing task j is element c_{ij} in the cost matrix C.
- The overall "cost" is calculated by taking the sum of all the products of corresponding elements of the decision variable matrix and the cost matrix.

The spreadsheet formulation of assignment problems is somewhat different than in previous chapters.
- The SUMPRODUCT formula needs to be used for the objective function.
- Constraints typically involve row and column totals from the decision variable matrix.

Terms

Assignment Problem An assignment problem arises whenever a number of agents must be paired with a number of tasks

Agent In an assignment problem, the agents are the ones able to perform the tasks

Binary Decision Variable A decision variable that can take on only two possible values, zero or one.

Binary Indicator Coefficient A coefficient that takes the value of one if the quantity meets a given condition or zero if it does not.

Cost The "cost" depends on the context and units of the problem, but it represents the amount of the objective quantity required for an agent to perform a task

Cost Matrix For an assignment problem with m agents and n tasks, the cost matrix C will be an $m \times n$ matrix, and element c_{ij} will represent the cost of agent i performing task j

Summation Notation

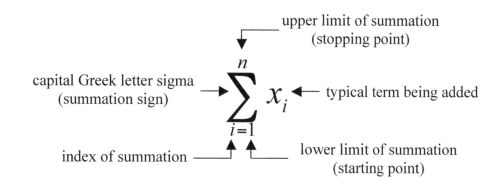

$$\sum_{i=1}^{n} x_i = x_1 + x_2 + x_3 + \ldots + x_{n-1} + x_n$$

Task In an assignment problem, the tasks are those things needing to be accomplished

Chapter 6 (Binary Programming) Objectives

You should be able to:

- Identify the objective of the problem

- Identify and define the binary decision variables

- Write the objective function, including using summation notation and double summation notation

- Identify the constraints involved in the problem

    o Standard assignment constraints

    - row totals ≤ 1 indicating each agent can be assigned to at most one task

    - column totals = 1 indicating that each task must be completed by exactly one agent

    o Extra constraints (e.g., modifications to totals, binary indicator coefficients)

- Write the functional constraints as inequalities, including using summation notation

- Use binary indicator coefficients in writing constraints

- Formulate the problem using a compact matrix

    o Each row represents an agent

    o Each column represents a task

- Enter the problem formulation into Excel

- Use SUMPRODUCT formula in Excel

- Set up Solver Parameters and Options, including the constraints that decision variables are binary

- Interpret the optimal solution in the context of the problem

- Analyze the Answer Report

Chapter 6 Study Guide

1. What is the objective function?

2. What are decision variables?

3. What is different about the decision variables in a binary programming (BP) problem compared to an integer programming (IP) problem?

4. What are functional constraints?

5. Besides functional constraints, describe two other types of constraints found in BP problems.

6. Consider a BP problem whose decision variables are of the form x_i, and one of the functional constraints is $\sum_{i=1}^{12} x_i \leq 7$.
   a. Explain how to determine the number of decision variables in the problem?
   b. Interpret the meaning of the constraint.

7. What information is found in the Answer Report for a BP problem?

8. In a BP problem, is it possible for a binary constraint to be nonbinding? Explain why or why not.

9. To what can the "Final Value" on the Answer Report refer?

10. Define *slack*.

11. In the Answer Report, what information is in the "Cell Value" column?

12. Explain how Excel would execute the command "=SUMPRODUCT(A1:B3,D1:E3)."

13. If an assignment problem has nine agents to assign to seven tasks, why is it convenient to use matrices to formulate the problem?

14. Assuming that each agent can perform at most one task and each task must be performed, what is the relationship between the number of agents and the number of tasks in an assignment problem? Why?

15. If an assignment problem has six agents, four tasks, decision variable matrix X, and cost matrix A, use double summation notation to write the objective function.

16. Consider an assignment problem with more agents than tasks. Explain why it is typical that the constraints based on row (agent) totals are inequalities while the constraints based on column (task) totals are equations.

17. Explain the "maximize the minimum" technique that was used in Problem 6.3 (assigning students to teams).

18. Do all assignment problems have a unique solution? Explain why or why not.

Appendix A: Summation Notation

Summation notation is often used as a simpler and more convenient way to write out a long sum in which all of the terms in the sum have some common feature. For example, in Section 6.1: Flipping Houses, the sum

$$z = P_1 \cdot x_1 + P_2 \cdot x_2 + P_3 \cdot x_3 + P_4 \cdot x_4 + P_5 \cdot x_5 + P_6 \cdot x_6 + P_7 \cdot x_7 + P_8 \cdot x_8 + P_9 \cdot x_9 + P_{10} \cdot x_{10}$$

could instead be written as:

$$z = \sum_{i=1}^{10} P_i \cdot x_i$$

In this sum, the common feature is: a product. The factors that are multiplied are an estimated profit (P_i) and a binary decision variable (x_i).

Summation notation, such as this, involves five main elements, as shown below:

$$\underset{\text{Index of summation} \quad \text{Starting point}}{\underset{i=1}{\overset{10}{\sum}}} P_i \cdot x_i$$

- Summation sign → Σ
- Stopping point (above): 10
- Typical element (right): $P_i \cdot x_i$
- Index of summation and Starting point (below): $i = 1$

The *summation sign* is the capital Greek letter Σ (sigma). To the right of the summation sign is a *typical element* of the sequence of terms being summed. In the Flipping Houses example, the typical element being summed is the estimated profit for a particular House i (P_i) multiplied by the binary decision value for that particular House i (x_i). That pattern can be seen when all of the terms in the sum are completely written out.

The index i plays two roles. First, it specifies a particular term in the sum. In the Flipping Houses example, a particular term can be thought of as a particular house. The second role of the index is to determine how many terms are in the sum. In the above example, this index is i. The *index of summation* is found below the summation sign, along with the starting point. The *starting point* (sometimes called the "lower limit of summation") refers to the first value that i takes on. Then, the *stopping point*, found above the summation sign, refers to the last value that i takes on. In the Flipping Houses example, the sum begins with House 1 ($i = 1$), ends with House 10 ($i = 10$), and therefore includes 10 terms.

Consider the following examples.

- The sum $\sum_{i=1}^{5} x_i$ can be rewritten as $\sum_{i=1}^{5} x_i = x_1 + x_2 + x_3 + x_4 + x_5$.
- The sum $\sum_{i=4}^{7} x_i$ can be rewritten as $\sum_{i=4}^{7} x_i = x_4 + x_5 + x_6 + x_7$.
- The sum $\sum_{i=1}^{5} x_i^2$ can be rewritten as $\sum_{i=1}^{5} x_i^2 = x_1^2 + x_2^2 + x_3^2 + x_4^2 + x_5^2$

- The sum $x_1 + x_2 + x_3 + x_4$ can be rewritten as $\sum_{i=1}^{4} x_i$.

- The sum $x_3 + x_4 + x_5$ can be rewritten as $\sum_{i=3}^{5} x_i$.

Double Summation Notation

If $C = \begin{bmatrix} c_{1,1} & c_{1,2} & c_{1,3} & c_{1,n} \\ c_{2,1} & c_{2,2} & c_{2,3} & c_{2,n} \\ c_{3,1} & c_{3,2} & c_{3,3} & c_{3,n} \\ c_{m,1} & c_{m,2} & c_{m,3} & c_{m,n} \end{bmatrix}$ and $X = \begin{bmatrix} x_{1,1} & x_{1,2} & x_{1,3} & x_{1,n} \\ x_{2,1} & x_{2,2} & x_{2,3} & x_{2,n} \\ x_{3,1} & x_{3,2} & x_{3,3} & x_{3,n} \\ x_{m,1} & x_{m,2} & x_{m3} & x_{m,n} \end{bmatrix}$, then

$$\sum_{i=1}^{m}\left(\sum_{j=1}^{n} c_{ij} \cdot x_{ij}\right) = \text{the sum of the elements of } \begin{bmatrix} c_{11} \cdot x_{11} & c_{12} \cdot x_{12} & c_{13} \cdot x_{13} & c_{1n} \cdot x_{1n} \\ c_{21} \cdot x_{21} & c_{22} \cdot x_{22} & c_{23} \cdot x_{23} & c_{2n} \cdot x_{2n} \\ c_{31} \cdot x_{31} & c_{32} \cdot x_{32} & c_{33} \cdot x_{33} & c_{3n} \cdot x_{3n} \\ c_{m1} \cdot x_{m1} & c_{m2} \cdot x_{m2} & c_{m3} \cdot x_{m3} & c_{mn} \cdot x_{mn} \end{bmatrix}$$

$$= \text{SUMPRODUCT}(C, X)$$

CHAPTER 7:

Find Optimal Locations with Algorithms

Chapter 7 — Finding the Best Place – Location Problems

`Section 7.0: Introduction

This chapter focuses on different types of location problems. Location decisions arise in many contexts. All fast food companies, oil companies, drugstore chains or other retail outlets routinely evaluate locations for new facilities. Similar decisions are made in the public sector with regard to the location of libraries, fire stations, school buildings, and health care clinics. In some instances a simple measure of travel distance suffices to guide the decision. In other instances multiple criterion are used as in Chapter 1 of this text.

Section 7.1 presents a simplified example involving two smoothie stands located along a single stretch of road. This example introduces the concept of minimizing the average distance traveled. Section 7.2 explores where to locate a small warehouse to store excess inventory for a downtown store. The goal is to minimize the total number of truck shipment miles. In section 7.3 the decision involves the location of a warehouse along a major interstate. Trucks from this warehouse are to make deliveries to different cities along the interstate. In this example the number of deliveries to each city is not the same. The example in section 7.4 involves a two-dimensional location problem. As in the previous section, customers are not uniformly distributed throughout the surrounding area. Section 7.5 introduces a new measure instead of average travel distance. In some contexts, the preferred measure involves ensuring that all potential users of a service are within a fixed distance of the nearest facility. Any set of users that are within the prescribed distance are said to be **covered** by that facility.

Section 7.1 Cooperative Location

Two mobile smoothie stations are located on Main Street in downtown Raleigh, NC: Ellie's Eco-Smoothies and Fran's Freshtastic Smoothies. Ellie and Fran offer similar prices and menus of fruit smoothies and fresh juices to their customers. Most of their customers are employees of nearby companies in the downtown area. In a **competitive location** situation, both Ellie and Fran would seek to maximize their market share and profits while minimizing their *individual* costs. In this case, "cost" represents the round-trip walking distances their customers must travel to their respective stations. However, since Ellie and Fran have noticed that customers on Main Street tend to visit the smoothie station that is closest to their office, they have decided to work together to maximize both of their profits while minimizing their customers' average walking distance. This is known as a **cooperative location** problem, in which the two stations operate as one overall system to minimize the overall *system* costs.

Assume the following:
- The segment of Main Street that Ellie and Fran serve is approximately one mile in length,
- Customers are equally distributed throughout the mile, and
- Customers will walk to the station that is closest to their office or company.

Using the line below as a depiction of Main Street, where do you think Ellie and Fran will position their stations along the street to minimize their customers' average walking distance?[1]

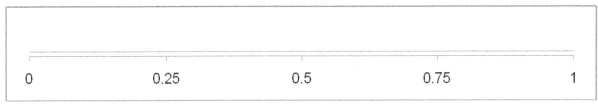

Figure 7.1.1: Line representing Main Street

Suppose Ellie and Fran position their stations next to each other at the halfway point of the line. Draw figures to represent their stations on the line provided in Figure 7.1.1. Notice that they each would have an equal market share of 0.5 miles of customers to serve. The maximum distance a customer would need to travel is 0.5 miles, and the minimum distance would be 0 miles. Thus, the *average* distance that a customer would travel to one of the stations is $(0 + 0.5) / 2 = 0.25$ miles.

Let's try another scenario: suppose Ellie positions her station at the 0.25-mile mark, while Fran positions hers at the 0.75-mile mark. Again, each woman has an equal market share of customers along a 0.5 mile segment of Main Street.

1. What will be the maximum distance a customer would need to travel? What would be the minimum?

2. What will be the customers' *average* walking distance?

3. Which positions of the stations minimize the *average* walking distance for customers?

Section 7.2: Supply Chain Exploration

Consider the supply chain of Spike's Sports Emporium, a small store located on Main Street in downtown Raleigh. Spike would like to construct or rent warehouse space to store excess inventory (basketballs, golf clubs, kayaks, etc.). His sports equipment supplier in Durham, which is 30 miles from Spike's, will send one large truckload per week to the warehouse. Spike will require two smaller truckloads of equipment from the warehouse to be delivered to the store each week to meet his customers' demands. Spike uses the smaller trucks to transport his goods since renting an 18-wheeler truck and trailer would be too costly for him. Also he would not be able to park it in the parking lot nearby! Figure 7.2.1 illustrates the decision scenario. The supplier is located at mileage marker 0 and the customer, Spike, is at mileage marker 30.

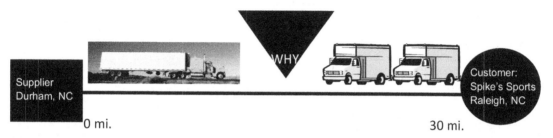

Figure 7.2.1: Distance between Spike's Sports Emporium and its supplier

Let's explore what happens if we locate the warehouse at the geographical center of the supply chain at Mile Marker 15. What would be the total *one-way* cost of the system in terms of the distance traveled? The large truck from Durham would travel 15 miles to get to the warehouse and the two small trucks from Raleigh would also travel 15 miles each. Disregard the size of the trucks for a moment and only consider the number of trips that they make each week. We see that the supplier makes 1 trip to the warehouse whereas Spike makes two trips. Thus the cost of this scenario (in terms of one-way distance traveled) would be:

Durham: 15 miles × 1 trip = 15 trip miles
Raleigh: 15 miles × 2 trips = 30 trip miles
Total = 45 trip miles

What happens if we locate the warehouse one mile closer to the supplier in Durham at Mile Marker 14? What happens if we locate the warehouse one mile closer to Spike in Raleigh at Mile Marker 16? For each of these scenarios, what would be the total one-way cost of the system in terms of the distance traveled? Following the same calculation methodology as before:

Proposed Warehouse Location	Weekly One-Way Distance Traveled by the Supplier	Weekly One-Way Distance Traveled by Spike	Total Trip Miles
Mile 14	14 miles × 1 trip = 14	16 miles × 2 trips = 32	46
Mile 16	16 miles × 1 trip = 16	14 miles × 2 trips = 28	44

Table 7.2.1: Evaluating warehouse location one mile from midpoint

Notice how the total cost (i.e. one-way distance) of the system increased when the warehouse location shifted one mile closer to Durham, but decreased when shifted one mile closer to Durham. Thus, we should keep shifting toward Raleigh until we find the location with the smallest total trip miles.

1. Calculate the costs if the warehouse were located at the Mile Markers given in Table 7.2.2.

Proposed Warehouse Location	Weekly One-Way Distance Traveled by the Supplier	Weekly One-Way Distance Traveled by Spike	Total Trip Miles
Mile 15	15 miles × 1 trip = 15	15 miles × 2 trips = 30	45
Mile 20	20 miles × 1 trip = 20	10 miles × 2 trips = 20	40
Mile 25			
Mile 27			
Mile 29			
Mile 30			

Table 7.2.2 Converging on the best location for Spike's warehouse

2. What trend do you notice in the total cost?

It turns out that it is optimal to locate the warehouse in Raleigh at Mile Marker 30 near the customer.

3. Why is this case?

Let's ignore the distances momentarily and think only about the weights of the supplier and customer locations. The number of trips that each entity of the supply chain (i.e., the supplier and Spike) makes is known as their **weight**. Keep in mind that the term "weight" does not refer to the physical weights of the trucks and cargo. Rather, it is a representation of that entity's contribution to the overall system (i.e. Spike requires 2 trips to meet his customers' demand). What if we position the warehouse to balance the systems such that no more than one-half of the total weight of the system is to the right and to the left of the warehouse's location? Let's explore.

> Total weight of system = 1 + 2 = 3 truckload trips
> One half of the total weight = 3 / 2 = 1.5

The value 1.5 is known as the **median location**. It represents the location at which no more than one half of the total weight of the system is to the right and no more than one half of the total weight is to the left.

Would the geographical center location at Mile Marker 15 satisfy this condition? This would mean that there would be less than 1.5 truckload trips to the left and more than 1.5 truckload trips to the right of Mile Marker 15. However, at Mile Marker 15, a weight of one would be located to the left of the warehouse and a weight of two would be located to the right, so the condition has not been satisfied.

Recall how you noticed that the total cost of the system decreases as you locate the warehouse closer and closer to Raleigh at Mile Marker 30. It turns out that it is optimal to locate the warehouse in Raleigh near the customer (Spike's Sports Emporium) to minimize the overall system costs. It is only in Raleigh that we find no more than 1.5 truckloads both to the left and to the right. In fact, there is one truckload to the left and 0 truckloads to the right. The system costs (distance x weight) are calculated as follows:

Durham: 30 miles × 1 trip = 30 trip miles
Raleigh: 0 miles × 2 trips = 0 trip miles
Total = 30 trip miles

In actual practice, there would be other considerations such as the cost of purchasing or renting a warehouse. Since these costs are typically higher, often much higher, within a city, they would need to be considered. The practical solution might be to locate the warehouse as close to Raleigh as possible, taking the cost of land into consideration.

The concept of median location will be explored further in the next section.

Section 7.3: One-Dimensional Location: North Carolina I-40

Agros Fresh Farms is located in Greenville, South Carolina. It is a leading supplier of fresh strawberries to supermarkets in North and South Carolina. Agros would like to expand its operations and has decided to purchase one large plot of farmland along Interstate-40 in North Carolina to better serve customers in the following cities: Asheville, Statesville, Winston-Salem, Greensboro, Durham, Raleigh, and Wilmington.[2] It wants to have access to a large labor pool to staff the farm. Agros has decided to purchase the land in or just outside of one of the aforementioned cities. Figure 7.3.1 gives a map of North Carolina showing the potential cities along I-40 where the supplier will purchase the plot of farmland. It also shows the number of highway miles from the beginning of I-40 at the state's western border to each city. The farm should be located to minimize the distance trucks travel to serve Agros' customers.

This analysis assumes that I-40 will be used for all travel. The weekly demand in truckloads to each city is in parentheses: Asheville (10), Statesville (8), Winston-Salem (10), Greensboro (14), Durham (12), Raleigh (20), and Wilmington (6). Determine where the supplier should purchase farmland and locate his new strawberry farm.

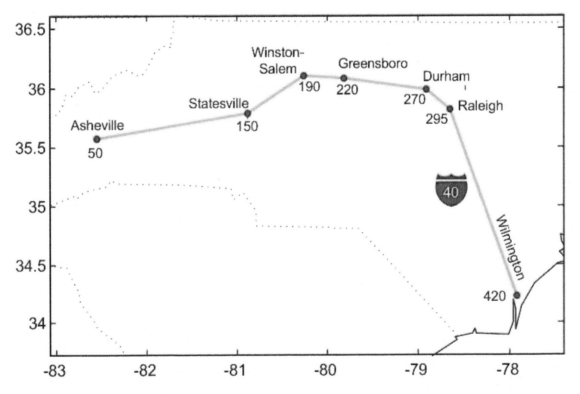

Figure 7.3.1: Mile-marker for each city along I-40 in North Carolina

Chapter 7   Finding the Best Place – Location Problems

Building upon the procedure from the Spike's Sports Emporium problem:

1. Label each city's "weight" on the map (i.e. their contribution in truckload trips to the overall system).
2. Determine the total weight of the system:
3. Identify the median location by cutting the total weight in half:

In step 2, we created the total weight of the system by simply adding all the weights together. To find the optimal solution, we need to identify the *cumulative sum of the weights* of the system. A cumulative sum is found by identifying the sum of the weight at a specific location and the weights of all the locations to the left of that location. For example, the cumulative sum for Winston-Salem is 10 + 8 + 10 = 28. The optimal solution will be the location where the cumulative sum just surpasses the median location found in step 3.

1. Starting at Asheville, create a cumulative sum of the weights for all the cities along I-40.

2. Locate the farm at the city where the cumulative sum exceeds 40 for the first time.

This represents the location where no more than one-half of the total weight of the system is to the right and to the left of the farm's location. You can check your work by finding the cumulative sum of the cities to the left of the strawberry farm and to the right of the city. (Do not include the weight for the city in which the farm is located).

3. What is the number of truckloads to the left of the city (i.e., the median)?

4. What is number of truckloads to the right of the city (i.e., the median)?

Check your work by starting at Wilmington and creating a cumulative sum of the weights that are to the right of the strawberry farm. This, too, should be less than 40, but does it necessarily have to be the same as the cumulative sum on the left.

5. Starting at Wilmington, create a cumulative sum of the weights for all the cities along I-40.

6. Explain how the median location is the same but the cumulative total is different.

As we did with the Spike's Sports Emporium problem, the total cost of the system will be the sum of each of the distances from the farm multiplied by the respective weights of each city. It is crucial to note that the distance from a fixed point is measured in absolute value and our distances will be measured in reference to the city where the farm is located. Finally, recall that the weight represents the weekly demand, in truckloads, of the customers in each city (referred to as "trips" in Spike's problem).

7. What are the total truckload miles for the best location?

8. What are the total truckload miles for the cities just east and just west of the optimal?

Lead Author: Amy Craig Reamer, supplemented by Kenneth Chelst

Section 7.4 Two-Dimensional Location: Ellie's Eco-Smoothies Store

All of the previous examples measured distance in only one dimension. For example, Ellie and Fran's mobile smoothie stations and Spike's Sports Emporium focused on moving only left or right along the same street or highway. Similarly, the location of the strawberry farm that supplies Ellie and Fran used only one dimension, traveling west or east on I-40. If we expand our location problems to include two dimensions, we encounter the difficulty of measuring distance.

Given two points $P_1 = (x_1, y_1)$ and $P_2 = (x_2, y_2)$, there are a number of different ways you can calculate the distance between points P_1 and P_2, depending on the context of the problem. One such distance is the **Euclidean distance**, which is defined as the straight-line distance between two points. For example, an airplane would travel a Euclidean distance between two points P_1 and P_2. Euclidean distance between points P_1 and P_2 is given by the following equation.

$$d(P_1, P_2) = \sqrt{(x_2 - x_1)^2 + (y_2 - y_1)^2}$$

However, in the context of some of our problems, Euclidean distance will not work. Another type of distance is **rectilinear distance**, which is defined as the distance between two points measured along axes at right angles.[3] For example, a taxicab would travel a rectilinear distance along the streets in the downtown area of a city between two points P_1 and P_2. The taxicab cannot just drive along the straight-line path given by Euclidean distance; it must drive along the streets of the city. This assumes that streets are parallel or perpendicular to one another. Rectilinear distance is given by the following equation.

$$d(P_1, P_2) = |x_2 - x_1| + |y_2 - y_1|$$

To see how rectilinear distance will be a necessary component of solving a two-dimensional location problem, let's revisit Ellie's mobile smoothie station. Recently, Ellie formulated a new smoothie with an anti-aging compound that is all the rage among her customers. Sales have increased so dramatically that she has decided to close her mobile smoothie station and open a smoothie, fruit, and juice store in the downtown area. Patrons that work in ten local companies are expected to make daily trips to her store according to Table 7.4.1.

Chapter 7 Finding the Best Place – Location Problems

Company	Location (x, y)	Number of customer visits/day from …
1. Dee's Dirt Bikes	(0.5, 7)	8
2. Sid's Cell Phones	(7, 9)	20
3. Paula's Purses	(6, 9)	12
4. Pam's Paper Company	(2.5, 7)	19
5. Spike's Sports Emporium	(1.5, 2.5)	3
6. Fred's Financial Group	(5, 1)	16
7. Anita's Animal Shop	(5, 6)	5
8. Michelangelo's Mechanics	(9, 7)	21
9. Jill's Junk Antique Store	(6, 1)	9
10. Tronda's Toys	(2.5, 2.5)	17

Table 7.4.1: Companies with employees that will visit Ellie's new store

Figure 7.4.1 shows the locations of the ten companies from which Ellie will draw her customers. Our goal is to determine where Ellie should open her new store. She wants to minimize the total rectilinear distance that all customers must travel from their respective offices to the store.

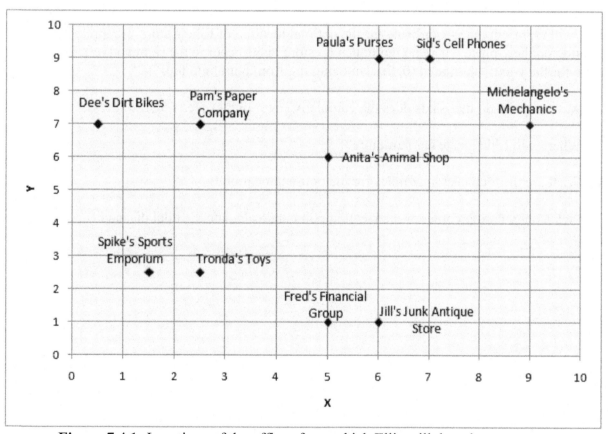

Figure 7.4.1: Locations of the offices from which Ellie will draw her customers

This problem is very similar to the North Carolina I-40 problem, except now we must work in two dimensions instead of one. The first step is to label each company's "weight" (number of customers) on the coordinate plane shown in Figure 7.4.1. Recall that we started the North

Carolina I-40 problem by identifying the median location of the system. We must do the same here in the two-dimensional problem. To find the median location of the system we must find the total weight of the system and divide it by two.

$$\frac{8+20+12+19+3+16+5+21+9+17}{2} = 65$$

Therefore, the median location will be at a site where in the x direction the cumulative value first exceeds 65. Similarly, the median location will be at a site where in the y direction the cumulative value first exceeds 65.

In the one-dimensional problem, our next step was to start at a fixed endpoint and calculate the cumulative sum of the weights of each location until we just surpass the median location. The two-dimensional problem continues in the same way, except that we need to consider two separate cumulative sums—one along the x-axis and one along the y-axis. Starting at the origin (0, 0), create the cumulative sum of the weights as you approach each plotted point, while moving from left to right.

1. At what point on the x-axis does the cumulative sum first exceed 65 customers?

This will represent the x-coordinate for the optimal location of Ellie's store. It represents the location on the x-plane where no more than 65 store visits occur to the right and left. Do the same for the y-axis, starting at (0, 0), while moving from bottom to top.

2. At what point on the y-axis does the cumulative sum first exceed 65 customers?

3. Where will Ellie locate her new store?

4. What are the total customer miles for the optimal location?

5. Pick a neighboring corner or intersection point and calculate the total distance.

Section 7.5 Set-Covering Problem: Multiple Local Disaster Response Facilities

Now consider six cities in central Oklahoma. These are some of the most tornado-ravaged cities in the United States. The governor would like to determine in which cities to construct disaster response facilities to house just a few emergency response vehicles at each site (no such facilities currently exist). Due to budgetary constraints, the governor wants to build the minimum number of facilities needed to ensure that at least one facility is within a 25-minute drive of each city.

The time in minutes required to drive between the six cities is shown in Table 7.5.1. In addition, the governor does not want agencies in both Moore and Oklahoma City. We will formulate a 0–1 integer programming problem that will inform the governor how many agencies need to be constructed and in which of the six cities they should be located.[4]

(minutes)		To					
		Edmond	Midwest City	Moore	Mustang	Norman	Oklahoma City
From	Edmond	0	27	31	36	42	20
	Midwest City	27	0	22	33	30	12
	Moore	31	22	0	29	17	16
	Mustang	36	33	29	0	40	24
	Norman	42	30	17	40	0	28
	Oklahoma City	20	12	16	24	28	0

Table 7.5.1: Driving time in minutes between six cities in central Oklahoma[5]

1. What decisions about this problem situation must the governor make?

2. Are any of these "yes/no" decisions?

3. Define the binary decision variables.

4. Since our goal is to identify how many facilities to build and their optimal locations, what is the equation for the objective function?

5. Is this a maximization or minimization problem?

6. In constructing the constraints for the IP formulation, it is helpful to think about which locations can reach each city in 25 minutes or less. Using the data from Table 7.5.1, fill in the cities that are within 25 minutes of each given city in Table 7.5.2.

City	City(ies) within 25 minutes
Edmond (x_1)	
Midwest City (x_2)	
Moore (x_3)	
Mustang (x_4)	
Norman (x_5)	
Oklahoma City (x_6)	

Table 7.5.2: Cities within 25 minutes driving time

7. How can the information in Table 7.5.2 translate into viable constraints?

Let's take a look at Edmond. According to Table 7.5.1, Oklahoma City is the only city that is within a 25-minute drive of Edmond. And, of course, Edmond is within a 25-minute drive of itself. Thus, the first constraint must require that there be at least one agency constructed in either Edmond or Oklahoma City or both to satisfy the 25-minute driving distance requirement for Edmond. The constraint for Edmond would be:
$$x_1 + x_6 \geq 1.$$

Similar logic can be followed for the other cities.

8. Use the information in Table 7.5.2 to identify the constraints for this problem.

9. Formulate the 0-1 integer programming problem in Excel.

10. Use the Excel Solver to find the optimal solution.

Chapter 7 (Location Problems) Homework Questions

1. The E. G. Cream products company is planning to establish a processing facility along I-90 in Wisconsin. This facility will make high value milk products. It has contracted to receive regular truckloads of raw milk from dairy farms situated along the I-90 corridor. The data in Table 1 summarizes the number of truckloads of milk from farms near each of the exits. The exit numbers correspond to mileage markers. For example, E.G. Cream will receive 4 truckloads of milk from farms near Onalaska by exit #3. In contrast, they have contracted with farms for 9 truckloads of milk from the Mauston area around exit #69. The distance between Onalaska and Mauston is 66 miles. The V.P. for operations, Alfred Swiss, wants to identify the location that will minimize the distance the milk carriers will travel to make deliveries to the new facility.

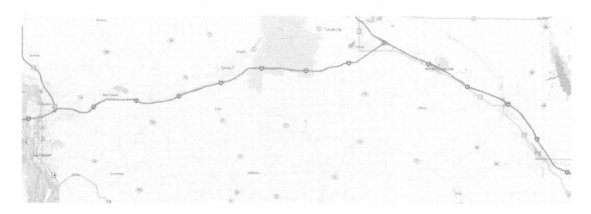

	City						
	Onalaska	West Salem	Sparta	Tomah	Camp Douglas	New Lisbon	Mauston
I-90 Exit Number	3	12	25	41	55	61	69
Truckloads/week	4	3	4	6	7	4	9

Table 1: Truckloads and Exit Numbers along I-90 in Wisconsin

   a. Near which exit should E.G. Cream build a facility?

   b. What is the average travel distance from farms to the processing facility? (Use the exit number as the farm location.)

   c. Mr. Swiss is thinking ahead. He believes there may be opportunities to purchase additional quantities of milk from farms in the eastern part of the state. These farms are in the New Lisbon area. Would the optimal location change if 3 more truckloads were to come from the New Lisbon area?

   d. A group of farm operators in the western part of the state have approached E.G. Cream management with a proposal to deliver a total of six more truckloads per week. These extra truckloads would be evenly split between Onalaska, West Salem, and Sparta. (Ignore part b above.) Would this affect the optimal location of the processing facility?

e. Refer back to part c. It was stated that the truckloads are evenly divided among the three exits. Would the answer to part c change if all six truckloads came from Onalaska? Would the answer to part c change if all six truckloads came from Sparta?

f. Review your answers to parts b, c, and d. Generalize your findings as to the sensitivity of the optimal locations to additional truckloads from different parts of the I-90 corridor.

2. A number of service stations and convenience stores along the I-94 corridor in Michigan have formed a purchasing cooperative. They plan to share rented storage space for canned goods they sell in their stores. Each of these stations has limited storage space on site. As a result they periodically receive truckloads of goods from a regional warehouse. Table 2 lists the locations of the stations and stores, the mile exit marker and the annual number of truckloads.

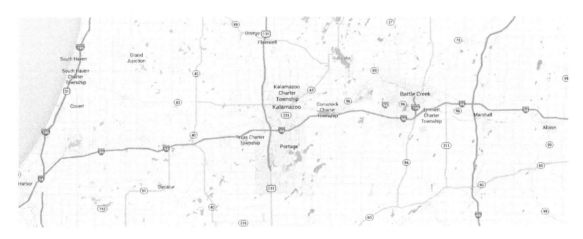

City	Michigan I-94 Mile Exit	Truckloads per year
Watervliet	41	6
Hartford	46	9
Lawrence	52	4
Paw Paw	60	12
Mattawan	66	7
Galesburg	85	7
Marshall	110	16
Albion	121	18

Table 2: Michigan Cities, I-94m exit, and Truckloads

a. Near which city-exit should the warehouse be located to minimize delivery distances?

b. What is the average travel distance from the warehouse to the stations and stores? (Use the exit number as the station and store location.)

c. If the cooperative is successful, other businesses in the far western part of the state may also decide to rent space. If these businesses need four truckloads a year, would this affect the optimal location? What if they require ten truckloads per year?

d. In part b, we simply referred to these business as in the far western part of the state. How would the answers in part b change, if the stores were located near exit 24 or Exit 16?

e. If the cooperative were to expand their group, what strategy would keep the optimal location unchanged from that found in part a?

3. The Road-Aid Company has a contract to provide road assistance along a section of the Garden State Parkway in northern New Jersey. Its contract covers the section of road from Garfield, just north of Clifton, all the way down to Iselin. Minnie Driva was interested in knowing where to keep his vehicle in between calls for assistance. She planned on getting permission to stay at a gas station just outside an exit. Road-Aid had the exact GPS location for each call last winter. Driva decided to group the data by assigning each call to the nearest exit. Table 3 summarizes the number of calls for assistance in the month of January. In Minnie's experience, the geographic pattern of calls is different on weekdays than on weekends. The data are group by weekdays, weekends and total.

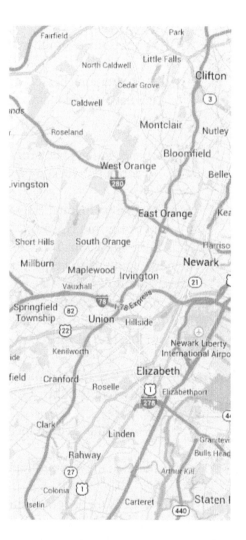

January Calls for Assistance				
City	Exit #	Week-days	Week-ends	Total
Garfield	157	36	11	47
Clifton	154	32	13	45
Montclair	151	40	15	55
Bloomfield	148	16	8	24
E. Orange	147	24	5	29
Irvington	143	20	9	29
Union	141	40	5	45
Cranford	137	48	7	55
Clark	135	40	8	48
Iselin	131	48	10	47

Table 3: Garden State Parkway calls for assistance in January

a. Where should Minnie stay so as to minimize the average travel distance to a call for assistance on a weekday?

b. What is the average travel distance to a call for assistance on a weekday?

c. Where should Minnie stay so as to minimize the average travel distance to a call for assistance on a weekend?

d. What is the average travel distance to a call for assistance on a weekend?

Minnie thought that it might be better to stay at the same gas station all week long. That way she would only have to ask one station manager for permission.

e. What is the preferred location using the total call for service data?

f. What is the average travel distance to all calls for assistance?

g. If she stations herself at the optimal location for the week, what is her average travel distance on weekdays? How much longer is this average than the answer found in part b?

h. If she stations herself at the optimal location for the week, what is her average travel distance on weekdays? How much longer is this than the answer found in part d?

4. USC University Hospital (H_1), Cedars Sinai Medical Center (H_2), Olympia Medical Center (H_3), and California Hospital (H_4) are all located in the Los Angeles area. They are cooperating to establish a centralized blood-bank facility that will serve all four hospitals. The new facility is to be located such that distance traveled is minimized. The hospitals are located as follows: $H_1 = (5, 10)$, $H_2 = (7, 6)$, $H_3 = (4, 2)$, and $H_4 = (16, 3)$. The number of deliveries to be made per year between the blood-bank facility and each hospital is estimated to be 450, 1,200, 300, and 1,500, respectively. Assuming rectilinear travel, determine the optimum location of the blood-bank facility.[6]

a. Draw a graph and plot the locations of the four hospitals. Label each hospital with its respective weight.

b. What is the median location?

c. What is the x-coordinate for the optimal location of the blood-bank facility?

d. What is the y-coordinate for the optimal location of the blood-bank facility?

e. Calculate the distance from each hospital to the blood-bank facility and using these distances and the respective weights, calculate the total travel cost of this system.

f. What assumption must be made concerning the deliveries of blood?

g. Is it reasonable to expect delivery trips will always be to only one hospital when a delivery is made?

5. Mrs. Williams has to pick up several items for her daughter's sweet sixteen party at various locations. She wants to locate a parking space downtown such that she can get to all the stores in a minimum amount of time without reparking. Each block is square, 100 feet on a side. Streets running north to south are numbered consecutively. Those running east to west are lettered consecutively. The bakery is at 6th and E; she must walk half as fast as normal from the bakery so that she won't drop the birthday cake. At 10th and D is the grocery store. The dress shop is at 12th and G. Mrs. Williams picks up her daughter, Cecilia, from the hair salon at 10th and G and they walk twice as fast as normal back to the car so that the wind doesn't mess up her hair. It is assumed she must stay on the sidewalks that enclose each block- distance used crossing streets is considered negligible. It is also assumed that she must return to the car after visiting each store before visiting the next one.[7]

   a. Construct a grid to represent this problem situation.

   b. Determine the location of the parking space that satisfies her objective taking note of all assumptions you make in formulating your decision.

   c. What is the distance from each shop to the parking space?

   d. How did the walking speeds influence your solution?

6. Refer back to the scenario posed in Section 7.5. Suppose the minimum driving time was set at 30 minutes instead of 25. How does that change the problem formulation? Does it have an effect on the optimal solution?

7. Referring back again to the scenario posed in Section 7.5, take away the constraint limiting the locations of agencies in both Moore and Oklahoma City and find a solution. Is this solution unique? Interpret your findings in the context of the problem.

8. In metropolitan areas it is common for high schools to share football stadiums since each school may not have enough land adjacent to it to construct a large football stadium complex. Figure 1 shows a representation of a collection of high school districts. Each shape contains one high school and the surrounding area in which students live who attend that high school. The nodes represent candidate locations for new football stadiums that will be constructed in the region in 3 years. Each football stadium can serve all schools that are adjacent to it. For example, potential football stadium 1 could serve high schools 2, 3, and 4. Your objective is to determine the minimum number of football stadiums that need to be constructed to cover each and every high school. Also identify which high schools each stadium would serve.

Chapter 7 — Finding the Best Place – Location Problems

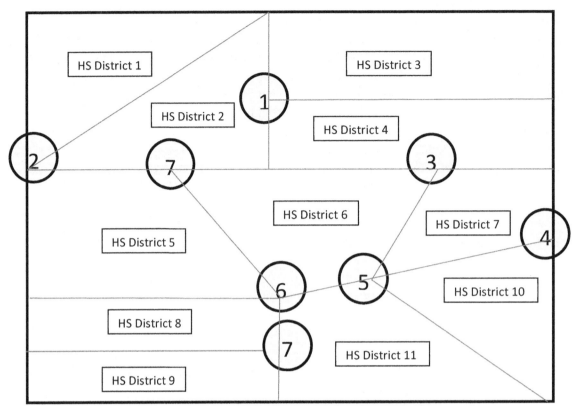

Figure 1: Metropolitan area high school districts

a. What kind of mathematical programming problem is this?

b. Define the decision variables and objective function.

c. Complete Table 4. Identify for each high school, the list of potential stadium sites that would cover that high school.

High School District	Adjacent candidate football stadium locations
1	
2	
3	
4	
5	
6	
7	
8	
9	
10	
11	

Table 4: Football stadium locations adjacent to high schools

d. What are the constraints?

e. Formulate and solve the problem using Excel and Solver.

f. Are there any high schools that are covered by more than one stadium?

9. Rapid Response Inc. has a contract to provide rapid emergency medical services to 15 neighborhoods. The contract specifies that a Rapid Response vehicle must be within five minutes travel time of each neighborhood. There are eight possible places where Rapid Response can station one of its vehicles. Table 5 has the travel time in minutes from each location to each neighborhood.

		Neighborhoods														
		A	B	C	D	E	F	G	H	I	J	K	L	M	N	O
Potential Locations	1	7.4	2.0	6.4	7.1	0.3	2.5	3.5	8.6	6.3	9.6	1.4	9.3	4.3	1.3	0.5
	2	0.5	1.6	9.3	8.7	0.2	8.7	9.3	5.7	3.6	4.9	9.9	9.1	2.7	4.5	6.7
	3	3.8	9.7	8.1	4.8	0.3	7.4	2.2	2.0	4.4	3.2	3.0	9.0	3.3	1.5	5.0
	4	9.6	5.9	8.6	2.4	7.5	4.0	3.2	3.1	6.7	6.8	9.3	6.0	8.2	7.9	0.2
	5	6.5	2.9	2.4	9.0	3.9	3.9	3.7	0.1	8.8	5.8	5.9	8.8	0.7	2.3	1.7
	6	2.3	3.5	8.5	6.7	7.4	8.1	3.6	6.9	5.6	7.5	3.0	6.3	8.0	8.6	1.2
	7	8.1	2.4	6.4	3.9	8.0	4.6	7.4	2.0	0.5	0.9	6.5	7.4	4.7	1.0	3.8
	8	3.9	5.3	9.7	3.4	7.4	3.5	6.5	5.4	1.2	8.5	5.2	2.1	1.4	0.2	0.5

Table 5: Travel times between potential vehicle locations and neighborhoods

In order to apply the concept of a covering, each cell in the above table would need to be compared to five minutes. If the travel time were less than five, you would record a 1 to signify that location covers that neighborhood. If the travel time were greater than five minutes, you would record a 0 to signify that the location does not cover the neighborhood. There are 120 travel time cells in the table. This process would be tedious. However, Excel ha a function that can automate this task.

Imagine that the above table were in an Excel file. The command is the IF command. This command operates as follows. You specify the relationship you are checking for. You also specify the value to return if the relationship is true. You also specify the value to return if the relationship is false. Imagine that the column letters and row numbers in Table 5 were cell locations in a spreadsheet.

Here is an example of the command that would be used in this context.

=If(A1<=5,1,0) → 0

This command states if the value of cell A1 is less than or equal to 5, then return a value of 1. If it is greater than 5, return a value of 0. Since cell A1 contains the value 7.4, the function returns the value of 0. Here are the values to be returned for cells A2, A3 and A4.

=If(A2<=5,1,0) → 1
=If(A3<=5,1,0) → 1
=If(A4<=5,1,0) → 0

a. To check your understanding, determine each of the following values.
=If(B4<=5,1,0)
=If(C7<=5,1,0)
=If(F11<=5,1,0)

You can use the concept of copy and paste to generate the entire matrix of 0 and 1 values.

b. Define your decision variables. How many are there?

c. Write one constraint. How many are there?

d. Use Excel Solver to determine the minimum number of locations needed to meet the five minute travel time constraint.

e. What would be the impact on the minimum number of locations needed if the travel time constraint were reduced to four minutes?

f. What would be the impact on the minimum number of locations needed if the travel time constraint were increased to six minutes?

Chapter 7 Summary

What have we learned?

In this chapter we have studied location decisions. The primary objective function was to minimize the total distance that individuals travel to reach the facility. The optimal location in this case is called the median. We have presented step-by-step procedures for identifying the median in both one-dimensional and two-dimensional situations. In addition we have presented an alternative optimal location model that minimizes the number of facilities needed to provide coverage. This last model applies the concept of binary decision variables presented in Chapter 6.

Terms

Cooperative Location Situation in which two or more entities in a supply chain (example: supplier, retail store, distribution center) operate as one overall system to minimize the overall system costs, while maximizing profit.

Competitive Location Situation in which two or more entities in a supply chain seek to maximize their individual market share and profits while minimizing their individual costs.

System A collection of organized things; as a solar system.[8]

Supply Chain
(aka Logistics Network) The system of organizations, people, technology, activities, information and resources involved in moving a product or service from supplier to customer. Supply chain activities transform natural resources, raw materials and components into a finished product that is delivered to the end customer.[9] [Hint: Ask students how it is that you can go to McDonalds and they are never out of hamburgers?]

Euclidean Distance The straight line distance between two points (see also appendix 1).

$$d(P_1, P_2) = \sqrt{(x_2 - x_1)^2 + (y_2 - y_1)^2}$$

Rectilinear Distance The distance between two points measured along axes at right angles (see also Appendix 1).[10]

$$d(P_1, P_2) = |x_2 - x_1| + |y_2 - y_1|$$

Chapter 7 (Location Problems) Objectives

You should be able to:

- Determine the optimal location for a new facility in one- or two-dimensional problems.

- Distinguish between a cooperative and competitive location problems.

- Use linear programming techniques to determine the optimal location for a new facility in a set-covering problem.

Appendix 1: Theory & Equations

Single Facility Location Problems

The single facility location problem involves determining the location of one new facility (e.g. industrial plant, hospital, school, etc.) with respect to M existing facilities[11]. The new facility's location is the location which minimizes the sum of the weighted distances between the new facility and the M existing facilities[12]. The problem can be represented as follows, which is often referred to as the "cost" of locating a facility in a given location:

$$\min f(X) = \sum_{i=1}^{M} w_i \cdot d(X, P_i)$$

where:

$f(X)$ = the "cost" of locating the new facility at point X

w_i = weight of locating new facility at point X. To calculate the weight, find the product of cost per unit distance traveled and number of trips made per unit time between the new facility and existing Facility i ($/mile)

X = point location of new facility

P_i = point locations of the M existing facilities

$d(X, P_i)$ = distance between points X and P_i

Assuming the cost per unit distance is a constant, the objective of the single facility location problem is the determination of a location that minimizes the sum of all the distances traveled[13]. One example of a single facility location problem is a large clothing retailer deciding where to open a new retail store to serve its customers.

Solving 1-Dimensional & 2-Dimensional Single Facility Location Problems

Recall, a median is the middle number in a list of numbers arranged in order from smallest to largest (or vice versa). In 1-dimensional and 2-dimensional facility location problems, *median location* is used to calculate where a new facility should be located given the weights of the existing facilities. An example of a "weight" could be the number of trips a student makes to the snack machine from their classroom each day. If the student makes two trips to the snack machine per day, then the "weight" (w) is 2.

A median location is the first existing facility location where the cumulative weight of the existing facilities up to that point is at least half of the total weight of all existing facilities. In other words, it is the existing facility location that splits the total weight into equal halves.[14] The procedure is as follows:

1. Order the existing facilities along a dimension x (e.g. gas stations located along an interstate, where the dimension is the interstate mile marker by which they are located):

   $$|x_1| \le |x_2| \le \ \le |x_M|$$

2. Locate the new facility at the existing facility where:

Chapter 7 Study Guide

1. What is the difference between a cooperative location problem and a competitive location problem?

2. How are "weights" calculated in location problems?

3. How is the "median location" determined in a one-dimensional location problem?

4. How is the "median location" determined in a two-dimensional location problem?

5. Why are set-covering problems formulated similarly to binary programming problems?

References

[1] Adapted from Kay, M. ISE 453 & ISE 754 Course Notes. North Carolina State University, Raleigh, NC. 2008.

[2] Adapted from Kay, M. ISE 453 & ISE 754 Course Notes. North Carolina State University, Raleigh, NC. 2008.

[3] "Rectilinear." NIST. 17 Dec. 2004. 27 May 2008 <http://www.nist.gov/dads/HTML/rectilinear.html>.

[4] Adapted from Winston, W. Operations Research Applications and Algorithms (4th ed.). Belmont, CA: Brooks/Cole Thomson Learning, 2004.

[5] http://www.mapquest.com. Mapquest Driving Directions. 27 May 2008.

[6] Francis, R. & White, J. Facility Layout and Location: An Analytical Approach. Englewood Cliffs, New Jersey: Prentice-Hall, 1974.

[7] Francis, R. & White, J. Facility Layout and Location: An Analytical Approach. Englewood Cliffs, New Jersey: Prentice-Hall, 1974.

[8] "System." Wikipedia. 25 Sept. 2008. 30 Sept. 2008 <http://en.wikipedia.org/wiki/System>.

[9] "Supply Chain." Wikipedia. 25 Sept. 2008. 30 Sept. 2008 <http://en.wikipedia.org/wiki/Supply_chain>.

[10] "Rectilinear." NIST. 17 Dec. 2004. 27 May 2008 <http://www.nist.gov/dads/HTML/rectilinear.html>.

[11] Francis, R. & White, J. Facility Layout and Location: An Analytical Approach. Englewood Cliffs, New Jersey: Prentice-Hall, 1974.

[12] Kay, M. ISE 453 & ISE 754 Course Notes. North Carolina State University, Raleigh, NC. 2008.

[13] Francis, R. & White, J. Facility Layout and Location: An Analytical Approach. Englewood Cliffs, New Jersey: Prentice-Hall, 1974.

[14] Kay, M. ISE 453 & ISE 754 Course Notes. North Carolina State University, Raleigh, NC. 2008.

CHAPTER 8:

Waiting in Line with Polynomials – Non-linear functions and Queueing

Section 8.0: Queueing Theory Introduction

"Queues" is the British word for waiting lines, and we have all experienced them. It is estimated that Americans spend over a billion hours per year waiting in queues. Queueing theory is the mathematical study of queues that operations researchers and industrial engineers use to increase the efficiency of queueing systems. The term **queueing system** refers to the people currently waiting in the queue and the person(s) being served. Inefficient queues not only hurt customer satisfaction, but they also have an impact on a nation's economy. If the time we wasted waiting in queues could instead be spent productively, it would amount to the equivalent of half a million additional workers.

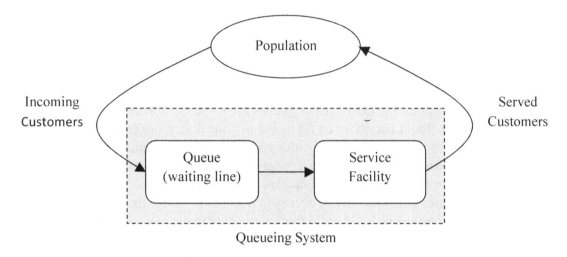

Figure 8.0.1: Basic queuing model

Figure 8.0.1 represents the key components of queueing systems. The system begins with customers seeking service and entering the queuing system. They are characterized as an arrival process with an average rate of λ. The second critical element is the service facility. This can involve one individual, a single server or multiple servers. The basic building block is the rate, μ, at which service is completed by a continuously busy server. The number of servers is simply, s. The final element is the queue of customers waiting to be served. When there is more than one server, a decision has to be made as to how to organize the queues. The customers can be organized to form a single queue as most banks do. Alternatively, there may be a separate queue for each server as occurs with cars waiting to pay tolls. Once a customer is served, the customer leaves the queueing system.

The sequence in which customers are served is called the queue discipline. The simplest procedure is FIFO, first-in-first-out. The next customer served is the one at head of the line. However, the customers seeking service may fall into distinct categories that are treated differently. For example, in an emergency room, patients experiencing a heart attack will be given priority over someone with a broken finger. Most airlines give priority boarding to frequent flyers. Some amusement parks allow patrons to pay a premium to reduce their waiting time at rides.

Chapter 8 — Waiting in Line with Polynomials - Non-linear Functions and Queueing Theory

Before discussing the mathematics, we would like the reader to explore his or her own queueing experiences.

1. Describe two different queues in which you have spent time waiting in an actual line.

2. Describe one situation in which you were waiting for service but the people waiting were not in a physical line that you could see.

3. Estimate how much time you spend waiting in queues during a typical week.

4. Reflecting on the time you spend in queues, propose two strategies for reducing the amount of time customers spend waiting.

5. From a manager's perspective, what are the downsides of your suggestions?

Not all queues are created equal. Some queues are designed to include distractions to make the time you spend waiting more pleasant. Other queues are quite boring and may lead you to frustration as you think about ways you could be using your time more productively. The experience of waiting in line is influenced by the waiting area environment and our expectations as to the length of the wait. Imagine having to wait standing up in a dentist's office for twenty minutes, while a patient is screaming in an adjacent examination room. Now imagine an alternative wait in comfortable chairs with access to the "latest" magazines for a variety of customer tastes. For your ten-year old child there is a video game console, and the area is sound proof.

Disney is one example of a company that has become expert in understanding the psychology of waiting. Waiting in a line that is moving seems less boring than standing still in the same spot. TV monitors with engaging pictures help keep visitors' minds off the clock. In addition, if they can see and hear some of the excitement of those who have completed their wait and enjoyed the attraction, anticipation increases and waiting seems worthwhile. Lastly, expectations are a major factor in determining customer satisfaction. If customers approach a line and are told the wait will be fifteen minutes, at least they have the information to make an informed judgment as to join the line or not. If it turns out to be less than the quoted fifteen minutes, they are pleasantly surprised.

Another dimension to the psychology of waiting relates to fairness. It can be very upsetting to see someone arrive after you in line and end up being served before you. This can happen if there are two separate lines. You might get stuck behind a customer who has a complicated request that takes a long time to service. As a result, people who have joined the other line even after you might ending up waiting less time. Many businesses have addressed this potential inequity by creating one line which all arriving customers enter. Thus, anyone who arrives after you must be further back in line and cannot begin service before you do.

6. Propose two strategies for making the time spent in a queue more pleasant.

7. Compare your responses to numbers 4 and 6; which approach seems more cost effective?

Chapter 8 — Waiting in Line with Polynomials- Non-linear Functions and Queueing Theory

All of us have experienced the annoyance of having to wait in line. Unfortunately, this phenomenon continues to be common in congested, urbanized and "high-tech" societies. We wait in line in our cars in traffic jams or at toll booths; we wait on hold for a help-line professional waiting for assistance; we wait in line at supermarkets to check out; we wait in line at fast-food restaurants; and we wait in line at banks and post offices. As customers, we do not generally like these waits. The managers of the establishments at which we wait also do not like us to wait, since it may cost them business. Why then is there waiting?

The answer is relatively simple: There is more demand for service than there is facility for service available. Why is this so? There may be many reasons; for example, there may be a shortage of available servers; it may be infeasible economically for a business to provide the level of service necessary to prevent waiting; or there may be a limit to the amount of service that can be provided. Generally, this limitation can be removed with the expenditure of capital.

A key contributor to queue formation is the randomness inherent in the arrival process and the service time. For example, if customers were scheduled every 15 minutes and each customer required exactly 14 minutes of service, no line would form. However, arrival and service of randomness will result in long lines as will be discussed later in this chapter.

The chapter contains a number of formula used to calculate queueing statistics. All of the mathematical models and examples in this chapter have three critical assumptions
- The randomness in the arrival process is described by the **Poisson distribution**.
- The randomness in the service time distribution is re represented by the **exponential distribution**.
- The statistics apply to long-term averages which in the queuing literature is called **steady state**.

To know how much service to make available, a manager would need to answer such questions as, "How long will a customer wait?" and "How many people will form in the line?" Queueing theory attempts to answer these questions through detailed mathematical analysis. The word "queue" is more commonly used in Great Britain and other countries than in the United States, but it is rapidly gaining acceptance in this country. However, it must be admitted that it is just as unpleasant to spend time in a queue as in a waiting line.

A queueing system can be simply described as customers arriving for service, waiting for service if it is not immediate, and if having waited for service, leaving the system after being served. The term customer is used in a general sense and does not imply necessarily a human customer. For example, a customer could be a ball bearing waiting to be polished, an airplane waiting in line to take off, a computer program waiting to be run, or a telephone call waiting to be answered.

Queueing theory can trace its origins back to a Danish mathematician named A. K. Erlang. In 1909, Erlang published *The Theory of Probabilities and Telephone Conversations* based on work he did for the Danish Telephone Company in Copenhagen. In the long gone days, all telephone calls passed through a telephone exchange operator who made the connection and calls were backed-up when all operators were busy. Work continued in the area of telephone applications and was a major factor in managing the first 911 systems in NYC.

There are many valuable applications of the theory, including traffic flow (vehicles, aircraft, people, communications), scheduling (patients in hospitals, jobs on machines, programs on a computer), and facility design (banks, post offices, amusement parks, fast-food restaurants).

Chapter 8 Waiting in Line with Polynomials- Non-linear Functions and Queueing Theory

Section 8.1: Arm-and-a Leg Tickets: Does This Line Ever Move?

Mr. I. M. Boss is vice president in charge of operations for Arm-and-a-Leg Ticket Sales. He is concerned about complaints regarding long waits at the ticket windows on Friday afternoons at many of the malls. To reduce the number of complaints, Mr. Boss has hired Dr. Hye I. Cue, an expert on queueing theory. Before we begin to analyze Mr. Boss's problem, which is called a single-server model, we will make the following assumptions:

- Individual customers arrive at random to purchase tickets.[1]
- The time to complete a purchase is also random. This might be due to the number of tickets the customer purchases or the customer asking for information about dates and seat locations.[2]

The arrival process is represented as an average rate of λ. The service time is also represented as a rate, the number of customers served per unit time when the server is continuously busy. For example assume the time to service a customer were 12 minutes. This is converted to a rate of five customers per hour. The Greek letter μ is the standard symbol for the service rate.

To use queueing theory, Dr. Cue needed to collect data about the customers. She spent several Friday afternoons observing the situation and collecting the data. She found that the average number of customers arriving per hour is 18. She also determined that it takes on average 3 minutes to process a customer. This means the single ticket agent can serve 20 customers per hour when he is continuously busy.

1. Dr. Cue observed that $\lambda =$ ____ and $\mu =$ ____. (Be sure to include units.)

2. If 18 customers arrive per hour, on average, how much time is there between successive arrivals?

3. If λ customers arrive per hour, on average, how much time is there between arrivals?

4. If μ customers can be served per hour, on average, how long does it take to serve one customer?

To develop a mathematical model of our queue, we must have an idea of the utilization, ρ. This is the ratio of the average rate of customer arrivals per unit of time, λ, to the average rate of customers who can be served per unit of time, μ. In order for this ratio to make sense, the time units of λ must be the same as those of μ.

Let $\rho = \dfrac{\lambda}{\mu}$.

5. What is the utilization on Friday afternoons at the Arm-and-a Leg Ticket counter?

---

[1] The pattern of random arrivals is assumed to follow the Poisson distribution.
[2] The pattern of random duration of service is assumed to follow the exponential distribution.

The average number of customers in the system is represented by L. This includes those in line and the customer at the ticket window. Experts in queueing theory have determined the following relationship between L and ρ.[3]

$$L = \frac{\rho}{1-\rho}$$

For example, there are times during the week when the number of arriving customers is much less than 18. For example, on Tuesday evenings the arrival rate is only 12 per hour. In that case the utilization, ρ, is equal to 0.6. Then

$$L = \frac{0.6}{1-0.6} = 1.5 \text{ customers}$$

There would be on average 1.5 customers in the line including a customer being served. (This is an average. Therefore, there is nothing wrong with this average not being an integer value.)

6. Use the value of ρ for Friday afternoons, to calculate L for Friday afternoons.

Customer satisfaction actually is more dependent upon the length of time it takes to get a ticket than the length of the line.

Let W = the average time a customer waits in a system including the time to be served by the ticket agent.

There is a well-known equation known as Little's formula, that relates W to L.

$$L = \lambda W$$

This equation is easily manipulated to determine W in terms of L.

$$W = \frac{L}{\lambda}$$

When the arrival rate was only 12 per hour, L was equal to 1.5.

$$W = \frac{1.5}{12} = 0.125 \text{ hours}$$

---

[3] The calculation of L assumes the system is approaching steady state and the statistics are long-term averages. This calculation involves the following steps. Create of an infinite series of state rate balance equations. These are solved by recursion. Then L is calculated with a formula for an infinite geometric series. F.S. Hillier and G. J. Lieberman (2012) Introduction to Operations Research, McGraw Hill.

Chapter 8   Waiting in Line with Polynomials- Non-linear Functions and Queueing Theory

Thus, the average time in the system is 0.125 hours. The unit of measurement is hours because the arrival rate was given in arrivals per hour. This number is multiplied by 60 to translate the waiting time into an average of 7.5 minutes. This total waiting time to complete service can be split into its two components. The first component is the time waiting to begin service. This is referred to as the wait in queue, W_q. The second component is the service time.

W = total waiting time

    = average time waiting to begin service + average service time

    $= W_q + \dfrac{1}{\mu}$

Thus to determine W_q we simply subtract the average service time from W.

$$W_q = W - \dfrac{1}{\mu}$$

Earlier it was calculated that W was 7.5 minutes and the average service time was three minutes. Thus, the average wait in queue on Tuesday evenings is 4.5 minutes.

$$W_q = W - \dfrac{1}{\mu}$$
$$= \left(0.125 - \dfrac{1}{20}\right) \text{ hours}$$
$$= (7.5 - 3) \text{ minutes}$$
$$= 4.5 \text{ minutes}$$

7. What is the average waiting time, W, on Friday afternoons? (The first calculation will be in units of hours. Convert this to a more readily understandable unit of minutes.)

8. What is the average waiting time in queue, W_q, on Friday afternoons?

Dr. Cue was interested in understanding waiting time on other days. She asked the person in charge of the computer system to show her what data was recorded with each ticket sale. She found that the system kept track of the actual time the ticket was sold. She decided that she could use those times to estimate the number of customers arriving per hour.

9. Why is the time stamp on the ticket not a perfect indication as to when the customer came to purchase a ticket?

Dr. Cue found that the arrival rate varied by day of week and time of day. The average arrival rate ranged from a low of 4 per hour on Monday afternoons to a high of 18 per hour on Friday afternoons. She also noticed that around the holiday season, the rate could even exceed 19 customers per hour.

10. Complete the table below to determine the average utilization (ρ), queue length (L), waiting time (W), and waiting time in queue (W_q) for different arrival rates (λ). (The initial calculation of W will be measured in hours since both the arrival and service rates are defined in units per hour. Convert this number into minutes.)

λ	$\rho = \dfrac{\lambda}{\mu}$	$L = \dfrac{\rho}{1-\rho}$	$W = \dfrac{L}{\lambda}$	$W_q = W - \dfrac{1}{\mu}$
4				
7				
10				
12	0.6	1.5	0.125 hrs. = 7.5 min.	4.5 min.
14				
16				
18				
19				
19.5				

Table 8.1.1: Queues for different arrival rates

Dr. Cue asked her assistant to graph L as a function of ρ using the data in Table 8.1.1. After reviewing the graph, she noticed there appeared to be two asymptotes.

11. Describe the vertical line that appears to be an asymptote.

12. Describe the horizontal line that appears to be an asymptote.

Now let's consider other values of ρ not mentioned in Table 8.1.1.

13. Suppose $\rho = -0.5$. What would be the value of L?

14. Why do values of ρ that are negative make no sense in this problem context? Why do the corresponding values of L also have no meaning?

In the Table 8.1.1 the highest value of λ was 19.5 customers per hour. It is possible that during the busiest periods the number of arrivals might be a as high 25 customers per hour. If that occurs

$$\rho = \frac{25}{20} = 1.25$$

When the utilization is greater than 1, the arrival rate exceeds the service capacity. In this instance, there are five more customers per hour than can be serviced. This is 25% more than the service capacity. If you insert 1.25 into the equation for L, the equation yields the value -5.0.

$$L = \frac{1.25}{1-1.25} = -5 \text{ customers}$$

There can be no such thing as negative five customers waiting in line. In other words this equation is not meaningful if ρ is greater than one.

What happens to the equation when the arrival rate equals 20, the same as the service rate? If ρ were equal to 1, the denominator would be 0 and the expression for L would be undefined. In fact ρ equal to 1 is an asymptote. As ρ approaches 1 from below, L grows infinitely.

Thus, the equation for L provides usable and meaningful values over the domain $0 \leq \rho < 1$. To represent this fact, we provide the domain as well as the equation when specifying L as a function of ρ.

$$L = \frac{\rho}{1-\rho} \quad \text{for} \quad 0 \leq \rho < 1$$

Why is it that the equation does not work for values of ρ greater than or equal to 1? The mathematics used to develop the equation of L applies when the system fluctuates around a long-term average value, L. However, when the arrival rate exceeds the service rate, the lines simply continue to increase hour after hour. For example if λ is 25, then each hour there will be five more people than the system can handle. As a result, the queue would continually increase.

8.1.1 Reducing Waiting Time at Peak Hours

Dr. Cue and Mr. Boss began discussing ways to reduce the waiting time during peak hours. One short term solution is to provide a minimum wage assistant to help the ticket agent speed up processing during peak hours. With the assistant, the average processing is estimated to decrease by 25% to two minutes and 15 seconds.

15. On average, what is the maximum number of customers the agent can process in an hour with the assistant's help?

16. With this solution, what is the total waiting time, W, when customers are arriving at rate of 18 per hour on Friday afternoons?

During the holiday season, the arrival rate can increase to as much as 19.5 customers per hour.

17. With this solution would the waiting times be reasonable even during the holiday season?

A longer term solution is to provide a self-service terminal alongside the ticket booth. By placing it near the booth, people who are having difficulty can always buy from the ticket agent. It is estimated that 20 percent of the customers will be able to use the terminal to make a simple purchase. The other customers will buy from the ticket agent.

18. If this solution is used, what will be the total waiting time, W, for customers on Friday afternoons?

19. With the addition of a terminal, would the waiting times be reasonable even during the holiday season?

Chapter 8 Waiting in Line with Polynomials- Non-linear Functions and Queueing Theory

Section 8.2 Post Office in Britton, MI (single server)

The village of Britton, in southeast Michigan, has a population of 4,500 people. Due to its small size, there is only one postal clerk working at Britton's post office. On Saturdays, the post office counter is open only four hours in the morning.

Dave Goldsman, a USPS supervisor, has received complaints about long waiting times. He decided to collect data. Every transaction the postal clerk performs is logged into the computer. The transaction includes a date and time. Dave gathered and organized the data for the last 20 Saturdays. The data are reported in the Table below.

Saturday	Customers	Saturday	Customers	Saturday	Customers	Saturday	Customers
1	40	6	33	11	29	16	37
2	31	7	25	12	44	17	30
3	29	8	37	13	30	18	29
4	29	9	39	14	28	19	30
5	22	10	31	15	35	20	32

Table 8.2.1: Number of customer on Saturdays

1. On average how many customers are served on a Saturday morning?

2. What is the average rate of arrivals per unit of time? Include appropriate units in your answer.

3. What is the average interarrival time (i.e., amount of time between arrivals)?

The second set of data Dave needed was the average time to service a customer. He hired a local high school student to collect data over four weekends. The student determined the average to be 6 minutes per customer. But he told Mr. Goldsman there was a great deal of variability in the time to service a customer. The standard deviation was approximately 6 minutes as well. Mr. Goldsman explained that this pattern was consistent with the exponential probability distribution.

4. If the average time to service a customer is six minutes, at times when the postal clerk is continuously busy, what is the average rate of service in customers per hour?

Dave Goldsman was specifically interested in the total waiting time for a customer, W. In general the formula for L is simpler to calculate directly than the corresponding formulas for W. The reason is that L does not have units of measurements. In contrast W can be in seconds, minutes, hours or even days depending on the context. The key input for all queuing models is ρ, the utilization. Recall from the previous section that ρ is the ratio of the arrival rate, λ, to the service rate, μ.

In other words $\rho = \dfrac{\lambda}{\mu}$.

5. What is the utilization on a Saturday morning at the Britton Post Office?

The formula for determining L as a function of ρ when there is just one server is

$$L = \frac{\rho}{1-\rho}.$$

Dave calculated is the average number of customers in queue on Saturday mornings.

$$L = \frac{0.8}{1-0.2} = 4 \text{ customers}$$

Little's formula defines the relationship between L and W.

$$L = \lambda W$$

The total time spent in the system, W, includes both the wait time and the service time.

$$W = \frac{L}{\lambda} = \frac{4}{8} = 0.5 \text{ hours} = 30 \text{ minutes}$$

Dave thought that perhaps the better measure is the wait time to begin service, W_q. This is calculated by removing the service time component of waiting to complete service on Saturday mornings.

$$W_q = W - \frac{1}{\mu}$$
$$= 30 - 6 = 24 \text{ minutes}$$

John Q. Henry, a friend of Dave's had seen the formula for L_q, the average number of people waiting to begin service.

$$L_q = \frac{\rho^2}{1-\rho}$$

He wondered if he started with this formula, would he obtain the same value for W_q after using Little's alternative formula.

$$L_q = \lambda W_q$$

6. What is L_q when ρ is 0.8?

7. Calculate W_q.

Chapter 8 Waiting in Line with Polynomials- Non-linear Functions and Queueing Theory

8.2.2 Options to Improve Performance

After reviewing the results, Dave was deeply troubled. The USPS has a standard that the average waiting time should be less than 10 minutes. Currently, the average wait in queue was more than double this standard. He is considering three options he has to improve service.

A. A 40-hour retraining program designed to improve the efficiency of the postal worker. Typically, participants can reduce their average service time by 1 minute after completing the program. That will cost the postal service $2,500.

B. A redesign of the workplace and the addition of some specialty equipment will reduce the average service time by 1.5 minutes. The cost of these changes is estimated to be $5,000.

For Option A, the average service time is reduced from six minutes to five minutes.

1. What are the new values of μ and ρ?

2. Calculate the values of L_q and W_q and complete Table 8.2.2

For Option B the average service is reduced from 6 minutes to 4.5 minutes. The average service rate is now 13.3 customers per hour.

3. What is the new value of ρ?

4. Calculate the values of L_q and W_q and complete Table 8.2.2.

5. Consider the cost and the impact on waiting. Which option would you recommend and why?

	Options		
	Current	A - Training	B - Equipment
Added Cost ($)	0	2500	5000
λ (customers arriving per hour)	8		
μ (customers served per hour)	10		
ρ (utilization rate)	0.8		
L_q (customers waiting for service)	3.2		
W_q (minutes waiting for service)	24		

Table 8.2.2: Evaluate post office options

6. What justification can be provided for considering Option C (i.e., making no change)?

8.2.3 Merge Two Offices – 2 servers

Dave decided to consider a broader view of the services provided in the region. There is another small post office in Macon Township, a ten minute drive from Britton. It services approximately the same number of people. It too provides only limited hours on Saturdays. He is considering recommending that on Saturdays the Macon office close and that the worker from Macon Township join the worker at the Britton office. This will save the post office an estimated $45 per Saturday in reduced utility and cleanup expenses. There will be no labor savings as the worker will simply work in a different location but the same number of hours.

As a result, there will now be two servers in the Britton office. For simplicity's sake, we will assume that all of the customers who would otherwise use the Macon office on Saturday will make the short drive to Britton instead. Thus, the arrival rate will double to 16 customers per hour at the merged branch in Britton. For the individual post office the average utilization was 0.8. The formula for average utilization for the merged post offices is adjusted to recognize the fact that there are now two clerks who service twice as many customers per hour.

$$\rho = \frac{\lambda}{2\mu}, \text{ and } \lambda = 16 \text{ customers per hour}$$

Dave looked up the formula for a two server queuing system. It is given below.

$$L_q = \frac{2\rho^3}{1-\rho^2}$$

7. What is the average number of customers waiting in line to begin service at the combined facility?

8. Is this better or worse than the two separate post offices? Explain.

Little's formula for relating L_q and W_q applies to all queueing systems, even those with two servers. Thus the formula for W_q is

$$W_q = \frac{L_q}{\lambda}.$$

9. Is W_q better or worse than the two separate post offices? Explain.

Let's think a little more deeply about the situation from the perspective of people in Macon Township who would on average have to travel ten minutes further.

10. If you lived in Macon, would you prefer the current situation or a Saturday merger with a longer drive?

11. This alternative comes with reduced cost. Would you recommend implementing this?

8.2.4 Merge Two Offices and Provide Training

Dave reexamined the options and realized that combining two options might even be better. He considered merging the two offices on Saturday and sending both postal workers for retraining. The savings from the merger would pay in part for the extra training for two workers.

12. What is the net cost of this combination of options?

Recall that retraining reduces the average service time to five minutes.

13. What is the average utilization for the two server system after training?

14. What is the average number of people waiting to be served and average waiting time for the two server system after training?

Consider the cost and the impact on both groups of potential customers.

15. Now, what would you recommend?

Section 8.3 Airport Security Screening and a Queueing Approximation

8.3.1 Minimum Number of Screeners

The Transportation Security Agency (TSA) is concerned about long back-ups in the security screening area at Charlene York International Airport (CYA). The area has 7 x-ray screening devices. TSA can vary the number of staff to operate these devices up to a maximum of seven. In the morning, the number of screeners can be adjusted every hour. Any unneeded screeners are assigned other tasks.

As customers go through the screening area, they move through a series of security screening steps. First, their boarding pass and ID are checked. Then, they wait to join one of the screening lines. There they wait to approach a long counter where they begin unloading their personal items into bins to be passed through an x-ray detection device. As their personal items are screened, they pass through a metal detection device. The primary backlog is created by the bins passing through the x-ray equipment. Each bin is looked at carefully. There is significant variability as the number of bins and contents vary from customer to customer. On occasion people have to be specially checked or pass through the metal detector a second time. Ralph Waldo is in charge of personnel staffing for CYA. Ralph Waldo's data collectors found that the average time for this x-ray and personal screening was 45 seconds.

Customer Arrivals		Number of Screeners	
Interval	Total	Total workload (man_hours)	Minimum
6-7 am	315	3.94	4
7-8 am	375	4.69	5
8-9 am	470		
9-10 am	500		

Table 8.3.1: Airport arrivals and number of screeners

His staff has collected data on the total number of arrivals during one-hour intervals from 6 am until 10 am. After that time, there was a significant decrease in the number of arrivals until late in the afternoon. They found, for example, between 6 am and 7 am on average 315 customers arrive daily. The average service time per customer is 45 seconds. The total screening workload per hour is 315 multiplied by 45 seconds.

Total Workload per hour = 315 * 45 seconds = 14,175 seconds of work

This number was then divided by the number of seconds in an hour, 60 * 60=3,600.

Total Workload per hour = 14,175 / 3,600 = 3.94 hours of work

That means there is a need for at least four screeners between 6 am and 7 am.

Similarly, between 7 am and 8 am on average 375 customers arrive daily. The average service time per customer is the same 45 seconds. The total screening workload per hour is 16,875 seconds or 4.69 hours.

Total Workload per hour = 375 * 45 seconds = 16,875 seconds = 4.69 hours.

Ralph saw an opportunity to simplify the calculation of total workload. The workload is proportional to the arrival rate. Once he had an estimate for 6 am to 7 am, he could simply use ratios to determine the workload for every other hour.

Let A_1 = number of arrivals in the first hour, 6 am to 7 am
A_2 = number or arrivals in the second hour, 7 am to 8 am
W_1 = workload in the first hour, 6 am to 7 am
W_2 = workload in the second hour, 7 am to 8 am

Then, $\frac{W_2}{W_1} = \frac{A_2}{A_1}$.

This ratio can be rearranged to determine W_2:

$$W_2 = \left(\frac{A_2}{A_1}\right) W_1$$
$$= \left(\frac{375}{315}\right)(3.94)$$
$$= 4.69 \text{ hours}$$

1. Complete Table 8.3.1 to determine the minimum number of screeners required each hour.

Ralph Waldo is concerned that staffing to the minimum could result in long waits to be screened. He therefore, asks Dr. N. Queue to determine the wait in queue for the minimum number of screeners and one more than the minimum number.

Little's law relates the average number in the queueing system to the average waiting time. It relates L_q and W_q as well as L and W. Ralph Waldo is interested in W_q, the number of people waiting to begin the scanning process. The relevant formula is below.

$$L_q = \lambda W_q$$

Dr. N. Queue applied the appropriate formula for L_q as a function of ρ, the average utilization. This utilization is determined by dividing the arrival rate by the total service capacity. This service capacity is the rate of service by one server multiplied by the number of servers, s. Thus the average utilization can be calculated using the equation below.

$$\rho = \frac{\lambda}{s\mu}$$

Next, Dr. Queue determined the average utilization with four screeners. The arrival rate is specified in customers per hour. We need to define the service rate in the same units of measurement. The average service time of 45 seconds is equal to $\frac{3}{4}$ minutes. This is equal to $1/\mu$. Thus the service rate μ is $\frac{4}{3}$ customers per minute or equivalently 80 per hour.

$$\rho = \frac{315}{4 \cdot 80} \approx 0.984.$$

If there were four screeners, they would be busy on average 98.4% of the time. If there were five screeners, the corresponding value is 78.8% of the time.

$$\rho = \frac{315}{5 \cdot 80} \approx 0.788$$

2. Help Dr. N. Queue fill in the shaded cells in Table 8.3.2 by determining the utilization for the minimum number of screeners and one more than the minimum number of screeners.

		Average Screening Service Time = 45 seconds							
		Minimum				Minimum Plus 1			
Interval	Arrivals	Screeners	ρ	L_q	W_q	Screeners	ρ	L_q	W_q
6-7 am	315	4	0.984	59.3	11.3	5	0.788		
7-8 am	375	5	0.938			6	0.782		
8-9 am	470								
9-10 am	500								

Table 8.3.2: Airport screening and waiting in line

8.3.2 Queueing Formula and Waiting Time

Dr. N. Queue found the formula for L_q with four servers in *Modeling Random Processes for Engineers and Managers* by James J. Solberg (2008).

$$L_q = \frac{32\rho^5}{3 + 6\rho + 3\rho^2 - 4\rho^3 - 8\rho^4}$$

She applied this formula to the first time period, 6 am to 7 am when $\rho = (0.984)$.

$$L_q = \frac{32(0.984)^5}{3 + 6(0.984) + 3(0.984)^2 - 4(0.984)^3 - 8(0.984)^4}$$
$$L_q = 59.3 \text{ customers}$$

Chapter 8 Waiting in Line with Polynomials- Non-linear Functions and Queueing Theory

She then calculated W_q using Little's formula.

$$W_q = \frac{L_q}{\lambda} = \frac{59.3}{315} = 0.188 \text{ hours or } 11.3 \text{ minutes}$$

In summary, assume there are four screening machines fully staffed and operating this first hour. If John Q. Public arrived at 6:47 am, he would see on average more than 60 customers waiting in line ahead of him. His average time in queue would be less than 12 minutes. This did not seem unreasonably long for Ralph Waldo.

3. Explain why it will take on average less than 12 minutes to service 60 customers in line with four screeners.

Ralph Waldo was also interested in determining the impact of staffing one more x-ray machine.

4. What is the effect on the utilization of adding of one more screener between 6 and 7 am?

Next, Dr. N. Queue found the formula for L_q with five servers.

$$L_q = \frac{625\rho^6}{24 + 72\rho + 84\rho^2 + 20\rho^3 - 75\rho^4 - 125\rho^5}$$

5. Calculate the value of L_q when there are five screeners.

6. Now use Little's Law to find the value of W_q.

The addition of just one more screener using an x-ray machine would eliminate almost all of the waiting time. However, the screeners would be busy less than 80% of the time. More than 20% of the time they would have nothing to do.

7. Ralph Waldo needed to make a tough decision. The cost of staffing and operating one additional x-ray machine was $45 per hour. Was it worth spending that much money to almost eliminate waiting time for customers arriving between 6 am and 7 am?

8. What would you recommend and why?

8.3.3 Queueing Approximation

It was obvious to Dr. Queue that the formulas were getting more and more complicated. Dr. Queue wondered if there was some approximate formula that would be easier to use. She found the following approximation in Solberg's book.

$$L_q = \frac{\rho^{\sqrt{2(s+1)}}}{1-\rho}$$

This formula looks complicated but easily computed with a calculator or computer. The utilization, ρ, is raised to the power of $\sqrt{2(s+1)}$. For example, if s equals four, this exponent is the square root of 10. There is nothing wrong with the power term not being a rational number. We are familiar with squaring a term, and usually this can be carried out without the use of a calculator. However, when the exponent is an irrational number, such as $\sqrt{10}$, ease in calculation requires technology.

Dr. Queue decided to see how good the approximation was by first applying it to the four server example. Here are her calculations using the approximation.

$$L_q = \frac{(0.984)^{\sqrt{2(4+1)}}}{1-0.984} \approx 59.4 \text{ customers}.$$

Dr. Queue was excited to find that the result was less than 0.2% different from the exact value. The nice thing about the formula is it does not get more complicated as the number of servers increases. Before proceeding to use the formula for all of the analysis, she decided to check out the approximation for five servers between 6 am and 7 am.

The approximate formula yielded an estimate of 2.1 customers.

$$L_q = \frac{(0.788)^{\sqrt{2(5+1)}}}{1-0.788} \approx 2.1 \text{ customers}$$

Although, this approximation was 5% higher than the actual value, Dr. N. Queue felt this number was close enough to justify using the approximate formula for the rest of the analysis. However, she asked her staff to repeat the comparison one more time. She asked them to apply both formulas for the time period between 7 am and 8 am. Recall that five screeners are needed in that time frame, and the average utilization would be 0.938.

Dr. N. Queue hired a local high school student, Matt Wannabee, to complete the table. Matt was also curious about the approximation. He therefore decided to check it out for himself using the 7am to 8 am data and five servers. Matt was also going to time how long it took him to do the exact and approximate calculations and obtain the correct answer. (With complex calculations such as these, mistakes are common and the calculations may need to be redone. If the answers differ by more than 5%, you have made a mistake. Determine the total time it took you obtain the correct answer.)

9. Determine L_q by using the exact formula for five servers and $\rho = 0.938$. How long did it take to obtain this number?

10. Determine L_q by using the approximate formula for five servers and $\rho = 0.938$. What is the percentage difference? How long did it take?

Matt let Dr. Queue know that the formula for five servers was much more accurate when the utilization was above 0.9 as compared to before when it was less than 0.8.

11. Use the approximate formula to determine the queueing statistics for 6 and 7 screeners during the time periods specified in Table 8.2.3.

		Average Screening Service Time = 45 seconds							
		Minimum				Minimum Plus 1			
Interval	Arrivals	Screeners	ρ	L_q	W_q	Screeners	ρ	L_q	W_q
6-7 am	315	4	0.984	59.3	11.3	5	0.788	2.1	0.4
7-8 am	375	5	0.938	12.9	2.1	6	0.782		
8-9 am	470		0.977			7	0.840		
9-10 am	500		0.893			8	--	--	--

Table 8.3.3: Airport Screening and Waiting in Line for 6 or more screeners

The completed results were presented to Ralph Waldo. Now he had to decide how many x-ray machines to staff each hour. Should he staff the minimum each hour or should he sometimes add one more screener at a cost of $45 per hour.

He reviewed his budget and decided he could go above the minimum required staffing for only one or at most two hours.

12. What would you recommend and why?

8.3.4 Heightened Security Screening – More Time Processing

At least once a month, CYA is ordered on heightened security either because of an unspecified threat to airline safety. During this time, screeners are expected to more carefully review the images on the x-ray screen and hand check more carry-on packages. As a result the average time to screen a customer increases to 52 seconds during these time periods.. Although this does not seem like much of an increase, Ralph Waldo is concerned about the possibility of long back-ups. Dr. Queue has told him the waiting time is a steep function of the average utilization and is not a straight line.

13. He asked Matt Wannabee to start with the basic calculations to determine the minimum number of screeners. Complete the Minimum column in Table 8.3.3. How many more would he need each hour when compared to the minimum for a regular day?

	Average Screening Service Time = 52 seconds				
		Minimum			
Interval	Arrivals	Screeners	ρ	L_q	W_q
6-7 am	315				
7-8 am	375				
8-9 am	470				
9-10 am	500				

Table 8.3.3: Airport Screening and Heightened Security

Mr. Waldo planned to use the minimum number of screeners. He nevertheless wondered if there would be any hours with extremely long waiting times.

14. Matt Wannabee completed the Table 8.3.3 using the approximation to determine L_q and W_q.

Ralph took a closer look and noticed CYA was going to have a problem between 9 am and 10 am.

15. What problem did Ralph Waldo find?

Ralph Waldo was considering asking supervisors to help during this one hour period. They would assist passengers in organizing the stuff to be x-rayed so as to increase the efficiency of the processing. He believed this would reduce the overall average service time to less than 50 seconds.

16. Would these extra personnel reduce the required minimum number of screeners between 9 am and 10 am?

Chapter 8 (Queueing Theory) Homework Questions

1. Identify the customers and the servers in the queueing system in each of the following situations:

    a. The self-service checkout in a large supermarket

    b. At a bank

    c. Online assistance with computer problems

    d. Oil change service

    e. A bicycle repair shop

    f. An airport runway

    g. A college admissions office

    h. An urgent care facility

2. Donald Clark handles all checkouts at the Oasis County Public Library. On average, 40 people per hour arrive at the counter to check out books, CDs, and DVDs. It takes Donald an average of one minute to checkout a person.

    a. On average, how many people are in line including the person being checked out?

    b. On average, how much time will a person spend checking out?

    c. On average, how much time will a person spend in line before handing Donald the stuff to be checked out?

    d. In the last hour before the library closes, the number of people checking out increases. During this time, the average increases to 50 people per hour. How much does the average waiting time increase during this final hour? In your opinion is this wait excessive?

3. The customer service helpline at Koala Foods handles special requests from businesses placing orders. There is only one customer service representative staffing the helpline at any one time. There is call-waiting that allows .an unlimited number of customers to be placed on hold. On average, the customer service representative takes seven minutes to process an order. There are, on average, five calls per hour to the customer service helpline.

    a. How busy is the customer representative handling orders?

    b. Calculate W, a customer's expected total time in the queueing system.

c. What percent of customers who call in to the helpline are placed on hold?

d. On average how long do customers wait to speak to the customer service representative?

e. If you are the manager of the customer service department, do these results concern you? Why or why not?

4. The Hy Life pharmacy has a drive through window. In the early afternoon, on average, 12 customers arrive per hour. Jimenez Nova, the pharmacy assistant for the window, takes on average three minutes to process a customer at the drive-through window.

   a. What percent of time is Jimenez processing drive through customers?

   b. What is the average length of the line including any car being at the window?

   c. What is the average total time a car spends picking up a prescription?

   d. What is the average waiting time in line before moving up to the window?

5. Marianne Rivero, the Hy Life manager, noticed that in the early evening, the lines were backing up into the parking area. She was concerned whenever the average queue length exceeded three cars. She asked Jimenez what the problem was. Jimenez informed her that the peak times were weekday evenings. The number of customers can increase to 15 per hour. Worse yet, on Friday evenings the number is about 17 per hour. At those times Jimenez feels he is working nonstop and can barely catch his breath.

   a. How busy is Jimenez on weekday evenings?

   b. What is the average length of the line including any car being at the window?

   c. On average how many minutes is he busy per hour during these peak times?

   d. What is the average waiting time in line before moving up to the window?

   e. How busy is Jimenez on Friday evenings

   f. On average how many minutes is he busy per hour during this peak time?

   g. On Friday evenings, how long are the queues of cars including the one at the window?

6. Jimenez discussed the situation with Marianne. He told her that if he could have an aide during peak hours, he could reduce the customer processing time from three minutes to an average of two minutes.

a. Would this change solve the problem for weekday evenings?

b. Would this change solve the problem for Friday evenings?

7. At an outdoor concert, there are two portable toilets for women (one at each end of the field), and one portable toilet for men. Assume that the average customer arrival rate at the men's toilet is 30 per hour, and the average customer arrival rate at "each" of the women's toilets is 15 per hour. Assume also that the average service time for men is 1.5 minutes and for women is 3 minutes.

   a. What is the proportion of time the men's portable toilets is in use?

   b. What is the average number of men waiting in line outside the toilet?

   c. What is their average waiting time in queue?

   d. What is the proportion of time the women's portable toilets are in use?

   e. What is the average number of women waiting in line to use a toilet?

   f. What is a woman's average waiting time?

   g. Are the portable toilets distributed fairly? Use queueing theory to support your answer.

   h. Would your sense of fairness be different if men walk an average of six minutes to get to the only men's toilet. Women have to walk only three minutes to get to one of the two women's toilets.

8. There is a proposal to place the two women's toilets together, next to the men's toilet.

   a. What would the waiting time in queue, W_q, for women become?

   b. What are the advantages and disadvantages of this proposal?

9. The Foxwood Stadium has six food stations spread evenly around the stadium. An estimated 210 customers per hour, or 3.5 per minute, are going to get food at one of the stations. Each station handles one-sixth of the customers. Each station services one customer on average in 1.5 minutes. The average time for a sports fan to walk to the nearest station is six minutes. It takes another six minutes to return to his seat.

   a. What is the average time a customer spends getting food including walking to and from the station?

Chapter 8 — Waiting in Line with Polynomials- Non-linear Functions and Queueing Theory

10. Bill Eates, food manager for the stadium, has proposed combining stations into just three larger stations. Each station would have two servers. He believes that this would be more efficient use of space and personnel. With just three food stations, the average walk time for a customer would increase to eight minutes each way.

    a. Would this change increase or decrease the total time it takes a customer to get his food and return to his seat?

11. The Hugh Holloway Trucking Company is building a large warehouse on the east side of Gotham City. Mr. Holloway forecasts that eight trucks an hour will arrive at the warehouse to unload their goods. It takes an average of 25 minutes to unload a truck.

    a. What is the minimum number of unloading docks that are needed to handle the eight trucks an hour?

    b. Assume the company builds the minimum number of docks. What is the average number of trucks that will be waiting in the parking area for an unloading dock to become available? (Use both the exact formulas and the approximation. Assess the accuracy of the approximation.)

    c. What is their average waiting time for a truck to begin unloading?

    d. If the company were to build one extra dock, what would the average wait be?

    e. What would you recommend? What additional information would you want to help make the decision?

    f. Use the approximation formula to redo parts c and d above.

    g. How close are the estimates for W_q for each number of unloading docks?

    h. Would the change in your estimates of W_q make any difference in what you would recommend?

12. Mr. Holloway has been offered another property on the west side of the city. Hugh estimates that trucks coming from west of the city would take almost 40 minutes to drive across the city to get to the east side warehouse. He is considering building two smaller warehouses one on each side of the city. Assume that equal numbers of trucks approach the city from both the west and east directions. If he builds a second warehouse, four out of every eight trucks would save 40 minutes of travel time. He is considering building two warehouses. Each warehouse would have just two loading docks. Each would unload an average of four trucks per hour.

    a. What is their average waiting time for a truck to begin unloading at either warehouse?

b. Would you recommend two separate warehouses or one bigger warehouse? Justify your answer.

13. Al Waze has launched the Always There rapid response car service, in Unbroken, NJ. He estimates demand for his service will average seven calls an hour at peak times during the day. The average ride lasts 22 minutes. He is trying to recruit independent drivers.

    a. What is the minimum number of drivers that he needs?

    b. What is the expecting waiting time for an available car to be sent? (Use the approximation formula for all calculations)

Al believes he will have difficulty keeping repeat customers if the average wait time for a car to be sent is more than 10 minutes.

    c. How many drivers does Al need to meet or exceed the ten minute standard?

14. Tatianna Dobric is the owner of a small cozy coffee house. It has only five tables for couples. During the evening, couples arrive at a rate of four per hour. Couples on average spend one hour and five minutes at the table. However, there is significant variation around this 65 minute average. Tatianna has noticed the lines can get long and is concerned that she may be losing customers. She was wondering what would happen to waiting time if she squeezed in one more table. Customers would feel a little bit cramped but at least they will not have to wait as long.

    a. What percent of the time are the five tables occupied?

    b. What is the average number of couples waiting to be seated with five tables?

    c. What is the average wait time to be seated with five tables?

    d. With six tables, what percent of the time would the tables being occupied?

    e. What is the average number of couples waiting to be seated with six tables?

    f. What is the average wait time to be seated with six tables?

15. Nancy Chicila is a customer service manager at MONEYMARK. It is a local financial planning and advising company. Nancy has six highly trained staff who handle the customer service hot line. She has recently received complaints from customers about being placed on hold for a long time. Nancy decided to analyze the problem to determine whether MONEYMARK needed to hire more employees in order to reduce waiting time.

After studying historical data, she found that the arrival rate of calls followed the Poisson distribution. There were on average 34 calls per hour. The time to service each call averaged 10 minutes. The average cost, including salary and training per employee per hour, was $18.

A typical employee works 2,000 hours per year. The annual cost of each staff person is $36,000.

The six staff are sufficient to handle the average call volume. However, calls arrive at random and the service time is random. As a result, at times there will be more customers seeking advice than the six staffers can immediately handle. If Nancy hires additional staff, customers will spend less time on hold. However, employing these additional servers will increase her costs. The problem she is facing is to find the best balance between cost and waiting time. She is considering adding one or even two more staff. Complete the table below. Then make a recommendation on staffing and provide a justification.

Staff	Annual Cost	Average wait time on hold
6		
7		
8		

Chapter 8 Summary

What have we learned?

In this chapter we have studied queueing systems and used mathematical models to predict key measures of system performance. There are four related average measures of performance:
- The average number of customers in the queueing system (L)
- The average time customers spend in the queueing system (W)
- The average number of customers waiting to begin service (L_q)
- The average time customers spend waiting to begin service (W_q)

The key inputs to the models are the average arrival rate of customers (λ) and the average service rate of customers (μ). We have explored the relationship between these parameters and the customer waiting experience. The examples demonstrated the importance of defining the appropriate domain and range of a nonlinear function.

This chapter is fundamentally different from earlier chapters. The first seven chapters present prescriptive models. These models help the decision maker identify the best decision from a set of alternatives. In contrast, queueing models provide only estimates of system performance under different operating conditions. While they are useful in helping investigate specific changes to the queueing system, they do not identify the best decision.

The most common queueing system decisions involve specifying the number of servers to provide at different times and where they should be located. There is always a tradeoff between the number of servers and customer waiting times. The cost of service is borne by the service provider. However, there is generally no explicit assessment of the cost of waiting. It is, therefore, not possible to directly compare the cost of service and the cost of waiting. However, queueing system models do enable the manager to understand the relationship between investment in service and the customers' experience while waiting in the system. It is then up to the manager to decide on the level of investment in service and the acceptable waiting time.

Terms

L	Average number of customers in system including customers being served
L_q	Average number of customers waiting in queue to begin service
W	Average time customers spend in the system including service time
W_q	Average time customers wait until service begins
λ	Average arrival rate of customers per unit time.
μ	The average rate at which customers are served. $1/\mu$ is the average service time.
ρ	The proportion of time a server is busy.
Exponential distribution	A continuous distribution that is often used to represent the randomness in the time it takes to service a customer.
Poisson distribution	A discrete distribution that is often used to represent the random arrival pattern of customers.

Chapter 8 (Queueing Theory) Objectives

You should be able to:

- Determine minimum number of servers required to service all customers

- Calculate the average utilization or fraction of time that servers are busy with customers

- Use the appropriate nonlinear function to calculate L_q for different numbers of servers

- Starting with L_q, determine all of the other performance measures, L, W, and W_q.

- Use a complex function to approximate L_q for any number of servers

- Compare the performance of different plans for the queuing systems

Chapter 8 — Waiting in Line with Polynomials- Non-linear Functions and Queueing Theory

General Formulas and Relationships

Utilization rate or fraction of time servers are busy	$\rho = \dfrac{\lambda}{s\mu}$	ρ = utilization rate λ = arrival rate s = number of servers μ = service rate
Little's Law for L	$L = \lambda W$	L = number of customers in system W = time in system
Determine W from L	$W = \dfrac{L}{\lambda}$	
Little's Law for L_q	$L_q = \lambda W_q$	L_q = number of customers waiting for service W_q = time waiting for service to begin
Determine W_q from L_q	$W_q = \dfrac{L_q}{\lambda}$	
Relationship between L and L_q	$L = L_q + \dfrac{\lambda}{\mu}$	average number of customers in the system (L) = average number in queue (L_q) plus the average number in service (λ/μ)
Relationship between W and W_q	$W = W_q + \dfrac{1}{\mu}$	average time customers are in the system (W) = the average time spent in queue (W_q) plus the average service time ($1/\mu$)

Number of servers	Equation for L_q
1	$L = \dfrac{\rho}{1-\rho}$ and $L_q = \dfrac{\rho^2}{1-\rho}$
2	$L_q = \dfrac{2\rho^3}{1-\rho^2}$
3	$L_q = \dfrac{9\rho^4}{2 + 2\rho - \rho^2 - 3\rho^3}$
4	$L_q = \dfrac{32\rho^5}{3 + 6\rho + 3\rho^2 - 4\rho^3 - 8\rho^4}$
5	$L_q = \dfrac{625\rho^6}{24 + 72\rho + 84\rho^2 + 20\rho^3 - 75\rho^4 - 125\rho^5}$
Approximation	$L_q = \dfrac{\rho^{\sqrt{2(s+1)}}}{1 - \rho}$

Lead Authors: Kenneth Chelst and Thomas Edwards

Chapter 8   Waiting in Line with Polynomials- Non-linear Functions and Queueing Theory

Chapter 8 Study Guide

1. What is a queueing system? Give several examples.

2. What are the parameters needed to calculate the performance of a queueing system?

3. How do you determine the minimum number of servers for a queueing system to function?

4. How do you find the expected value for the average number of people in a single server queuing systems?

5. What is the relationship between L and W?

6. Explain the difference between L and Lq?

7. Explain the difference between W and Wq? When do you think W is the better measure and when is Wq the better measure of performance? Provide examples.

8. What can be a disadvantage of putting all of the service facilities in one location?

9. How do you find the expected value for L_q using the exact equations for $s = 1, 2, 3, 4$ and 5?

10. After having determined L_q, how do you go about determining the other measures, L, W and W_q?

11. How do you find the expected value for L_q using the approximation formula for any value of s? How do you measure the accuracy of the approximation formula?

Made in United States
North Haven, CT
20 September 2022